PENGUIN BOOKS
The New Map

Named 'Energy Writer of the Year' by the American Energy Society

'There are many stories in this wonderful book, all of them directed at the transformation of the global map of power and wealth that has happened in the 21st century . . . capacious and well-written . . . We are in a disordered period of rising tensions, but, then again, we usually are. Human competence races ahead, but wisdom remains as rare a commodity as it ever was' Bryan Appleyard, Sun*day Times*

'[Yergin] has turned his considerable talents to explaining how the world continues to be shaped by oil in his latest book . . . supremely readable' *Wall Street Journal*

'As Daniel Yergin writes in his recent book *The New Map*, which I highly recommend to you, climate change will have enormous impact on how energy is produced, transported, consumed, and in strategies and investment, in technology and infrastructure, and in relations between countries.' Hon. Scott Morrison, Prime Minister of Australia

'The book breezily takes the reader through the developments of the past few decades in the oil business and energy more generally, with an eye to political repercussions. With sections on America, Russia, China, the Middle East, the car industry and climate politics and policy, at its best it is both brisk and authoritative, an impressive combination' *The Economist*

'The US author is the energy guru par excellence. . . . *The New Map* comes after a decade of rapid transformation in energy, with renewables and other sources having broken through technological and financial barriers to become genuine challengers to the supremacy of fossil fuels (as Yergin had predicted). . . . Sweeping thematic narrative gives way to the fascinating detail garnered by someone who has sat at the top table with presidents, kings and chief executives for decades, but who still brings a journalistic eye to the proceedings . . . rich in dramatic detail' Frank Kane, *Arab News*

'Fans of the author's previous books will appreciate the snappy prose and plethora of well-told anecdotes . . . revealing and apposite . . . brings the general reader admirably up to date'
Edward Lucas, *The Times*

'Yergin . . . knows the issues inside and out.' *Nikkei Asia*

'Energy and geopolitics . . . If anyone has a front row seat and can explain their unfolding development and can interpret the implications, it is Mr Yergin.' *Channel News Asia*

'The latest on global energy geopolitics from the pen of an expert . . . Yergin delivers a fascinating and meticulously researched page-turner . . . Required reading. Another winner from a master' *Kirkus Reviews*

'A master class on how the world works' National Public Radio

'At a time when solid facts and reasoned arguments are in retreat, Daniel Yergin rides to the rescue. . . . Yergin provides an engaging survey course on the lifeblood of modern civilization – where the world has been and where it is likely headed. By the final page, the reader will feel like an energy expert herself' *USA Today*

'*The New Map* earned energy's highest literary prize for its ambitious survey and realistic assessment of energy and how it shapes all of human affairs. It is also an exceptional literary triumph in its narrative and in the quality of writing that we have come to expect from Dan Yergin' The American Energy Society, in awarding Daniel Yergin 'Energy Writer of the Year'

ABOUT THE AUTHOR

Daniel Yergin is one of the most influential voices on energy, international politics and economics in the world. He is the vice chairman of IHS Markit and a recipient of the United States Energy Award for 'lifelong achievements in energy and the promotion of international understanding' and was named 'Energy Writer of the Year' by The American Energy Society in 2020. Yergin is the author of *The Quest* and received the Pulitzer Prize for *The Prize: The Epic Quest for Oil, Money and Power*, which became a number one bestseller and was made into an acclaimed eight-hour PBS/BBC series.

DANIEL YERGIN

The New Map

Energy, Climate, and the
Clash of Nations

PENGUIN BOOKS

PENGUIN BOOKS

UK | USA | Canada | Ireland | Australia
India | New Zealand | South Africa

Penguin Books is part of the Penguin Random House group of companies
whose addresses can be found at global.penguinrandomhouse.com.

First published in the United States of America by Penguin Press,
an imprint of Penguin Random House LLC, 2020
First published in Great Britain by Allen Lane, 2020
Published with a new Epilogue and Appendix in Penguin Books 2021

008

Pages 505 and 506 constitute an extension of this copyright page

The Appendix was originally published, in slightly different form, as 'The World's
Most Important Body of Water' in *The Atlantic* on 15 December 2020

The moral right of the author has been asserted

Printed and bound in Great Britain by Clays Ltd, Elcograf S.p.A.

The authorized representative in the EEA is Penguin Random House Ireland,
Morrison Chambers, 32 Nassau Street, Dublin D02 YH68

A CIP catalogue record for this book is available from the British Library

ISBN: 978–0–141–99463–5

www.greenpenguin.co.uk

MIX
Paper from
responsible sources
FSC® C018179

Penguin Random House is committed to a
sustainable future for our business, our readers
and our planet. This book is made from Forest
Stewardship Council® certified paper.

To Angela, Rebecca, Alex, and Jessica

Contents

Introduction   xi

AMERICA'S NEW MAP

1. The Gas Man   3

2. The "Discovery" of Shale Oil   14

3. "If You Had Told Me Ten Years Ago":
   The Manufacturing Renaissance   25

4. The New Gas Exporter   31

5. Closing and Opening: Mexico and Brazil   41

6. Pipeline Battles   46

7. The Shale Era   52

8. The Rebalancing of Geopolitics   58

RUSSIA'S MAP

   9. Putin's Great Project   69

  10. Crises over Gas   78

  11. Clash over Energy Security   84

  12. Ukraine and New Sanctions   90

  13. Oil and the State   99

  14. Pushback   102

  15. Pivoting to the East   115

  16. The Heartland   120

CHINA'S MAP

  17. The "G2"   129

  18. "Dangerous Ground"   136

  19. The Three Questions   142

  20. "Count on the Wisdom
       of Following Generations"   147

  21. The Role of History   152

  22. Oil and Water?   155

  23. China's New Treasure Ships 161

  24. The Test of Prudence   165

  25. Belt and Road Building   177

MAPS OF THE MIDDLE EAST

  26. Lines in the Sand   193

  27. Iran's Revolution   206

  28. Wars in the Gulf   210

  29. A Regional Cold War   220

  30. The Struggle for Iraq   229

  31. The Arc of Confrontation   236

32. The Rise of the "Eastern Med"   253

33. "The Answer"   259

34. Oil Shock   272

35. Run for the Future   291

36. The Plague   311

ROADMAP

37. The Electric Charge   327

38. Enter the Robot   347

39. Hailing the Future   358

40. Auto-Tech   366

CLIMATE MAP

41. Energy Transition   377

42. Green Deals   388

43. The Renewable Landscape   394

44. Breakthrough Technologies   403

45. What Does "Energy Transition" Mean in the Developing World?   407

46. The Changing Mix   411

Conclusion: The Disrupted Future   423

Epilogue: Net Zero   431

Appendix: The Four Ghosts Who Haunt the South China Sea   452

Acknowledgments   465

Notes   468

Illustration Credits   505

Index   507

Introduction

———————

This book is about the new global map that is being shaped by dramatic shifts in geopolitics and energy. It is also about where this map is taking us. Geopolitics focuses on the shifting balance and rising tensions among nations. Energy reflects far-reaching alterations in global supply and flows, driven in major part by the remarkable change in the energy position of the United States, and by the growing global role of renewables and the new politics of climate.

Different kinds of power are in play. One is the power of nations that is shaped by economics, military capabilities, and geography; by grand strategy and calculated ambition; by suspicion and fear; and by the contingent and the unexpected. The other is the power that comes from oil and gas and coal, from wind and solar, and from splitting atoms, and the power that comes from policies that seek to reorder the world's energy system and move toward net zero carbon in the name of climate.

This is no simple map to follow, for it is dynamic, constantly changing. It has been made even more complicated by the coronavirus that swept out of China and across the planet in 2020, bringing grief and vast human suffering and disarray. It also shut down the world economy, disrupted commerce both local and global, destroyed jobs and businesses and impoverished many, plunged the world economy into the deepest recession since the Great Depression, added enormously to public debt, accentuated the tensions among countries, and created vast turmoil in global energy markets.

This book seeks to illuminate and explain this new map. How the shale revolution has changed America's position in the world. How and why new cold wars are developing between the United States on one hand, and Russia and China on the other, and energy's role in them. How swiftly—and potentially perilously—the overall relationship between the United States and China is changing from "engagement" to "strategic rivalry" and what begins to look like an emerging cold war. How unsteady are the foundations of a Middle East that still supplies a third of the world's total petroleum and a significant amount of natural gas. How the familiar ecosystem of oil and autos, which has held for more than a century, is now being challenged by a new mobility revolution. How climate concerns are reshaping the map of energy, and how the much-discussed "energy transition" from fossil fuels to renewables may actually play out. And how has the coronavirus changed the energy markets and the future roles of the Big Three—the United States, Saudi Arabia, and Russia—which now dominate world oil.

"AMERICA'S NEW MAP" TELLS THE STORY OF THE UNANTICI-
pated shale revolution that is transforming America's place in the world, upending world energy markets, and resetting global geopolitics. Together, shale oil and shale gas have proven to be the biggest energy innovations so far in the twenty-first century. Wind and solar are both innovations of the 1970s and 1980s, though they came into their own only over the last decade. The United States has surged ahead of Russia and Saudi Arabia to become the world's number one pro-

ducer of both oil and gas, and is now one of the world's major exporters of both.

Though targeted for bans by some politicians, the shale revolution has fueled America's economic growth, enhancing its trade position, generating investment and job creation, and lowering utility bills for millions of consumers. The supply chains supporting shale reach all across the United States, into virtually every state, creating jobs even in New York state, which prohibits shale development within its borders, owing to environmental opposition.

Starting with the energy crises of the 1970s, Americans became accustomed to thinking the country was vulnerable because of U.S. dependence on imported sources. But the geopolitical consequences for the United States, now that it is almost self-sufficient, are apparent in new dimensions of influence, increased energy security, and greater flexibility in foreign policy. Yet there are limits to this newfound self-assurance, for energy remains a globally-interconnected industry and these consequences are still only part of the overall nexus of relations among nations. Moreover, shale was already in search of its next "revolution" when coronavirus sent it spinning into a new crisis.

"RUSSIA'S MAP" IS ABOUT THE TINDER CREATED BY THE IN-teraction of energy flows, geopolitical competition, and the continuing contention over the unsettled borders that resulted from the collapse of the Soviet Union three decades ago—and from Vladimir Putin's drive to restore Russia as a Great Power. Russia may be an "energy superpower," but it is also economically dependent on oil and gas exports. Today, as in Soviet times, those exports are stoking fierce debate about the possible political leverage over Europe that may come in their wake. Yet, any potential leverage has been dissipated by changes in both the European and global gas markets.

The consequences of the abrupt transformation of the Soviet Union into fifteen independent countries remain uncertain, nowhere more so than between Russia and Ukraine, where conflict over natural gas has been central. Following the 2014 Russian annexation of Crimea, the struggle moved to the battlefield in southeastern Ukraine. In the strange way that history works, that war—and, specifically, the matter of U.S. weapons to resist Russian tanks—triggered the impeachment of Donald Trump by the House of Representatives, followed by his acquittal by the Senate.

U.S.-Russian relations have sunk to a level of hostility not seen since Soviet days in the early 1980s. At the same time, Russia has "returned" to the Middle East and is "pivoting to the east," to China. Moscow and Beijing are united in asserting "absolute sovereignty" and their opposition to what they decry as American "hegemony." There are also practical considerations to their burgeoning relationship: China needs energy, and Russia needs markets.

"CHINA'S MAP" IS ROOTED BOTH IN WHAT IT CALLS THE "CENtury of Humiliation" and in its tremendous gains in global economic and military power over the last two decades, and by the energy needs of what will become the world's largest economy (and, by some measure, already is). China is expanding its reach in all dimensions: geographically, militarily, economically, technologically, and politically. The "workshop of the world," it now seeks to move up the value chain and become the global leader in the new industries of this century. China is also asserting its own map for almost the entirety of the South China Sea, the most critical oceanic trade route in the world, and now the sharpest point of strategic confrontation with the United States. Energy is an important part of that claim.

China's Belt and Road Initiative is designed to redraw the economic map of Asia and Eurasia and beyond, putting what was once the "Middle Kingdom" in the middle of a reordered global economy. The initiative seeks to assure that China will have markets for its goods and access to the energy and raw materials that it needs. But to what degree is the Belt and Road mainly an economic project, or, as critics assert, a geopolitical project aimed at creating a new Chinese order in world politics?

The "WTO consensus" that goes back to the beginning of this century has broken down. Criticism of China is one thing that unites divided Democrats and Republicans in the United States, and the national security establishments in both countries increasingly focus on the other as the future adversary. Yet the two countries are more integrated economically and more interdependent than many recognize, as the 2020 coronavirus outbreak unhappily demonstrated; and they are mutually dependent on global prosperity. But that reality counts for less as calls grow louder for "decoupling" between the world's two largest economies, accompanied by growing mistrust, which has been amplified by the coronavirus crisis, one of the lasting consequences of which will be greater tension between the two countries.

THE MIDDLE EAST'S GEOGRAPHIC BOUNDARIES WERE CON-tinually redrawn throughout antiquity, with the rise and fall of so many empires. Though the Ottoman Turkish Empire ruled for six centuries, its borders were often shifting. The map of the modern Middle East was laid down during and after the First World War, in the vacuum resulting from the collapse of the Ottomans and yet based on the provincial lines left behind by the Ottomans. The maps have been challenged ever since—by pan-Arab nationalism and political Islam, by opposition to the state of Israel, and then by jihadists such as ISIS, who want to replace the very idea of a "nation-state" with a caliphate. One of the biggest challenges in the region today comes from the rivalry between Sunni Saudi Arabia and Shia Iran for preeminence, now made more complex by Turkey's new bid for that role, reclaiming a lineage going back to the

Ottomans. But also defining for the region are the four-decade confrontation between the United States and Iran and the prevalence of weak governance in many countries.

The Middle East has been shaped, of course, not only by the maps of frontiers but by different kinds of maps—of geology, of oil and gas wells, of pipelines and tanker routes. The oil and gas, and the revenues and riches and power that flow from them, remain central to the identity of the region. Yet the oil price collapse that began in 2014 has fed into a new debate about the future of oil. Not much more than a decade ago, the world worried about "peak oil," the idea that oil supplies would run out. The focus has shifted to "peak demand": how long consumption of oil will continue to grow and when it will begin to decline. Will oil lose its value and importance in the decades to come? The demand collapse for oil in 2020 has further fueled the urgency for oil exporters to diversify and modernize their economies, which Abu Dhabi had begun in 2007 with its Vision 2030, and which Saudi Arabia is now trying to do in double time.

If there is one major factor leading to the idea that demand, not supply, is the future constraint, it is related to the junction of climate policies and technology. The one market that seemed to be guaranteed for oil for a very long time was transportation and, specifically, the automobile. No longer, not on the "Roadmap" to the future. For oil now faces a sudden challenge from the New Triad: the electric car, which uses no oil; "mobility as a service," ride-hailing and ride-sharing; and cars that drive themselves. The result could be a contest for dominance in a new trillion-dollar industry: "Auto-Tech."

THE DEBATE OVER HOW RAPIDLY THE WORLD CAN AND MUST adjust to a changing climate, and how much it will cost, is unlikely to be resolved in this decade. But the endeavor will take on greater urgency as public opinion becomes more aroused and new policies seek to implement "net zero carbon." All this takes us to the "Energy Transition": the shift from the world of today, which depends on oil, natural gas, and coal for more than 80 percent of its energy—just as it did thirty years ago—to a world that increasingly operates on renewables. The Paris Agreement of 2015 galvanized the march toward a lower carbon future. Indeed, in terms of energy and climate, there are two distinct eras: "Before Paris" and "After Paris." Yet, while energy transition has become a pervasive theme all around the world, disagreement rages, both within countries and among them, on the nature of the transition: how it unfolds, how long it takes, and who pays. "Energy transition" certainly means something very different to a developing country such as India, where hundreds of millions of impoverished people do not have access to commercial energy, than to Germany or the Netherlands.

Solar and wind have become the chosen vehicles for "decarbonizing" electricity. Once "alternatives," they are now mainstream. Yet, as their share of generation grows larger, they confront the challenge of "intermittency." They can flood the grid with electricity when the sun shines and the wind blows, but then almost disappear when the day is cloudy or there is only a murmuring breeze. This points to major technological challenges: to maintain grid stability and find ways to store electricity at large scale for periods longer than a few hours.

"Climate" will be a profound determinant of the new map of energy. Here I build on the story I began in *The Quest*. In that book, I explored how "climate" went from being a subject of interest to a handful of scientists in mid-nineteenth century Europe, who feared the advent of another Ice Age that would obliterate civilization, to the consensus about warming that would bring 195 countries together in Paris in 2015 to forge a climate compact that has become the global benchmark. The focus in the pages of this book is on how the momentum of climate policies—powered by research and observation, by climate models, and by political mobilization and regulatory power, social activism, financial institutions, and deepening anxiety—will transform the energy system. "Net zero carbon" will be one of the great challenges of the decades ahead, not just politically but also in how people live their lives and in the costs of achieving it.

My first book, *Shattered Peace*, is about the origins of the Soviet-American Cold War. Now in these pages, readers will find the origins of new cold wars. *The Prize* is a sprawling canvas about geopolitics and oil over almost a century and a half, and that certainly is part of the narrative of *The New Map*. *The Commanding Heights*, which I coauthored, was about the world after the Cold War and the new age of globalization. Now the fragmenting of globalization becomes part of this story.

For the coronavirus has fueled a retreat that had already begun from globalization and from the international institutions and cooperation that have underpinned it. In 2008–2009, international collaboration was key to conquering the financial contagion. A dozen years later, such cooperation at the governmental and international level in fighting the contagion of the virus was notable by its absence. What had been talk of "decoupling" had turned into a rolling back of the supply chains that have been a foundation of a $90 trillion global economy. More broadly, borders go up, nationalism and protectionism rise, and the generally free movement of people becomes less free. One consequence of the global economic misery from the pandemic of 2020 could be greater prevalence of fragile and failed states, which would create new security challenges that, at some point, would reach beyond their borders. Yet governments would be hampered in responding to domestic and international needs, whether around security or health or energy and climate, by the huge debt and fiscal armor they have assumed in battling to preserve their economies.

But the journey on the road to the future had commenced well before the coronavirus crisis, not only with renewables and electric vehicles, but also with the shale revolution that has transformed the energy position of the United States, shaken global markets, and changed America's role in the world.

And it is on that road where we now begin.

AMERICA'S NEW MAP

Chapter 1

———

THE GAS MAN

I f you want to get to the beginning of the shale revolution, pick up Interstate 35E out of Dallas and head north forty miles and then take the turnoff for the tiny town of Ponder. Pass the feed store, the white water tower, the sign for the Cowboy Church, and the donut store that's closed down. Another four miles and you're in Dish, Texas, population about 400. You end up at a wire mesh fence around a small tangle of pipes with a built-in stepladder. You're there—the SH Griffin #4 natural gas well. The sign on the fence tells the date— DRILLED IN 1998.

That was not exactly a great time to be drilling a well. Oil and gas prices had cratered with the Asian financial crisis and the ensuing global economic panic. But SH Griffin #4 would change things more than anyone could have imagined at the time.

The well was drilled mainly with standard technology, but also with experimentation and ingenuity, despite considerable skepticism. The small band of believers working on the well were convinced that somehow you could extract natural gas from dense shale rock in a way that was commercially viable—something that the petroleum engineering textbooks said was impossible. More than anyone else, the unshakable conviction belonged to one man, their boss—George P. Mitchell. He had been a true believer for a long time.

To grasp the intensity of that conviction, you have to understand that the road to SH Griffin #4 really begins much longer ago, in a tiny village in Greece's Peloponnesian peninsula.

In 1901, an illiterate twenty-year-old shepherd named Savvas Paraskevopoulos decided that his only ticket out of a life of poverty was to emigrate to the United States. By the time he ended up in Galveston, Texas, he had been rechristened Mike Mitchell. He eventually opened a laundry and shoeshine shop that just barely supported his family. His son George enrolled at Texas A&M University, where he studied geology and the relatively new discipline of petroleum engineering. George was poor, and this was the time of the Great Depression. To pay his way through school, he sold candy and embossed stationery to the other students, waited on their tables, and did tailoring on their clothes. He also captained the tennis team and came top in his class.

After World War II, Mitchell did not want to work for anyone else. With a couple of partners, he opened an office as a consulting geologist atop a Houston drugstore. By the 1970s, he had built a sizable oil and gas company, though with ups and downs along the way. But he had an unusual proclivity. He favored natural gas over oil.

Around 1972, he came across *The Limits to Growth*, a book by an environmental group, the Club of Rome. It predicted that a soon-to-be overpopulated world would run out of natural resources. Intrigued, he became increasingly interested in environmental issues. Natural gas became for him not only a business but also a cause, for it was cleaner than burning coal. Sometimes he would call up people and berate them if he thought that they had said something nice about coal.

Fueled by his new environmental ethos, he launched a totally different business—creating a wooded, landscaped, forty-four-square-mile master-planned community north of Houston called The Woodlands. Its slogan was "the livable forest." (Today it has a population over one hundred thousand.) Mitchell involved himself in the decision making down to the details of the flower beds and trees and populating it with wild turkeys (until one got shot).[1]

Yet he could hardly ignore his energy business. He had a big problem. Mitchell Energy was contracted to provide 10 percent of Chicago's natural gas. But the reserves of gas in the ground to support that contract were running down. Mitchell Energy needed to do something. That is when Mitchell stumbled across a possible solution.

In 1981, he read the draft of a journal article by one of his geologists. The article offered a hypothesis that ran counter to what was taught in geology and petroleum engineering classes. It suggested that commercial gas could be extracted deep underground from very dense rock—denser than concrete. This was the source rock, the "kitchen" in which organic material was "cooked" for several million years and transformed into oil or gas. According to the textbooks, the oil and gas then migrated into reservoirs, from which it could be extracted.

It was thought at the time that oil and gas might still remain in the shale but could not be produced on a commercial basis because they could not flow through the dense rock. The draft article disagreed. Mitchell, beset by worries about the contract for Chicago, became convinced that here might be the road to his company's salvation. There had to be a way to prove the received wisdom wrong.

The test area would be the Barnett Shale, named for a farmer who had come out to the area by wagon train in the mid-nineteenth century—five thousand square miles in extent, a mile or more underground, sprawling out beneath the Dallas/Fort Worth Airport and under the ranches and small towns of North Texas. Year after year, the Mitchell team toiled away to break the shale code. Their goal was to open up tiny pathways in the dense shale so gas could flow through the rock and into the well. To do that, they applied hydraulic fracturing, later much better known as "fracking," which uses cocktails of water, sand, gel, and some chemicals injected under high pressure into rocks that would break open tiny pores and liberate the gas. Hydraulic fracturing is a technology that had been developed in the late 1940s and has been commonly used in conventional oil and gas drilling ever since.

But here the fracking was being applied not to a conventional reservoir but to the shale itself. Yet time was passing, and much money was being spent, with no commercial results. Criticism mounted inside the company. But when people dared to suggest to Mitchell that his idea would not work, that it was only a "science experiment," he would say, "This is what we're going to do." And since he controlled the company, Mitchell Energy went on fracking in the Barnett, but still with no good result.

By the mid-1990s, the company's financial position was precarious. Natural gas prices were low. Mitchell Energy cut its spending and slashed its workforce. The company sold The Woodlands for $543 million. When the announcement was passed to him for review, Mitchell jotted, "OK but sad." He later said, "I hated to sell it." But he had no choice. The company needed the money. But Mitchell would not bend on shale. One thing that characterized him, as his granddaughter once said, was "stubbornness." If he had doubts, he kept them to himself.[2]

BY 1998, THE COMPANY HAD SPENT A LOT OF MONEY ON THE Barnett—as much as a quarter billion dollars. When analysts did forecasts of America's future natural gas supplies, the Barnett did not even make the list. "All sorts of experienced, educated folks wanted to bail out of the Barnett," said Dan Steward, one of the believers at Mitchell. "They said we were throwing money away."[3]

Nick Steinsberger, a thirty-four-year-old Mitchell manager in the Barnett, was not among the skeptics. He was convinced that there had to be a technical solution to commercially produce from shale. Moreover, natural gas prices were low, and he was also trying to bring down the costs of drilling a well. To do that, he had to attack one of the biggest costs—that of guar.

Guar, mostly imported from India, is derived from the guar bean. It is used extensively in the food industry to assure consistency in cakes, pies, ice cream, breakfast cereals, and yogurt. But it has another major use—in fracking, in a Jell-O-like slosh that carries sand into the fractures to expand them. But guar and the related additives were expensive. At a baseball game in Dallas, Steinsberger ran into some other geologists who had successfully replaced much of the guar with water, but in another part of Texas and not in shale. In 1997, he experimented with their water recipe on a couple of shale wells, without success.

Steinsberger got approval for one final try. This was the SH Griffin #4 in Dish. The team was still using water to replace most of the guar, but this time they fed in the sand more slowly. By the spring of 1998, they had the answer. "The well," said Steinsberger, "was vastly superior to any other well that Mitchell had ever drilled." The code for shale had been broken.

The new technique needed a name. They didn't want to just call it "water fracking." That would have been too prosaic, even boring. So they called it "slick water fracturing."

The company quickly adapted the technique to its new wells in the Barnett. Production surged. Yet if it was going to develop shale on a large scale, Mitchell Energy needed a lot more capital, which it simply did not have. Reluctantly, George Mitchell started a process to sell the company. Personally, it was a difficult time for him. Although he could take great satisfaction that his intuition—and conviction—had been proved right after seventeen years, he was being treated for prostate cancer and his wife was slipping into Alzheimer's. There were no buyers. The sales process was called off, and the company went back to work.

———

OVER THE NEXT TWO YEARS, MITCHELL ENERGY'S GAS OUT-
put more than doubled. This caught the attention of Larry Nichols,
CEO of Devon Energy, one of the companies that had passed on Mitchell Energy during the earlier sales process. Nichols challenged his own
engineers: "Why was this happening? If fracking was not working, why
was Mitchell's output up?" Devon's engineers realized that Mitchell
Energy had indeed cracked the code. Nichols was not going to let the
company get away a second time. In 2002, Devon bought Mitchell for
$3.5 billion. "At that time," said Nichols, "absolutely no one believed
that shale drilling worked—other than Mitchell and us."

But shale drilling needed another technology to be economic. This
was horizontal drilling. It allowed operators to drill down vertically
(today, as much as two miles) to what is called the "kick-off point,"
where the drill bit turns and moves horizontally through the shale.
This exposes far more of the rock to the drill bit, thus leading to much
greater recovery of gas (or oil). While there was experience with horizontal drilling, the technology did not become more prevalent until
the late 1980s and early 1990s. This was the result of advances in mea-
surement and sensing, directional drilling, seismic analysis, and in special
motors that would do a remarkable thing—a mile or two underground,
they would propel the drill bit forward once it had made its ninety-
degree turn and started moving horizontally. And it required one other
thing—extensive "trial and error." Devon was now positioned to try to
meld horizontal drilling with fracking.[4]

IN THE HOT SUMMER OF 2003, A LARGE GROUP OF GOVERN-
ment officials, engineers, experts, and executives from the natural gas
industry were convening, 750 miles to the north, in a cavernous con-

ference room at the Denver airport Marriott. The objective was to review the results of a major study on the future of U.S. natural gas. The conclusions were deeply pessimistic. After languishing for years, natural gas prices had suddenly moved up sharply. Demand was rising, especially in electric power. Yet despite a doubling in the number of active drilling rigs, the report said, the "sobering" fact was that "sustained high natural gas prices" were not bringing the expected increased supplies of natural gas. In short, the United States was running out of natural gas.

New technologies and "non-conventional" or "unconventional" gas, the study chairman told the group, would hardly have any impact. Shale gas did not even get a fleeting mention on the list.

A professor from the University of Texas shot up to object. He noted that this estimate for "non-conventional" was only about a third of another projection. "That's a hell of a big difference," he caustically commented. The chairman disagreed. The dissenting projection of larger potential supply, he said, was flat wrong.

"Somebody is dead wrong here, aren't they?" retorted the professor.

Almost everybody in the room was convinced that it was the professor who was dead wrong and that the United States faced a permanent shortage of domestic natural gas. The main way to make up for the shortfall was to look overseas—to import liquefied natural gas (LNG). The United States would have to do something new in its history: increasingly depend on large imports of LNG from the Caribbean, West Africa, the Middle East, or Asia. The country, it was thought, was destined to become the world's largest importer of LNG, ever more dependent on global markets for its gas, as it already was for its oil.[5]

Yet that July 2003, while the natural gas study was being deliberated in the air-conditioned ballroom in Denver, Devon's crews were working away in almost-hundred-degree temperatures down in Texas, methodically drilling what eventually totaled fifty-five wells.

Larry Nichols, the Devon CEO, missed the Denver meeting because he was focused on Devon's drilling program. "As we drilled each well and as we saw the continuing production of the wells, we realized a little more each day that this was indeed a game changer," Nichols recalled. "There never was a single Eureka moment. There were lots of small Eureka moments as we gradually improved our technology."

By the end of that drilling program, they had the proof. Devon's engineers had successfully yoked together the two technologies—slick water fracturing with horizontal drilling—to liberate natural gas imprisoned in the shale. "The rest was history," Nichols would later say.[6]

IT WAS AS THOUGH A STARTING GUN HAD GONE OFF. NEWS OF the breakthrough set off a frenetic race among other companies to get their piece of that dense rock before anyone else.

These were not the very large companies whose logos are familiar at gas stations across the country. Those "majors" were still divesting from their on-land U.S. production because they thought it was a dead end. Instead, they were putting their money into the Gulf of Mexico's deep waters and into multibillion-dollar "megaprojects" around the world. As they saw it, the U.S. onshore was too picked over, too obviously in decline, to provide new resources of the scale they needed.

The onshore was left to the independents—companies focused on exploration and production, unburdened with gas stations or refineries, more entrepreneurial, faster-moving, and with the lower cost structures required to make money in the increasingly depleted onshore. "Independents" itself was a pretty broad term, ranging from companies with multibillion-dollar valuations down to small scrappy explorationists.

The race was to lease as much promising acreage as possible from ranchers and farmers and then begin the process of proving up the resource. All shales, it was soon learned, were not the same; some were more productive than others. One wanted to find the "sweet spots," the potentially most productive acreage, before anyone else. The advance men of this particular revolution were the thousands of "land men" who knocked on screen doors and left notes in rural mailboxes and got landowners to trade their heretofore worthless mineral rights in exchange for the possibility of future royalties—and maybe riches.

The independents carried the race to other shales in Louisiana and Arkansas, in Oklahoma, and then to what would prove to be the greatest shale gas play of all, the mighty Marcellus, a thick bedrock a mile or more underground that stretches beneath western New York down into Pennsylvania and Ohio and on into West Virginia. It also reaches into Canada. The Marcellus shale would turn out to be the second-largest gas province in the world—and possibly the largest. And another shale formation called the Utica lay below parts of the Marcellus. What particularly drove the independents to move as fast as possible was that great motivator known as price. "After decades of being cheap and plentiful," the Wall Street Journal reported, "U.S. natural gas is the most expensive in the industrialized world." High prices motivated a lot of experimentation, investment, and risk taking that would not have been undertaken at lower prices. It involved mastering a production profile that differed from that of conventional gas (and then oil). Initial output from a new well was high, but then declined much more rapidly than in a conventional well before leveling off. This created the need to continue drilling new wells in what came to be described as a manufacturing process.[7]

The year 2008 was the moment when the bell rang. That year, U.S. natural gas output went up instead of down, as had been the general expectation. That abruptly caught the attention of the majors, the big international companies. Some of the majors began to shift some of their investment back to the onshore United States. In some cases, they bought independents. And a number of international companies—from China, India, France, Italy, Norway, Australia, Korea—paid up to become partners of U.S. independents and provide them with the cash they needed to continue their frenetic advance.

With this new perspective, estimates of the U.S. gas resource base rose dramatically. By 2011, the Potential Gas Committee, which analyzes physical resources for the United States, projected recoverable gas resources 70 percent higher than it had a decade earlier. That year, President Barack Obama declared, "Recent innovations have given us the opportunity to tap larger reserves—perhaps a century's worth—in the shale under our feet."

The numbers have continued to go up. By 2019, the Potential Gas Committee's estimate for recoverable natural gas reserves was triple what it had been in 2002. Gas production was rising so fast that it became known as the "shale gale." As gas moved from shortage into oversupply, the inevitable happened: prices plummeted. The combination of abundant supply and low price changed the overall U.S. energy mix, with gas's share of total U.S. energy rising.[8]

The most decisive change was in the electric power sector. King Coal had long been the dominant source for electric power, a position that had been bolstered by government policies in the 1970s and 1980s, which promoted coal as a secure domestic source of energy and restricted the use of natural gas for electric generation (because at that time, too, the country was thought to be running out of gas). In the 1990s, before shale, gas never accounted for more than 17 percent of generation. But, with the arrival of shale, gas was highly competitive on price, and environmental opposition had made it virtually impossible to build a new coal-fired plant in the United States. As late as 2007, coal generated half of U.S. electricity. By 2019, it was down to 24 percent, and natural gas had risen to 38 percent. That was the main reason why U.S. carbon dioxide (CO_2) emissions dropped down to the levels of the early 1990s, despite a doubling in the U.S. economy.

Any thought of expensive LNG imports had been banished. The challenge was no longer how to eke out scarce new supplies, but rather how to find markets for the growing abundance of inexpensive natural gas. There was just so much of it.

Chapter 2

THE "DISCOVERY"
OF SHALE OIL

One morning in 2007, Mark Papa was getting ready for a board meeting in Houston. He was CEO of EOG, one of the leading independents in the Barnett Shale. Looking at the slides of how much gas EOG alone had found in the Barnett, his mind wandered down a troubling path. The magnitude was changing, Papa thought to himself. As a company, EOG used to think in terms of "bcf"—billions of cubic feet—for its natural gas reserves. Now, with all its shale gas in the Barnett, it was talking about units a thousand times larger—"tcf"— trillions of cubic feet. And "tcf," until the Barnett, was usually reserved for something like measuring the total gas reserves of the United States, not those of a single company!

Other companies were finding similar amounts. Papa added up the numbers in his head. The result was alarming. "This is going to affect the natural gas market," he realized.

Papa had the slightly surprised look of a chemistry professor who had just realized he was late to teach his class. He had grown up out-side Pittsburgh, and then, after coming across a brochure from an oil company, decided to study petroleum engineering at the University of Pittsburgh. "This will sound unscientific," he once said, "but most of the areas where petroleum is found are relatively warm. I like warm."

Over the course of his career, Papa had learned to keep an eye on the "macro"—the big picture. He had once worked for an oil economist who closely followed OPEC and the fluctuations in the oil market. "I learned that you'd better pay attention to supply and demand," said Papa. "I love the supply-demand mechanics, the ebb and flow."

Now, putting aside the slides he was reviewing, Papa visualized what the supply-demand mechanics were saying. "It was just absolutely obvious," he said. "Gas is a commodity, and the gas price was going to fall like a rock. And we would be heavily impacted."[1]

EOG had only three options. It could go international, but then it would be competing against the likes of Exxon and Shell and BP, and that would be very tough since EOG did not have the scale, resources, or experience. Or it could venture out into the deep water of the Gulf of Mexico. But it had no expertise there.

Or it could go where it did have some expertise—in shale—and see if it could extract oil from the dense rock as it did natural gas. But that would push Papa into a position similar to that faced by George Mitchell—climbing up a high wall of skepticism. "Industry dogma," in EOG's words, flatly asserted that shale rock was too dense, even with fracturing, for oil to flow. According to that dogma, oil molecules were much larger than gas molecules, and thus would not be able to fit through the tiny pores that fracking would create in the rock.

That was not the only reason for skepticism. There was also the almost universal conviction that America's day as a petroleum producer was fast ebbing away. By 2007, U.S. oil production would be down to 5.1 million barrels per day, little more than half of what it had been at the beginning of the 1970s. Meanwhile, net oil imports had risen to almost 60 percent of consumption. Politicians may have promised "energy independence." But the real question seemed to be at what rate imports would continue to rise.

EOG needed to answer the specific question: Were oil molecules too big to flow through shale that had been fracked? They were clearly bigger than natural gas molecules. But how much bigger?

"Let's look it up," Papa announced. Surely, there had to be some research papers. Yet, strangely, the EOG team could find no research that quantified the size of an oil molecule.

They would have to do the research themselves. How large was a natural gas molecule, how large was an oil molecule, and how big before and after fracking would be the pore spaces—the tiny spaces or holes, invisible to the naked eye, in the rock? After investigating the matter with electron microscopes and a CT scan and thin slices of cores, they had the answer—an oil molecule was anywhere from slightly bigger than a gas molecule to seven times bigger. But, crucially, oil molecules of even that size could slide through the "throat" of the pore.

Papa called together his senior managers. "These guys were all geared up for finding gas," he said. "We had been successful beyond our wildest dreams." So it was a shock when Papa said that the price of gas was going to collapse and could be low for many years. He told the stunned managers that the company was going to stop looking for shale gas. Instead they should start searching for shale oil.

The room was silent. Papa braced for the pushback. There could have been a rebellion. They could have said, "Mark, you're out of your mind." But instead, they said, "Okay, Mark, we'll do that."

Yet Papa was in no hurry to advertise the change publicly. Not long afterward, he went to an investors' conference in New York City and listened as the other CEOs all talked about how much gas they had discovered—and how much more they would find. Papa thought to himself, "Those guys are ignoring Economics 101." For his part, he was deliberately vague in public about EOG's new thinking.

But inside EOG, it was different. "We went 180 degrees in the other direction, looking for oil," said Papa.

EOG ended up focused on the Eagle Ford Shale, which underlies South Texas. The Eagle Ford was regarded as the source rock—the "kitchen"—for other Texas oil fields, but it was considered to have little commercial potential of its own. Yet in their research, the EOG geologists came across seismic logs from very low-production wells called "strippers" that had been drilled decades ago. As they examined the logs, they became more and more excited. The production profiles of those old wells matched up with how shale wells performed—high initial production, then declining to steady production at a much lower level. The play, said Papa, "was begging for horizontal drilling." The geologists and petroleum engineers at EOG suddenly visualized something that could not have previously been imagined—120 miles of pure oil.

Papa sent out orders to lease as much land as possible, but also as quietly as possible. By the time they were done, EOG's land men had acquired half a million acres at $400 an acre. EOG thought it had acquired almost a billion barrels of oil. But as it began to drill, it found that it had greatly underestimated the reserves. Papa broke the news at an investment conference in 2010. "We believe horizontal oil from unconventional rock will be a North American industry game changer," he said. Once it became apparent what EOG had done, other companies rushed into the Eagle Ford. The land price shot up from EOG's $400 an acre to $53,000. By 2014, EOG had become the largest onshore crude oil producer in the United States.

Within a couple of years, it would become clear that Papa had understated shale oil. It was not only a North American game changer. It would be a global game changer.[2]

THEN THERE WAS NORTH DAKOTA. AFTER DECADES OF DRY holes across the state, oil had first been discovered in 1951 in the

Williston Basin by a company called Amerada, which later became part of Hess Corporation.

The resulting boom led to a *Time* magazine cover story that described the state as an "El Dorado" for future oil production. A monument dedicated in 1953 at the site of the Amerada discovery said it "opened a new era for North Dakota." But it turned out that there was no El Dorado and no new era. Despite a lot of drilling, not much oil was found, and the boom petered out. Nevertheless, Amerada (and later Hess) stayed in the state, adding to its acreage. "We kept finding other geologic horizons, and that kept us at the table," said John Hess, CEO of Hess. "We thought that changes in technology would enable us to get more oil out. We held on because we continued to believe in it. There's an old theory in the oil business—if you have an oil province where you have multiple shots, that's something you want to keep."[3]

A few others also suspected that significant oil could be found in North Dakota. That included an Oklahoman named Harold Hamm. He was an oil man to his core. "The oil business grabbed my mind and my young imagination," he said. "I wanted to find oil."

Hamm had grown up dirt poor, one of thirteen children of an Oklahoma sharecropper. As a child, he would help his family pull cotton bolls. Because the harvesting season extended beyond the opening of the school year, he would often be months late joining his grade and would have to work hard to catch up. Instead of going to college, he went to work around oil fields. In the beginning, his main skills were his work ethic, his intelligence, and a fierce drive to succeed. He was, as he described himself years later, "a hungry young man."

One of his jobs was hauling diesel and lubricating oil to drilling sites, which is how he met oil men. He talked to them about the business, and was tutored by them on how to read maps and logs and how to drill and complete wells. At age twenty-five, in 1971, he scraped together the money to acquire rights to an oil field that a company was selling off. He had a different view of its prospects. "I had been in the business for five years," he said, "and I had a very strong conviction." He hit oil. He was launched. He also recognized that he had to catch up. He spent his evenings poring over books on geology and geophysics and took courses when he could at a local college. "I didn't go for the degree, but for the education," he said. He sold his first company in 1982. He started a drilling company. He also had his failures, including drilling seventeen dry holes in a row. But he persevered and built up a company that he named Continental Resources.

In the mid-1980s, he started looking for oil in the Williston Basin that straddles Montana and North Dakota. Continental discovered two fields on the Montana side using horizontal drilling. In 2003, Continental started acquiring acreage on the North Dakota side of the basin.

The shale revolution in natural gas was just getting going. Could this technology be applied in North Dakota? At a depth of two miles, sandwiched between a number of other strata, was a formation called the Bakken, named for a local farmer, and just below it the Three Forks. Though technically categorized as "tight sands," these are similar to shale and are usually called shale. Until the shale revolution, they were skipped over—no value. "People thought you could never produce from the Bakken," said John Hess. But what was happening in the Barnett in Texas suggested otherwise.

The answer in terms of technology was horizontal drilling in the form of "stages." Rather than trying to frack the length of the entire horizontal well all at once, the drillers would do so in stages, learning and experimenting and adjusting to the specific rock as they went. Doing this in stages two miles underground along a two-mile horizontal track took more time and cost more. But it *could* work. And by 2009, it *was* working.

The Bakken took off. In 2004, North Dakota had produced a grand total of 85,000 barrels per day. By 2011, it had more than quadrupled to 419,000 barrels. North Dakota overtook California as the third largest oil-producing state in the country, and then Alaska as the second largest, behind only Texas. By 2014, North Dakota was producing 1.1 million barrels per day—a fourteen-fold increase from a decade earlier. It turned out that *Time* magazine was absolutely right in predicting an "El Dorado" in North Dakota; it had just been sixty years too early.[4]

The oil boom in North Dakota gave a great boost to the state's economy and to state government revenues. Economic growth surged and so did incomes. Farmers operating on the margin but who owned mineral rights had an infusion of money. During the post-2008 years of high unemployment in the United States, North Dakota had the lowest rate, and the out-of-work migrated into the state.

But the rapidity and scale of the boom created its own problems— a shortage of housing, and overcrowded roads, schools, hospitals, and even courts. Also, North Dakota was not sufficiently connected to pipelines, and that meant large volumes of oil had to move by rail in trains with as many as one hundred cars. The amount of oil transported by rail in the United States went from 50,000 barrels per day in 2010 to more than 1 million in 2014. That business was much welcomed by railroads, whose coal-hauling business was in decline.

One of the most unusual challenges in the Bakken turned out to be birds. The U.S. Department of Justice, acting on a complaint from the U.S. Fish and Wildlife Service, brought a criminal indictment against Continental and two other oil companies for the death of twenty-eight migratory birds. In Continental's case, the entire death toll totaled one bird, of a species known as Say's phoebe, which the Cornell University Lab of Ornithology describes as "common around people, often nesting on buildings." By comparison, according to the Fish and Wildlife Service, half a million birds a year are killed by wind farms, sixty million birds a year by cars, and one hundred million by flying into windows. A federal judge finally threw out the case in 2012, saying a conviction would criminalize many everyday activities, including trimming and cutting trees, harvesting crops, driving a car, and owning a cat (estimated to be responsible for up to 3.7 billion bird deaths a year in the United States).[5]

AFTER THE BAKKEN AND THE EAGLE FORD CAME THE PERMIAN— the biggest of all. The Permian Basin sprawls across seventy-five thousand square miles in West Texas down into southeastern New Mexico. Much of it is characterized as a "featureless high plain." It draws its name from rocks that are characteristic of the Permian geologic age, which ended with the "great extinction" that wiped out most living creatures about 250 million years ago. The name itself was derived from the Russian city of Perm, where in the nineteenth century a British geologist had identified rocks of that geologic era.

At the beginning of the twentieth century, the parched Permian region was dismissed as a "petroleum graveyard." In 1920, the Permian was said to have "little to recommend it . . . as a potential oil province."

The first successful well came in 1923 on lands the state had endowed to the University of Texas—the "Santa Rita 1," named for the "Patron Saint of the Impossible." But subsequent wells were disappointing.[6]

Then, in October 1926, on a lease that had almost expired, a discovery opened up the Permian as a great oil province. The Permian would also become one of America's most important assets during World War II, as its production literally doubled to meet the wartime need for fuel. After the war, the Permian boomed yet again. The region and its oil business became a magnet for young men seeking their chance, among them a Navy veteran and Yale university graduate, George H. W. Bush, who moved there with his wife, Barbara, and baby, George.

Every day, the independents were rolling the dice. "If I hit," Bush said at the time, there was money to be made. "If I didn't hit, it's my hard luck." In 1974, the basin—really a collection of several different giant oil fields—reached its peak, providing almost a quarter of total U.S. oil supplies.[7]

But thereafter output in the Permian began a precipitous decline, hitting a low point in 2007. The patron saint of the impossible was no longer there to help out, and many were the last rites said over the region. "The role of the Permian Basin as a major oil producing province thus appears to be past," wrote one geologist in 2006, and its future "can thus be only one of continuing decline."

Yet by then, rising oil prices were starting to stimulate renewed activity in the Permian. The number of drilling rigs increased, and by 2011 it was getting harder to find a free table at the Wall Street Bar and Grill, a favorite eatery for oil people in Midland. But the new drilling was still the traditional vertical wells.

January 2011 marked the beginning of the "Arab Spring," which brought tumult to the Middle East and North Africa—and much uncertainty about that region's future. That same month, the title of a new report announced that the U.S. petroleum industry was changing—"The Shale Gale Goes Oily." The main case study was the Bakken. But it also called attention to a potential mega-shift—that "operators are taking a second look in their own backyard" in order to ascertain whether the new technologies could be applied in "existing fields" that were considered past their prime. The biggest backyard was the Permian Basin.[8]

In November 2011, the board members of Pioneer, a large independent, gathered in their conference room in Dallas to hear a three-hour presentation by the company's geologists. Pioneer's fortunes had mirrored the industry's. It had ventured into the deep waters of the Gulf of Mexico and internationally, developing projects in countries ranging from Argentina to Equatorial Guinea. In 2005, it decided to begin to sell off its international projects and come home. "The political arena and cost structure in our various assets outside of the lower forty-eight onshore in the United States were becoming too risky," said CEO Scott Sheffield. They could also see the success that other companies were having in the Barnett Shale. Better to plow money back into the United States, where contracts were generally observed and courts independent, than deal with foreign governments that could unilaterally change the terms under which a company operated.

For two years, Pioneer's geologists had studied the shales under Pioneer's nine hundred thousand acres in the Permian. Their conclusion was startling. Under it lay a potential bonanza—not just one layer of shale, but layer upon layer of tight rocks stacked on top of each other like pancakes a mile or two beneath the surface, whose oil could be made to flow in abundant volumes with hydraulic fracturing and horizontal drilling. "That," said Sheffield, "was the aha moment." Pioneer abruptly redirected its spending to that resource. In 2012, it drilled its first successful horizontal shale well in the Permian.[9]

Pioneer was only one of a host of companies that jumped on the new opportunity. Once again, the region was booming. The shortage now was not of oil, but of workers and housing and office space. Plans were made for a fifty-three-story office building in Midland that would be the tallest skyscraper between Houston and Los Angeles. Production skyrocketed. By 2014, Permian output zoomed from that low point of 850,000 barrels in 2007 to 2 million—almost 25 percent of total U.S. crude oil output.

Altogether, in a very short time the new technology transformed Texas, putting it on an extraordinary growth path. Between January 2009 and December 2014, the state's total crude oil output more than tripled. By this time, Texas was producing more oil than Mexico, and more than every OPEC country except Saudi Arabia and Iraq.

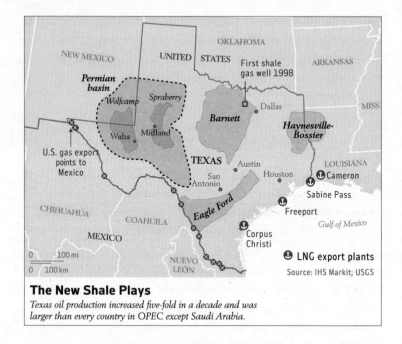

The New Shale Plays

Texas oil production increased five-fold in a decade and was larger than every country in OPEC except Saudi Arabia.

The unconventional revolution also transformed the map of oil resources. One area in the Permian—known as the Spraberry and Wolfcamp—was now deemed the second-largest oil field in the world, behind only Ghawar, Saudi Arabia's supergiant field. The Eagle Ford was ranked fifth, behind Burgan in Kuwait and another Saudi field, and ahead of the giant Samotlor field that is the foundation of Russia's oil might.

The United States was back, once again a major player in world oil.

Chapter 3

"IF YOU HAD TOLD ME TEN YEARS AGO": THE MANUFACTURING RENAISSANCE

St. James is a rural parish in Louisiana, on the banks of the Mississippi River. Its rich soil supports the sugarcane plantations that are the backbone of the local economy. It is known for the bonfires that are lit on Christmas Eve on the levees along the river, which, according to legend, are to welcome "Papa Noel," otherwise known as Santa Claus, and to help him avoid getting lost as he makes his way down the Mississippi bearing his satchel of gifts.

On an autumn Friday night in 2015, a different kind of ceremony was held in the local high school, this one to warmly welcome to the parish a new guest, who was carrying in his satchel a present of sorts—a large-scale investment in the parish, of a kind that had never before been seen. This visitor was Wang Jinshu, the chairman of Yuhuang Chemical Company, which is headquartered halfway around the world, in Shandong Province, China.

Wang had come to initiate the $1.9 billion first phase for a chemical facility that Yuhuang was building in St. James. Yuhuang had purchased not only thirteen hundred acres of sugarcane but also the adjacent high school in which the ceremony was being held, enabling the parish to build a new, more modern high school. Over time, the project would mean a lot of new jobs in St. James Parish and more income.

What brought Yuhuang to the parish was inexpensive natural gas. It was more economic for the company to take advantage of a pipeline that brought in shale gas, make the chemicals in Louisiana, and ship them to China, than to build a similar facility in China. A Yuhuang executive cited many reasons for the project—from the need for its product to the "beneficial" impact on U.S.-China relations and its alignment with the policies of Chinese president Xi Jinping. But the basis of it all was more down-to-earth—a twenty-year contract for inexpensive natural gas.[1]

In 2019, with the project 60 percent completed and a second phase planned but amid the U.S.-China trade war, Yuhuang prudently brought in a U.S. company as a joint-venture partner. But what had unfolded that night four years earlier, in 2015, in the high school in St. James Parish, was part of a much bigger story—America's manufacturing revival and its increased competitiveness in the world economy.

WITH THE UNCONVENTIONAL REVOLUTION, AMERICA'S EN-ergy position looked very different from what had been expected just a few years earlier. U.S. natural gas production was growing dramatically. The same was true for oil. Imports of oil were rapidly declining, as was the money that the United States spent on importing oil—all of which was reducing the U.S. trade deficit. Yet the impact of the shale revolution on the American economy went even further.

In 2014, Ben Bernanke, just retired as chairman of the Federal Reserve, described the unconventional revolution as "one of the most beneficial developments, if not the most beneficial development" in the American economy since the 2008–2009 financial crisis. This impact was amplified by the nature of the economic flows. The surge in economic activity stimulated by shale gas and oil, combined with the steep decline in imports, sent the benefits reverberating through supply chains and financial links right across the U.S. economy. This was very different from money flowing out of the country to support development elsewhere or ending up in the sovereign wealth funds of exporting countries. This domestic circulation of money would greatly multiply the impact.

Between the end of the Great Recession, in June 2009, and 2019, net fixed investment in the oil and gas extraction sector represented more than two-thirds of total U.S. net industrial investment. In another measure, between 2009 and 2019, the increases in oil and gas have accounted for 40 percent of the cumulative growth in U.S. industrial production.

In practical terms, that means money flowing into paychecks throughout the country. By 2019, the unconventional revolution was already supporting over 2.8 million jobs.* There were jobs in and around oil and gas fields, manufacturing jobs in the Midwest making equipment and trucks and pipes, jobs in California writing software and managing data, and jobs generated by increased income and spending, like real estate agents and car dealers. What is striking is that, owing to the linkages, the economic impact was felt across virtually all states. This was true even in New York state, where environmental activists and politicians succeeded in getting the state to ban hydraulic fracturing and prevent a new natural gas pipeline that would have carried inexpensive natural gas from the Marcellus in Pennsylvania to gas-short New England. The lack of new pipelines resulted in a prohibition in 2019 on gas hookups for new housing and small businesses in Westchester County, just north of New York City. Yet even New York registered over forty thousand jobs that were supporting shale activity in other states.[2]

---

* Altogether, before the COVID-19 shutdown of the economy, the entire oil and gas industry was responsible for 12.3 million jobs in the United States.

All of this incremental economic activity generates a lot of federal and state revenues, estimated to be $1.6 trillion between 2012 and 2025.

SHALE HAS GENERATED NOT ONLY REVENUES BUT ALSO ENVI- ronmental controversy and opposition as it grew. As with most major industrial activities, environmental issues around shale need to be prop- erly managed. In the early years of the shale revolution, the controversy was particularly focused on water contamination, either from the frack- ing process itself or the disposal of wastewater that comes out of the well. A decade later, as Daniel Raimi observes in his book *The Fracking Debate*, water contamination has proved not to be the systemic prob- lem that some feared. To begin with, the fracturing itself takes place several thousand feet below freshwater aquifers. There was also the view that shale was a "wild west" activity. But shale production, as with the rest of the oil and gas business, is highly regulated, in this case pri- marily at the state level. Some states needed time to ramp up their regulatory apparatus as shale development became significant in their area. Earthquakes were another concern, particularly after swarms were felt in Oklahoma. Follow-on studies attributed these quakes not to drilling but rather to disposing of wastewater in inappropriate locations, causing slippage of rock formations and thus quakes. With new regula- tion of where wastewater could be disposed and at what pressure, the number of earthquakes fell sharply. Much has been learned about man- aging the impacts on rural communities, including noise and the num- ber of trucks on local roads, while at the same time meeting those communities' needs for jobs and new sources of income.

The most significant question today concerns "fugitive" methane emissions—basically, natural gas leaking from equipment or pipelines— which is not limited to shale. The Environmental Defense Fund was among those at the forefront in directing attention to methane as a significant greenhouse gas. Reducing those emissions is now a priority for both regulators and industry and a particular focus of the thirteen-company Oil and Gas Climate Initiative. Moreover, the International Energy Agency notes, "Methane is a valuable product and in many cases can be sold if it is captured."[3]

THE EFFECTS OF THE SHALE REVOLUTION ON THE TRADE PO-sition of the United States are striking. Using 2007 as the baseline comparison, the U.S. trade deficit in 2019 was $309 billion lower than it would have been if there had been no shale revolution. Without shale, the United States would have continued to be the world's largest oil importer. It also would have become a large importer of LNG, competing for supplies with China, Japan, and other countries, adding greatly to the trade deficit.[4]

The shale revolution also dramatically improved the competitive position of the United States in the world economy. For years, industrial investment flowed out of the United States to countries that were lower-cost because of lower labor costs. But the tide turned. Over $200 billion is being spent on new and expanded U.S. chemical-related facilities.[5] Tens of billions of dollars more are going into steel fabrication and other manufacturing and processing plants, as well as refining and infrastructure. The primary reason is the abundance of low-cost natural gas. It is used both as a fuel and as a raw material for making chemicals. It also helps lower the cost of generating electricity.

For years, investment by the chemical company Dow had been outward bound, primarily to the Middle East, in the quest for access to cheap natural gas as a raw material for its products. But the advent of inexpensive gas in the United States pulled it back home. The company has since committed billions to expanding or building new petrochemical facilities in the United States. Announcing a $4 billion expansion in Texas in 2012, Andrew Liveris, Dow's then–chief executive, said in

2012, "Things change. We pivoted very fast." He added, "If you had told me ten years ago I'd be standing up on this podium making this announcement, I would not have believed you."

But it's not only U.S. companies. European manufacturers are escaping the burden of Europe's high energy costs to invest in the United States. In announcing a $700 million investment in Corpus Christi, Texas, the CEO of an Austrian steel manufacturer explained at the time that the low U.S. gas price compared to Europe "is the big economic advantage." The migrants include fertilizer companies from Australia and plastics companies from Taiwan. After decades of U.S. companies setting up factories in China, Chinese manufacturing companies were starting up new manufacturing facilities in the United States, of which Shandong Yuhuang, in the sugarcane region of Louisiana, is a case in point.

Inexpensive energy was not the only reason, of course. But for many companies—American and foreign—abundant low-cost natural gas—and the expectation that it will last for a long time—is decisive. All this makes shale gas a key contributor to what has been called the "manufacturing renaissance" in the United States and to the increased competitiveness of the United States in the world economy.[6]

Chapter 4

———

THE NEW GAS EXPORTER

t took two phone calls in 2009 to convince Charif Souki to turn
around the business he was trying to build. One was from the hard-
charging CEO of independent Chesapeake, one of the companies at
the forefront of shale gas development; the other, from one of the largest companies in the world, Shell. Both had the same question—could
Souki transform the facility he was building to import LNG into a
plant to export the growing supply of U.S. gas?

Souki was taken aback. He had bought into the consensus of the
early 2000s about the gas shortage and had raised hundreds of millions
of dollars and signed complex contracts on the premise that the United
States would have to import very large amounts of LNG. The calls suggested that he had made a big very bad bet.

Souki, with his longish hair, double-breasted suits, and traces of
an accent, did not exactly fit the profile of a wildcatter in the oil and
gas patch. He had grown up in Beirut, where his father was the well-
connected Middle East correspondent for *Newsweek*. Souki had begun

his career working for an investment bank in the Arab world, honing his skills of persuasion. Returning to the United States, he became an investment adviser, then opened restaurants in Aspen, Colorado, and Los Angeles before ending up in Houston, where he put together a company to explore for natural gas. He named his company Cheniere—a Cajun word for the raised ground in a swamp.

Cheniere got nowhere as an exploration company. But it convinced Souki, like many others, that America was running short of natural gas, leading him to the audacious idea of importing LNG from around the world. Audacious was actually an understatement. Souki had been a restaurateur, he had no money but would need billions of dollars, and he was going to try to make deals with the world's largest oil and gas companies and with major exporting nations. Though short of money, he was not short of confidence. Still, he was a novice trying to break into a big global business that was already more than forty years old.[1]

IN FEBRUARY 1959, THE *JOURNAL OF COMMERCE*, IN A STORY headlined "Cargo Ship with Methane on High Seas," announced that a converted World War II freighter, renamed the *Methane Pioneer*, had set sail from Louisiana for England. It carried a cargo that had never before been shipped over the seas—liquefied natural gas—LNG. Liquefied natural gas is the product of a complex process that refrigerates natural gas to extreme cold, down to minus 260 degrees Fahrenheit, thus compressing it into a liquid. Since in its liquid form the gas takes up only one six-hundredth of the space that it would in its gaseous state, it can be pumped into tanks on refrigerated ships and transported across oceans and then "regasified"—turned back into gas—at the other end and pumped into a pipeline system in the receiving country.

The technology had been developed during World War I. But it was only after World War II that experiments began to liquefy gas in order to transport it. The real spur was the killer fog that enveloped London in 1952. Burning cleaner gas instead of coal to generate electricity would help alleviate pollution, and LNG could be the source of that gas. It took time to work out the designs and find the right materials for the tanks. By 1959, the *Methane Pioneer* was ready to sail. This shipment, the head of the new company said, "is the prelude to a new era when natural gas, previously wasted or shut in for want of accessible markets in many parts of the world, will be liquefied and transported by tanker to countries where gas is not naturally available." That was a pretty good description of what would unfold over the next several decades.[2]

Yet things did not go quite as expected. The major market for LNG in Britain and in Europe largely evaporated with the discovery of the huge Groningen natural gas field in the Netherlands and then additional gas in North Africa and in the seabed off the east coast of Britain.

The growth market for LNG turned out to be on the other side of the world, in the East Asian "economic miracle"—Japan, South Korea, and Taiwan. To lower their dependence on Mideast oil for generating electricity and increase energy security, and to reduce pollution, those countries entered into complex contracts for LNG from Indonesia, Malaysia, and the sultanate of Brunei. Also, a small LNG facility in Kenai, Alaska, would intermittently ship supplies to Japan.

This new LNG business required very large investments—eventually billions of dollars—to find and develop and pump the gas; to construct the plants that, at one end, would liquefy the gas and, at the other, regasify it; and to build the specially constructed tanker ships that would ply the thousands of miles of ocean in between. Given the amount of money, participants in the market required confidence about the long term. Thus a highly interconnected business model developed, in which the various partners would coinvest up and down the supply chain and gain predictability via twenty-year contracts. Molecules from a particular field in Indonesia or Brunei or Malaysia would end up in specific power plants in Japan, Korea, or Taiwan. There was no buying and selling along the way, no redirection, no middlemen. Prices were indexed to the price of oil. If oil went up, the LNG price would go up. If the oil price went down, the gas price would also go down.

It was on this basis that the LNG industry turned into a big business. For a number of years, it was largely Asia-bound. Then the emirate of Qatar transformed it into a global business. Qatar is a flat, sandy peninsula that projects out into the Persian Gulf from the eastern side of Saudi Arabia. For much of the twentieth century, it was a poor country, eking out a living from fishing and pearl diving. That started to change when modest oil production began in the late 1960s. But the rapid development of the North Field, offshore of Qatar, would transform its economic position and its global importance. The North Field is considered the world's largest gas field. Separated only by a demarcation line on the map is Iran's huge South Pars field.

Qatar and the companies it partnered with introduced ever-greater scale into every phase of the LNG operations, including tanker size. The objective was to be able to competitively ship gas anywhere in the world. By 2007, Qatar had overtaken Indonesia to become the world's largest supplier of LNG. It was poised to begin large exports to the United States to help allay the anticipated domestic shortage of gas that had so gripped the U.S. energy industry.

This was the global business into which Souki wanted to jump. He set out to build a regasification facility—or several of them. They would take the natural gas, which had been liquefied in Qatar or Trinidad or somewhere else, and turn it back into gaseous form so that it could be put into a pipeline and sent on to U.S. consumers.

For his new terminals, Souki identified sites on the U.S. Gulf Coast. But he was still missing something very important—money. Two dozen investors showed him the door with varying degrees of politeness and incredulity; only one yelled at him. But he did know someone who had capital, another unusual entrepreneur—Michael Smith.

Smith had originally moved to Colorado to study veterinary medicine. Instead, he had ended up dabbling in Colorado real estate. Then he heard about an oil discovery and invested $10,000 in some nearby oil leases. The company he built was eventually sold for $410 million. He then went right back into the energy business, this time offshore in the Gulf of Mexico. Souki's pitch coincided with his own thinking— that a gas shortage was coming. U.S. gas, and Smith's gas production in the Gulf of Mexico, was, as he later said, "falling flat on its face."[3]

Souki and Smith worked out a partnership. Smith took controlling interest in one of the proposed sites, Freeport, about seventy miles south of Houston. Souki pushed ahead on a project at Sabine Pass in Louisiana, on the border with Texas. Two international majors signed twenty-year contracts to use Cheniere's Sabine Pass facility to regasify their LNG shipped in from the other parts of the world. The financial markets were now taking Cheniere seriously. Its stock price rose twenty-five-fold and then split. Michael Smith brought in major investors for his Freeport facility. Construction began at both sites. By 2007, dozens more regasification projects were being proposed by other groups. In 2008, natural gas prices reached a high point of almost $9 per thousand cubic feet, providing further "proof" of a shortage and thus increased urgency to import LNG. Yet by 2008, skepticism was emerging about the financial strength of Cheniere, and its stock price was falling. Souki himself was becoming depressed about the prospects for his business as he kept reading and hearing about more new gas discoveries in the United States, which could mean less demand for LNG imports.

Then, in the spring of 2009, came the call to Souki from Aubrey McClendon, CEO of Chesapeake, who was at the forefront of the shale gag boom and had built up a huge inventory of drilling sites.

"Hey, can you guys do liquefaction at Sabine Pass?" asked McClendon.

"Why are you asking?" replied Souki.

McClendon became more explicit—could Cheniere build an export terminal for Chesapeake so that it could find markets outside the

United States for its swelling volumes of gas and help relieve the growing pressure of oversupply?

Despite his growing worries, Souki was still stunned; Cheniere had
been going all-out on an import terminal, not an export terminal. And
building an export liquefaction terminal could be literally ten times as
expensive as an import regasification facility. Then Shell called to ask
the same question. This had to be taken seriously, for Shell was no
entrepreneur; it was a supermajor oil and gas company, and one of the
leaders in LNG. These calls sounded the alarm that U.S. supply was
growing much faster than the market could absorb—meaning there
would be no market in the United States for imported LNG.

The Cheniere team started working on the numbers. By the spring
of 2010, Souki presented to the Cheniere board a plan for turning Sabine Pass into a liquefaction facility—at least $8 billion for the first
phase. The board was surprised. It seemed too good to be true. But,
the board concluded, the numbers worked.[4]

Under the Natural Gas Act of 1938, the federal government had
to approve gas exports. Cheniere's application was thorough and did
not garner much attention. After all, the whole plan was regarded as
not realistic. The application went through smoothly, and, in 2011,
Cheniere received its approval. That same year, it got the first of several commitments to buy LNG from Sabine Pass. Buyers ranged from
Spain to India.

Michael Smith recognized the same stark change owing to the
buildup of shale gas. "The enthusiasm for regasification was gone," he
said. "The market was gone." Shortly after Cheniere, Freeport put its
application in to the government to transform its import facility into
an export facility. But unlike Cheniere, it did not get a quick approval.
Nothing seemed to be happening. Someone explained to a frustrated
Smith, "In Washington, the first application is an application. The second application is public policy." What had taken Cheniere nine months

would take Freeport almost four years. The same proved true for another first-mover project, Sempra LNG at Cameron, Louisiana, as well as other newer projects. They would all have to wait.[5]

After the fact, the Cheniere approval had set off a storm of criticism and opposition. Senators thundered against the decision. Some manufacturing companies, notably from the chemical industry, feared that the export facilities would divert gas supplies that they were counting on and drive up the cost of gas, threatening the billions of dollars they were investing in new plants. They were joined by unlikely allies—environmental groups that opposed the development of shale gas altogether. One environmental organization methodically registered an official objection to virtually every single export application going through the regulatory process.

But the opposition from the manufacturing companies dissipated in the face of the evidence—a continuing increase in natural gas supplies and the persistence of low prices. What finally quelled the controversy and alleviated industry fears was the enunciation of an export policy by the Department of Energy. While declaring that the market provides "the most efficient means of allocating natural gas supplies," it pledged "intervention" that would "protect the public" in the event of a shortage. Such intervention was highly unlikely, but the protection was there. Approvals were forthcoming, and Freeport could finally begin construction on its $13 billion project in 2014. Sempra started the same year.[6]

NO COUNTRY HAS BENEFITED MORE FROM THE GROWING global LNG business than Qatar. Today it has the highest per capita income in the world, and a sovereign wealth fund of $350 billion— all for a country with about three hundred thousand citizens (and more than two million foreigners who work in Qatar). The LNG wealth finances its Al Jazeera global television network. It also finances an educational hub and the Mideast campuses of several educational institutions—Weill Cornell medical school, Georgetown University, Texas A&M, Northwestern, and Carnegie Mellon, along with Canada's University of Calgary, University College London, and the University of Aberdeen.

The LNG business is becoming ever more global. New export projects are being developed in Egypt, Trinidad, Oman, Israel, Angola, Nigeria, Canada, Mozambique, and Russia. The much-expanded scale and the growing list of new buyers has changed the market. Long-term contracts remain, and new ones are being signed. But some LNG began to be traded differently—sold on a short-term basis. Cargoes would set out for one destination, and then, as a new bid came in, change course for another, and then change course again. LNG was now no longer only an integrated business; it was also becoming a competitive market of buyers and sellers.

In 2019, after an investment of more than a quarter trillion dollars, Australia overtook Qatar to become the largest LNG supplier. Qatar was not going to stay in second place. It lifted a self-imposed limit and announced plans to add more capacity to regain the number one position.

But a new era for LNG began in February 2016. After what had turned into an investment totaling $20 billion, the first shipment of U.S. LNG left Cheniere's Sabine Pass for Brazil. From then on, like clockwork, every few days tankers almost a thousand feet long were departing the dock at Sabine Pass for the rest of the world. Suppliers and consumers of natural gas in Asia, Europe, Africa, Latin America, North America, and Australia were now linked together in a global network of trade.

Freeport and Sempra Cameron started exporting in 2019. Several more plants are under construction in the United States. Altogether this will catapult the United States into becoming one of the top LNG exporters, along with Qatar and Australia.

———

WITH THE ARRIVAL OF THE TRUMP ADMINISTRATION, LNG
became a tool in conflict over trade. This administration—and Donald
Trump himself—was obsessed with deficits in the trade balance with
individual countries, and no deficit loomed larger than that with China.
The administration seized on U.S. energy exports, specifically LNG,
as a way to help reduce trade deficits. Previous administrations had also
promoted U.S. exports ranging from Boeing jets to corn and pork.

What was different, however, was that Donald Trump personally
turned himself into America's top LNG salesman. When India's prime
minister Narendra Modi visited Washington, Trump told him that he
was looking forward to "exporting more American energy to your
country," including "major long-term contracts to purchase American
natural gas, which are being negotiated right now." He added a laugh
line: "Trying to get the price up a little bit."

That was on a Monday. On Friday of the same week, Trump was
hosting the president of South Korea, Moon Jae-in, in Washington. South
Korea was yet another country whose trade surplus with the United
States rankled Trump. The meeting was meant to focus on the nuclear
threat from North Korea. But the president wasted no time in giving
top billing to renegotiating the U.S.-Korea trade relationship—"a rough
deal for the United States," he called it. "The United States has many,
many trade deficits with many, many countries and we cannot allow
that," the president said. "We will start with Korea right now." In fact,
that message had already gotten through. Five days earlier, Korea, one of
the world's biggest buyers of LNG, had signed a twenty-year contract
for U.S. LNG worth more than half a billion dollars a year.[7]

These foreign leaders might be a little confused by Trump's comments, since the U.S. government does not itself negotiate LNG contracts. But the message was clear: It behooved other governments to
push their companies to buy U.S. LNG.

Yet the stance of the Trump administration created some perplexity. "For many years, we have been arguing with the Russians and Chinese not to see energy trade in political terms," said the CEO of one of the major oil companies. "But now the U.S. president is doing exactly that, and the Russians and Chinese can say to us, 'We told you so.'"[8]

Chapter 5

———

CLOSING AND OPENING: MEXICO AND BRAZIL

The U.S. shale revolution extends into Mexico. But it is mostly not natural gas in liquefied form—LNG—that is being exported. Through 2019, U.S. exports of natural gas to Mexico via pipelines were larger than all of the LNG exports combined. The United States is delivering 60 percent of Mexico's total gas supply and 65 percent of its gasoline. This is part of the new map of North American energy integration.

Since the nationalization of the Mexican petroleum sector in 1938, that industry had been a government monopoly. Pemex, the state oil company, was responsible for everything from drilling to gas stations. Mexico had been one of the world's major petroleum producers, with revenues from petroleum providing 30 to 40 percent of the national budget. But output has been on a steep decline. The industry was suffering from a lack of technology, a huge shortfall in investment, an agonizing debt burden, bureaucratic rigidities, corruption, and the stranglehold of the powerful workers' union. Without major reforms, and absent an opening up with the world, the deterioration would continue.

Mexico simply could not come up with the investment required. At the same time, the country was emerging as a major global manufacturing export platform. Nissan announced that it expected to produce more cars in Mexico than in Japan. By 2018, Mexico was the seventh largest car manufacturer in the world, and the fourth largest auto exporter, after Germany, Japan, and South Korea. Trade with the United States rose from $248 billion in 2000 to $614 billion in 2019, making it the United States' largest trading partner. But unreliable and high-cost energy supplies hobbled Mexico's drive to be competitive in the global economy, damaging economic growth and job creation. In the U.S. waters of the Gulf of Mexico, a large offshore industry was producing substantial amounts of oil. In the adjacent Mexican sector, despite similar geology, there was virtually nothing. Pemex had neither the cash nor the technology to venture into the deep water or even to fully develop its shallow-water reserves.

The Eagle Ford geological shale trend continued down into Mexico. But there was no domestic capability to unlock Mexico's shale rocks. Yet the speed and scale of the shale revolution in the United States added further urgency to develop Mexico's resources.[1]

The Institutional Revolutionary Party (Partido Revolucionario Institucional, or PRI) had dominated Mexican politics for most of the twentieth century, and it was a PRI president who had nationalized the industry in 1938. It took a new PRI president—Enrique Peña Nieto—to hammer out a consensus for reform of the oil industry. The constitutional amendment of December 2013, while reasserting Mexico's sovereignty over the subsoil, opened its development to Mexican and international companies. Pemex's monopoly was over. Subsequent laws enabled companies to bid against one another for rights to drill and created electric power markets. The energy sector was now open to competition.

The result was an inflow of investment and technology from both Mexican and international companies. It was estimated that the 107 upstream contracts signed with Mexican and foreign companies would, if all worked out, generate over $160 billion of investment. New, modern gasoline stations sprouted across the country. New pipelines and power plants were built. Shale gas from the United States flowed into Mexican electric power plants, replacing higher-cost LNG and oil. The foundations were now in place to drive down the cost of electricity to consumers and industry and make Mexico more competitive in the world economy.

Yet not everyone supported the reform. The most outspoken critic was Andrés Manuel López Obrador—AMLO for short—a longtime activist who at one point had lived in a shack in the tropical state of Tabasco, working with the indigenous people. He later became the mayor of Mexico City and then in 2018, on his third try, won the Mexican presidency. His populist platform of "self-sufficiency" harked back to the oil nationalism of the 1970s and the once-popular "third-world" ideas of *dependencia*, which railed against integration with the global economy and resulted in decades of high inflation and low growth in Latin America—and back to the 1930s and the original nationalization of the Mexican oil industry.

AMLO canceled a new airport project for Mexico City, on which $5 billion had already been spent, after "consulting with the people" before he became president in an informal poll that involved 1.2 percent of registered voters. Instead he planned to build an $8 billion oil refinery in his home state of Tabasco, which is expected to end up costing a good deal more. Overall, his number one target is energy reform. "The technocrats," he said, "deceived us." For him, Pemex and its nationalization are the great symbols of national identity. He pointedly announced bids for the new refinery on Oil Expropriation Day, which celebrates the 1938 nationalization; and he promised that he would initiate a "transformation of Mexico"—the previous one, in his view, having been the petroleum nationalization of 1938. Existing investments by foreign and Mexican companies, he said, would be "reviewed," but he closed the valve on the flow of new projects. Instead, he wants to restore Pemex—the most indebted company in Latin America—as the monopolist national champion. To drive home the point, he insisted that Pemex add "for the recovery of sovereignty" to its logo. AMLO has also shut down the competitive bidding for new power plants, driving up the cost of electricity.[2]

Yet, despite his animus toward energy integration, AMLO cannot escape the reality of the shale revolution. Seventeen pipelines cross the border, carrying U.S. gas into Mexico, with more to come. This means that Mexicans will continue to use U.S. natural gas to generate growing amounts of their electricity, even as AMLO seeks to return to a past when Mexico was more isolated from the global economy.

IN SHARP COMPARISON TO MEXICO'S DECLINE, BRAZIL'S OIL production has more than doubled since 2000, primarily because of the "pre-salt"—the discovery and development of vast new offshore reserves hidden under thick layers of salt. But nationalist legislation in Brazil had required that Petrobras, the partly state-owned company, be the operator in all "pre-salt" projects. While Petrobras is highly experienced in deep waters, this was a very heavy burden to put on the shoulders of one company, made worse by the huge debt load it would accumulate in the process. In 2016 and 2017, after the impeachment of President Dilma Rousseff, the Brazilian government changed tack, initiating major reforms that allowed foreign companies to bid to be operators in the pre-salt. The result was a surge of new investment, technology, and ideas.

On January 1, 2019, just weeks after AMLO's inauguration in Mexico City, the populist Jair Bolsonaro, a former army captain turned legislator, was inaugurated as president of Brazil. He succeeded in reforming the bankrupt pension system, which allowed government employees to retire early—in some circumstances, at fifty-six for men and fifty-three for women—and further opened up Brazil to the global economy and international investment. These developments restored confidence in Brazil's prospects. But further reforms of the energy sector have been hampered by turbulent politics and a congress with more than twenty political parties. The coronavirus pandemic in 2020 hit Brazil hard, adding further to the political turbulence. Still, substantial new investment has been going into Brazil's highly prospective offshore, and those waters are now one of the most active areas in the global oil industry.[3]

AMLO and Bolsonaro, coming to the presidencies of the two largest Latin American countries within a few weeks of each other, are leading their nations in opposite directions. When it comes to energy policies, Brazil and Mexico are two ships of state passing in the night. Mexico's oil production has fallen back to the level of 1979, while Brazil's is now 80 percent higher than Mexico's—and growing.

Chapter 6

PIPELINE BATTLES

Pipelines are the connectors, the necessary lines on the map that tie supplies to markets. In previous decades, they hardly featured in public discussion. "Getting permits to build pipelines," it used to be said, "was about as exciting as watching paint dry." But that changed with a proposed pipeline called the Keystone XL, which would move oil from Canada's huge oil sands reserves down to refineries in the United States. Anti–fossil fuel activists seized on blocking pipelines that would connect new resources to markets, and the twelve-hundred-mile Keystone segment became their galvanizing and highly visible symbol. Less noticed was that the proposed pipeline length was equivalent to about one-half of 1 percent of the over two hundred thousand miles of oil pipelines that already lay beneath the soil of the United States.

Canada and the United States are highly integrated on the North American energy map. In Canada, technological advances had led to rapid growth in production from the oil sands, primarily centered in the province of Alberta. Between 2000 and 2019, Canada's crude oil output more than doubled, reaching 4.5 million barrels a day—more than Iraq or Iran, pre-sanctions. While some is consumed in Canada, most is exported to the United States. Canada supplied, in 2019, about 50 percent of total U.S. oil imports, a volume three times greater than all the oil the United States imported from OPEC countries. The growth of Canadian imports was seen as a major contribution to U.S. energy security. This trade is also a key element in the $600 billion-plus trading relationship between the two countries. The greenhouse gas (GHG) intensity of Canadian oil sands has fallen by more than 20 percent over a decade, and current trends indicate at least another 20 percent decline. Though not well recognized, the life cycle of GHG emissions from some newer projects are at or near the same level as average of crude oil processed in North America and within the same range as other globally traded crude oils.

Keystone, whose initial leg was proposed in 2005, is actually a network of existing and proposed pipelines developed by TC Energy, formerly TransCanada. In 2012, when gasoline prices hit $4 a gallon and seemed destined to go higher, President Obama flew to Cushing, Oklahoma, the major hub for pipelines and oil storage in the United States. Striding dramatically out of a labyrinth of huge pipes, he took to the podium to, in effect, dedicate the southern segment of the Keystone system. "A company called TransCanada has applied to build a new pipeline to speed more oil from Cushing to state-of-the-art refineries down on the Gulf Coast," the president said. "Today, I'm directing my administration to cut through the red tape, break through the bureaucratic hurdles, and make this project a priority, to go ahead and get it done." He added, "My administration has approved dozens of new oil and gas pipelines over the last three years."

By this time, the new segment of the Keystone system—Keystone XL—had already been proposed to shorten the northern part of the route from Canada down to Nebraska. Pipe was bought, workers hired, but the project was stymied by multiple legal and regulatory challenges, necessitating one new review after another. But it was not only in the courts. Protestors chained themselves to the fence around the White House. Among them was climate scientist James Hansen, who said approval of the pipeline—which would move less than 1 percent of world oil—"would be game over for the planet."[1]

Because the pipeline would cross an international border, law required that the State Department sign off on it, including on the environmental impacts. Two bureaus at State spent a total of seven years reviewing the proposed pipelines and eventually, in 2015, came out with a report large enough to fill a bookshelf—eleven volumes in all—saying that the project should be approved and that there was no environmental reason not to do so.[2]

Secretary of State John Kerry disagreed. Despite the State Department's finding, he nixed Keystone XL for fear, he explained, of the unfortunate impression that would result. Approval "would undercut the credibility and influence of the United States in urging other countries to put forward ambitious actions and implement efforts to combat climate change." His decision was greeted with dismay in Canada.[3]

The Trump administration overturned Kerry's decision at the beginning of 2017, but the project remained mired in legal and regulatory challenges. In the spring of 2020, the province of Alberta made a $7 billion commitment to Keystone XL. Alberta premier Jason Kenney called the project absolutely "essential" for "Alberta's economic future." TC Energy announced that it would begin construction at last. But new legal challenges quickly emerged. In the meantime, substantial amounts of oil are being transported to the Gulf Coast by railcar.

What Keystone has clearly demonstrated is that pipeline approvals are no longer like watching paint dry. They have become potent political dramas.

———

IT WAS IN NORTH DAKOTA, HOWEVER, THAT THE ANTI-PIPELINE
opposition rose to a new level of intensity.

The Dakota Access pipeline was designed to move almost 600,000
barrels per day from the booming Bakken in North Dakota to a terminal in Illinois. The battle that ensued encapsulated what has become a
central struggle over energy in the United States. The new energy map
of North America is incomplete without new pipelines to move the
new oil and gas supplies from wellhead to markets. But opponents of
oil and gas have focused on blocking pipelines as a way to choke off oil
and gas.

Dakota Access, a project of Energy Transfer Partners and other
companies, would supplant 740 railcars of oil a day. In early 2016, the
$3.8 billion pipeline was moving ahead, with almost every mile of the
1,172 miles completed. It had gone through its environmental reviews
and had received the approval of the U.S. Army Corps of Engineers,
which is required by law to sign off on parts of pipelines that cross or go
below rivers and waterways. The company had also consulted with about
fifty Indian tribes and made 140 revisions in the route as a result.

The last thing to be completed was a 1,320-foot segment—a quarter
of a mile—that would be a hundred feet below the bed of the Missouri
River. The Army Corps of Engineers gave the go-ahead in a 1,261-page
report—almost one page for every foot of pipeline. But then the Standing Rock Sioux tribe, whose eighty-two hundred members live on a
nearby reservation, objected, saying that it had not been consulted
and that the pipeline, though buried deeply under the riverbed, would
threaten their drinking water as well as violate both sacred tribal sites
and the Fort Laramie Treaty of 1868. Energy Transfer replied that its
efforts and those of the Army Corps of Engineers to consult had been
spurned by the tribe and that the pipeline passed under private and

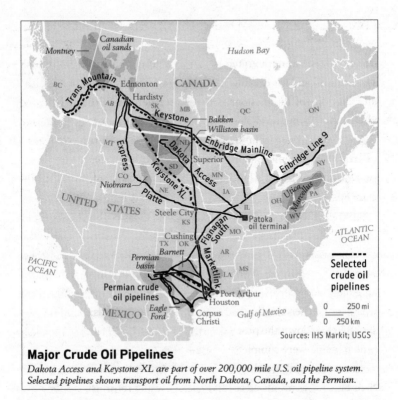

Major Crude Oil Pipelines

Dakota Access and Keystone XL are part of over 200,000 mile U.S. oil pipeline system. Selected pipelines shown transport oil from North Dakota, Canada, and the Permian.

federal lands, not tribal lands, and would be secured far beneath the riverbed. But then protestors on foot and horseback breached a barbed-wire fence into a construction area where six bulldozers were operating. Guards sought to push them back with pepper spray and a couple of guard dogs. An activist filmmaker was in position at the front line to record the event. The video went viral.[4]

Eventually as many as ten thousand demonstrators, rallied by the environmental group Greenpeace, converged near the site, creating a media spectacle that went on for more than two hundred days. Among the protestors was Alexandria Ocasio-Cortez, who would describe her experience as "transformational" and say that it had "galvanized" her.

(Two years later she would be elected to Congress and emerge as the lead author of the "Green New Deal.") The legal challenges against the last 1,320 feet were mounted by another group, Earth Justice, which says that "oil and gas drilling are destroying our air, water, and health" and that one of its missions is to "fight pipelines." Protests became violent, an improvised explosive device went off, Molotov cocktails were thrown, people were injured, and North Dakota authorities needed to call in law enforcement support from eleven other states. At one point protestors tried to get a herd of bison to charge law enforcement officers. Federal judges turned down Earth Justice's legal challenges.

But the Obama administration, in its last months, stepped in to halt construction of the last 1,320 feet of the pipeline, overruling both the courts and the Army Corps of Engineers' approval. A few months later, with the change in the White House, the Trump administration issued an executive order overturning the shutdown. Meanwhile, freezing weather and then the threat of melting snow and floods prompted the state to close down the protest camps. The last 1,320 feet of the 1,172-mile pipeline were completed.

But the battle was hardly over. The protestors and activists and their legal allies promised to fight both new and refurbished pipelines. In turn, Energy Transfer sued Greenpeace under RICO—the Racketeer Influenced and Corrupt Organizations Act, originally used to go after the Mafia. For its part, North Dakota was left with a $43 million bill that included the cost of cleaning up large amounts of debris left by the protestors, who, said a local emergency management official, had left "more garbage down there than anybody anticipated."[5]

By the end of May 2017, four months after the executive order giving the go-ahead, the first oil was flowing through the now-completed Dakota Access. But pipelines will continue to be at the center of the battle over new energy infrastructure in the United States, spurred by the clash between the shale revolution and environmental activism.

Chapter 7

———

THE SHALE ERA

In August 1946, exactly one year after the end of World War II, a tanker sailed into the port of Philadelphia laden with 115,000 barrels of oil for delivery to a local refinery. The cargo, loaded a month earlier in Kuwait, was described at the time as the first significant "shipment of Middle East oil to the United States." Two years later, Saudi oil was imported for the first time, in order, said the U.S. buyer, "to meet the demand for petroleum products in the United States."[1]

That year—1948—marked an historic turning point. The United States had not only been a net exporter of oil, but for many years the world's largest exporter, by far. Six out of every seven barrels of oil used by the Allies during World War II came from the United States. But now the country was becoming a net importer of oil. By the late 1940s, with a postwar economic boom and car-dependent suburbs spreading out, domestic oil consumption was outrunning domestic supplies.

Two decades later, by the beginning of the 1970s, the balance between supply and demand in the global oil market had narrowed and indeed had become taut. That set the stage for the oil crisis of

1973—the October War and the Arab oil embargo—that sent petro-leum prices spiraling up fourfold. Lines at the gas station and high gasoline prices fueled public outrage, and imports became a burning national issue in the United States. The oil crisis was a shock not only because of the economic and price impact. It also meant, many said, that the United States was weaker, dependent on OPEC, its foreign policy and its economy vulnerable to the decisions of oil exporters or to disruptions in supply.

In November 1973, President Richard Nixon proclaimed the goal of U.S. "energy independence" within ten years. That objective became the mantra of every president after him. But the track record never matched the rhetoric. By 2005, net imports had risen to 60 percent of total consumption, and seemed sure to rise in the decades ahead.

But the advent of shale oil over the next few years meant that the United States could once again even become an exporter of crude oil. This prospect ignited a new political battle, especially since crude oil exports were for the most part legally prohibited.

Why was there a ban on exporting crude oil? After all, the United States exports an almost endless list of things, from jet airplanes and chemicals to soybeans and movies. The ban was a relic, imposed dur-ing the oil crisis years of the 1970s. An angry public would have been even more angry were oil to have been exported at a time when prices were skyrocketing. The ban had also helped to protect the domestic system of price controls that the Nixon administration had instituted to fight inflation. A few small exceptions to the ban were allowed. But the ban did not really matter, as the United States was a larger importer, and domestically produced oil went directly into U.S. refineries.

In the late 1970s, President Jimmy Carter, despite opposition from the liberal wing of his own party, began phasing out a complex, dys-functional system of energy price controls, which discouraged invest-ment and set different prices for the same type of molecule. Ronald Reagan's first official act on becoming president in January 1981 was to completely abolish domestic price controls on oil. The following autumn, without any fuss or hullabaloo, the Reagan administration also ended restrictions on exporting petroleum products—gasoline, diesel, and jet fuel, and other products that had gone through the refining process. No one seemed to notice. But the ban on crude oil exports remained. There matters rested until the sudden emergence of shale oil. As the volumes rose so dramatically, so did a call to lift the export ban.[2]

At first glance, it might have seemed strange to be exporting crude oil when the United States was still a crude oil importer. The answer has to do with the quality of the shale oil and the nature of the U.S. refining system. Very simply, there is a big mismatch. Much of the U.S. refining system is "complex," meaning heavy-duty processing facilities designed to handle low-quality, heavier crude oils from Canada, Mexico, Venezuela, and the Middle East. Since the early 1990s, over $100 billion had been invested to reconfigure refineries along the Gulf Coast and in the Midwest to handle these heavy crudes.

As a result, these refineries are not well suited to handle the swelling volumes of lighter, higher-quality shale oil. As the volumes of light oils increase, the refineries become less efficient, imposing a financial penalty and, ultimately, higher costs on motorists and other consumers.

From an economic point of view, the most rational thing to do is let markets determine where the oil goes. That does not change America's net oil balance. If the United States exports a hundred barrels of light oil from one port on the Gulf Coast and imports a hundred barrels of heavier oil somewhere else on the same coast, the U.S. net oil balance is unchanged. But the system would be more efficient, and the economic benefits for all parties, including consumers, would be greater.

The politics, however, were quite different. In contrast to the placid response to the decontrol of product exports in 1981, crude oil exports blew up into a white-hot political issue. The most vigorous opponents were environmental groups and their congressional allies, who were against oil and gas development, and from those refining companies, mainly on the East Coast, whose systems were not "complex" and depended instead on lighter, higher-quality oil.

The result was a raucous public debate. The turning point came in April 2015, when Senator Lisa Murkowski, chairman of the Senate Energy Committee, observed that the export ban "equates to a sanctions regime against ourselves." Why, she asked, was the U.S. government lifting the "sanctions on Iranian oil" as part of the 2015 nuclear deal "while keeping sanctions on American oil"? She was joined by two other senators in arguing that exporting crude oil to "our friends and allies" would bolster both the security of U.S. partners and America's own international position. The European Union broadcast the same message, declaring that U.S. crude oil exports would, in the aftermath of Russia's moves on Ukraine in 2014, enhance European energy security.[3]

To get the ban lifted came down to a deal on Capitol Hill between Republicans and Democrats—removing the prohibition of crude oil exports in exchange for extending and expanding tax credits for solar and wind. The deal was signed into law on December 18, 2015. A week and a half later, a tanker cast off from the Texas port of Corpus Christi carrying a cargo of crude oil from the Eagle Ford, bound for France.

The ban on crude oil exports was now history. By 2019, U.S. net oil imports were down from the 60 percent of total supply they had been in 2008 to less than 3 percent. And even as the United States continued to import oil, it was also exporting almost three million barrels per day of crude oil—making it one of the largest crude oil exporters in the world—as well as over five millions barrels per day of petroleum products.

This was not just a matter of additional oil supplies. What was unfolding was an historic shift in both world oil and the global economy, and in power relations around the world.

———

AS LATE AS 2003, $20-$25 A BARREL WAS ASSUMED TO BE the long-term price for oil. But later in the year and into 2004, that price, along with that of other commodities, began to rise—and kept rising. This signaled a momentous change in the world economy, the entry into the era of the BRICs—Brazil, Russia, India, and China. What had been known as "developing countries" were transmogrified into "emerging markets" by powerful forces—high economic growth, world trade, more open markets, technology and communications, the collapse of the Soviet Union, the opening of China and India, global supply chains, and the overriding power of globalization.[4]

The growth of the BRICs was dramatic. Between 2003 and 2013, China's economy grew more than two and a half times over; India's more than doubled. The world economy grew by just 30 percent, the United States by 17 percent, Europe by 11 percent, and Japan just 8 percent.

This BRIC era was characterized by what became known as the "commodity supercycle"—high and rising prices for oil, copper, iron ore, and other commodities driven by strong economic growth in those countries. During the BRIC era, it was this growth in demand, particularly from China, that became the defining factor for the world oil market.

In response to the supercycle and rising commodity prices, companies dramatically stepped up their investments to produce more commodities. For surely China's demand would only continue to grow. How could one think otherwise? In June 2012, a major international bank convened its annual conference for hedge funds and other investors at a bucolic conference center called The Grove outside London. On the panel on "Natural Resources and the Commodity Supercycle," CEOs of the world's largest mining companies absolutely agreed—no end was in sight; the commodity supercycle could only continue and prices would continue to rise; and therefore their companies would continue to add capacity, buy assets, and spend ever-larger sums of money.

Yet by that time the supercycle was already ending. Economic growth in the BRICs was slowing, and that meant a slowing in the growth of consumption of oil and other commodities. Demand and the BRICs would no longer be the defining factors for world oil. The defining factor was now U.S. shale. Other countries were adding new supplies, notably Canada, Russia, Brazil, and Iraq. But dominating the growth was shale—by far.

The United States was now on track to become one of the Big Three of world oil, along with Russia and Saudi Arabia. This was significant not only for the world market, but also for geopolitics. Maroš Šefčovič, at the time the European Commission's vice president for energy, told an audience in Washington that from the perspective of Brussels, the United States was now an "energy superpower."[5]

But what did that mean?

Chapter 8

THE REBALANCING
OF GEOPOLITICS

S t. Petersburg is so far north that every June, during the White
Nights, the summer sun barely sets over what in tsarist times
had been Russia's capital, foreshortening darkness to barely a
few hours. In modern times, however, it is not only the magic of the
White Nights, nor the majesty of the city and the charm of its palaces
and canals, that pulls people from around the world. During those days
almost ten thousand people are drawn there by the St. Petersburg International Economic Forum, which is held under the sponsorship of the
hometown boy who made good, Russian president Vladimir Putin.

In 2013, Putin shared the stage with German chancellor Angela
Merkel. This was when the global impacts of the U.S. shale revolution
were just beginning to be recognized. The lack of chemistry between
the two leaders was all too evident; the interaction between them was
brittle and cold, and they kept their eyes fixed more on the audience
than on each other.

When the formal interview was over, questions came from the

audience. The first question—about diversifying Russia's economy, reducing its heavy dependence on commodities—was directed to President Putin. In passing, however, the questioner happened to mention shale gas. At that, the Russian president erupted. He launched a broadside, warning against possible shale gas development in Eastern Europe, denouncing shale gas as a grave danger, an environmental threat, a despoiler of land and water. The questioner sank back into his seat.

Putin reacted so vehemently because shale gas was also becoming a matter of geopolitics. For shale was a challenge for Russia, at the time the world's largest producer of natural gas, as well as the major supplier to Europe. Around the world, it was becoming clear that the unconventional revolution is about more than the flow of oil and gas. It is also about the relative positions of nations.

FOR FOUR DECADES, U.S. ENERGY POLICY WAS DOMINATED— and its foreign policy hobbled—by the specter of shortage and vulnerability, going back to the 1973 oil embargoes and then the 1979 Iranian Revolution, which toppled the shah and brought the Ayatollah Khomeini to power. But no longer. The shale revolution "affords Washington," observed Thomas Donilon, national security advisor to President Obama, "a stronger hand in pursuing and implementing its international security goals." Secretary of State Mike Pompeo would subsequently put it differently—that the shale revolution has provided the United States with a flexibility in international affairs that it had not had for decades.[1]

For more than a century, energy—its availability, access, and flows—has been intertwined with security and geopolitics. As a Brookings Institution study put it, "In the modern era, no other commodity has played such a pivotal role in driving political and economic turmoil, and there is every reason to expect this to continue."[2]

The Middle East has been central to world oil, the security of its supplies crucially important to the world economy and a top priority for U.S. foreign policy. At the beginning of the Cold War in 1950, with Saudi oil exports starting to flow, President Harry Truman extended an explicit American security guarantee to King Ibn Saud. "No threat to your Kingdom," the president wrote, "could occur which would not be a matter of immediate concern to the United States."[3] That commitment, at the time aimed at preventing those resources from falling into Soviet hands, continued after the Cold War. The current extensive U.S. security engagement with the Arab Gulf countries is represented in a multitude of agreements, arms deals, exchanges, and a series of bases and facilities for air, ground, and sea forces.

An important element in the world oil market is "spare capacity." This is production capacity—that is, oil wells—that are not actually in operation, but can be swiftly brought on line if prices spike or if a disruption knocks out supply elsewhere. Today, most of the world's spare capacity is in Saudi Arabia, with some in the United Arab Emirates and Kuwait. That, combined with the size of its oil reserves and its ability to quickly increase or decrease output, makes Saudi Arabia the balancer in the world market. Sometimes it is described as the "central bank" of world oil.

The nature of the U.S. commitment to Persian Gulf security, the scale of the U.S. engagement, and the size of the region's resources led to a widespread view that the United States itself was heavily dependent on the Mideast. Yet in 2008, even before shale oil, imports from the Gulf amounted to less than 20 percent of total U.S. oil imports. As already noted, oil sands in the province of Alberta had made Canada the largest supplier of U.S. imports by far. In 2019, only about 11 percent of U.S. imports came from the Persian Gulf. For their part, Gulf producers are focused on Asia as their most important market.

The U.S. commitment to the region has endured not because specific barrels of oil are departing Saudi Arabia or Kuwait or the UAE for

U.S. refineries, but rather because these resources are central to the overall world economy and critical for America's most important allies and trading partners. Disruptions of supply affect the global system into which America is so integrated—with almost 30 percent of U.S. GDP and close to 40 million jobs resulting from trade with the rest of the world. Even if the U.S. is not importing much Middle Eastern oil, a supply disruption would drive up global prices, including in the United States.[4]

How has the shale revolution changed geopolitics? Case study number one is Iran and the 2015 nuclear agreement. In 2012, sanctions were applied on Iranian oil exports and finance. The aim was to force Iran to the negotiating table, as we shall see later. But it wasn't obvious that these sanctions would work. The expected shortfall in world supplies would drive prices up, hitting oil-importing countries, causing the sanctions to crumble. Certainly that is what Tehran expected as it confidently proclaimed the new sanctions "doomed to fail." But increasing U.S. production offset the reduction in Iranian exports. As we shall see later, the oil sanctions held, buttressed by financial sanctions; and the economic pressure on Iran led finally to the 2015 agreement that constrained Iran's nuclear program in exchange for the removal of sanctions.[5]

Case number two is Europe and relates back to Putin's angry rejoinder at St. Petersburg. The rise of shale has been one of the keys to diversifying the European gas market and enhancing energy security. When European leaders talk about energy security, they are often less focused on oil and more on natural gas—and in particular the degree of reliance on Russian gas. As Europe's top supplier of gas, Russia had, in the minds of some in the European Union and many in Washington, the ability to use gas supply as leverage for political objectives. This concern was magnified by the reliance on pipelines with their inherent inflexibility.

Enter U.S. shale gas. First it eliminated the need for LNG in the

United States, leading exporters to redirect some of their LNG to Europe. Then the export of LNG from the United States reinforced the shift toward competition in Europe—with U.S. gas, along with other LNG supplies, competing head-to-head with Russian gas. European buyers now had multiple options and choices, which meant diversification of supply—the keystone of energy security. "We have had many historical challenges with Russia," said Lithuania's energy minister. But now, as a result of the opening of the country's LNG importing facility, he continued, "gas supply has been depoliticized."[6]

IN MARCH 2016, A SUPERTANKER FILLED WITH OIL LEFT THE U.S. Gulf Coast and crossed through the Panama Canal into the Pacific. Its destination was China. The customer was Sinopec, one of China's two major oil companies and one of the world's largest buyers of oil. "U.S. crude oil exports are positive news for the global market and make it possible for Asia-Pacific refiners to diversify their supply," said a Sinopec executive. A few months later, another tanker unloaded at Shenzhen the first shipment of U.S. LNG to China. These voyages demonstrated that the supposed zero-sum life-and-death competition between China and the United States for access to constrained energy, so vividly imagined just a few years earlier, was not going to happen. Global energy supplies are ample, and China and the United States can interact through the global marketplace to mutual benefit. The shale revolution removed at least one major area of contention in U.S.-Chinese relations, creating a new commonality of interests between the nations— trade wars and contention over the coronavirus permitting.

Because of shale, the United States is "present" in Asia in a new and strategically important way for many countries. It adds to diversification, moderating dependence on the Middle East and the Strait of Hormuz and providing options on LNG. While the United States is only one among several suppliers of oil and LNG to India, this growing trade has brought the two nations closer together and added an important positive new dimension to a relationship that had been more contentious in the past.

Japanese companies, and the Japanese government, are also keen to receive U.S. oil and gas exports. They see these supplies as important to reducing Japan's trade surplus with the United States and as major contributions to global energy security—as Japan imports 99 percent of its oil and 98 percent of its natural gas. Prior to the 2011 Fukushima nuclear accident, nuclear power provided 30 percent of Japan's electricity. By 2020, not much more than 5 percent of the country's electricity came from nuclear. LNG, already significant for electric generation, filled much of the void—in 2020 responsible for almost 40 percent of its electricity generation.

South Korea, at the time of this writing, is the largest buyer of U.S. LNG. Moreover, the option of U.S. gas, as a senior Korean official said, "helps us negotiate with our traditional suppliers." And the more gas South Korea buys from the United States, the lower its trade surplus—something that Seoul does not hesitate to point out.[7]

IN THE AUTUMN OF 2018, THOUGH IT WAS HARDLY NOTED AT the time, something historic occurred: The United States overtook both Russia and Saudi Arabia to regain its rank as the world's largest oil producer, a position it had lost more than four decades earlier.

For how long and how much more will U.S. production grow? Some say it is still early in terms of the ability to recover resources. As a petroleum engineer who ran the Permian for one of the large companies said: "In my thirty years in the industry, the Permian is the only place where every time I made a map of resources, it was bigger than the last one. Everywhere else, each time the map of the resources got smaller."[8]

Yet changes in government policies or in the economics or in terms of infrastructure or capital constraints could constrain growth.

Environmental opposition could become more potent. A Democratic administration intent on a "Green New Deal" could throw up regulatory and legal obstacles. "Banning fracking" has become a refrain among some politicians. Yet most new wells drilled in the United States have some element of fracking; such a ban would mean a flood of imported oil, and the negative economic impacts would be felt across the country. Geological constraints that are not currently foreseen could emerge. Or, although hardly thought of beforehand, a viral epidemic could shut down demand, drive down prices, and lead to a retreat in drilling.

WHAT HAS HAPPENED SINCE THE DRILLING OF SH GRIFFIN #4 is extraordinary. The world has never before seen anything like the speed or scale of the growth. It is as if, in in terms of volume, the United States had added, in little more than a decade, the equivalent of another major oil-producing country. Yet even as U.S. oil output continued to surge, a new challenge had emerged. The shale revolution was in search of a new revolution—this one based not on technological breakthrough, but on its economics.

Shale has been described as a manufacturing business. In contrast to traditional wells, the output of shale wells, as noted earlier, falls significantly over the first year or so before leveling out. Thus companies constantly drill new wells to compensate for the declines in their previous ones. For the independents, it has been all about growth, funded by investors and debt. But then growth was no longer enough. The investors who had previously cheered on the companies as they strove for ever-higher production now soured on them. The companies, improving their efficiency, had cut the cost of a well from perhaps $15 million to $7 million, but that was not enough. Investors wanted money back; they wanted a return on their investment. When investors looked at the shale companies, it was no longer growth at any cost but rather growth at what cost. As share prices declined, companies were forced to reset their businesses, get their spending under control, and live within their budgets—and thus deliver returns to investors either in the form of dividends or share buybacks. The need to reduce costs set the scene for mergers and consolidation.[9]

The shale industry has changed in another way. While the independents have scaled back, the majors have stepped up. ConocoPhillips, once a very international company, has shifted back to emphasize North America onshore. Two of the largest majors, ExxonMobil and Chevron, have both partly pivoted back to North America, directing significant investment to the Permian in order to make it a growing part of their global portfolios. To a lesser degree, BP, Shell, and Equinor have moved in the same direction.

When it was all added up, the growth in U.S. output seemed destined to slow to a much smaller annual increase—far less than the hectic pace registered in preceding years. Yet, even with the slowdown, the United States had become the world's number one oil producer. By February 2020, it had reached the highest level of production ever—thirteen million barrels per day—more than Saudi Arabia and Russia and on the way to tripling the level of 2008.

At that moment struck the calamity of 2020—the coronavirus pandemic and the shutdown of the globalized world economy, which slammed shale as it did most industries. As a result of a drastic cutback in investment, shale output will go in reverse and decline. When growth returns, it will be at a slower pace. But, whatever the trajectory, shale is now established as a formidable resource.

THE SHALE REVOLUTION HAS TRANSFORMED THE WORLD OIL market and is changing concepts of energy security. "OPEC versus non-OPEC," the arrangement that had defined the world oil market for decades, has been overtaken by a new paradigm, the "Big Three"—the United States, Russia, and Saudi Arabia. This was made starkly clear by the unprecedented interaction among Moscow, Riyadh, and Washington in the vast oil market crisis in 2020 that the virus unleashed, to which we will turn later.

Moreover, the struggle to adjust to the crisis demonstrated in a new way how energy continues to be so central to geopolitics. Certainly that is the way Vladimir Putin sees it.

RUSSIA'S MAP

Chapter 9

———

PUTIN'S GREAT PROJECT

From the Bolshevik Revolution in 1917 through the end of the Cold War in 1991, the name "Russia" was interchangeable with "Soviet Union."* The overlap worked because the map of the Soviet Union at that time roughly matched up with the Russian empire (save for Poland, which was part of the empire at its maximum), because Moscow was so central—and because Russian culture and the Russian language so thoroughly dominated the Soviet Union. In 1991, the Soviet Union collapsed and fifteen newly independent states emerged, ranging from the tiniest, Estonia on the Baltic Sea, to Kazakhstan, equal to India in geographic size.

Yet the Russian Federation—Russia—still looms over the newly independent states. It sprawls across the map, encompassing eleven time zones from Europe in the west to the tip of the Chukotka Peninsula in the Far East, just sixty miles across the Bering Strait from Alaska. Its

———

* For instance, the book by the distinguished diplomat and historian George Kennan on Soviet foreign policy was titled *Russia and the West under Lenin and Stalin*.

population is only half that of the Soviet Union, and in 2019, its economy was only slightly larger than Spain's—although Spain has only a third of the population and ceased being a great power in the eighteenth century. Yet Russia still has formidable accoutrements of power. It has scale. It has a huge arsenal of nuclear weapons and missiles and considerable cyber skills. It has the determination to project itself on the world stage. And it has natural resources—particularly vast amounts of oil and gas that underpin its place in the world.

Three decades after the collapse of the Soviet Union, a new global competition between the United States and Russia has emerged—not the Cold War of history and nuclear doomsday, but still a cold war. It is playing out in regional conflicts, information warfare, cyberspace, energy, and overall relations. Since its interference in the 2016 U.S. presidential election, "Russia" has become a toxic subject and source of great rancor in Washington and domestic American politics.

Over his two decades as president, Vladimir Putin's great international project has been to reassert Russia's sway over the rest of the former Soviet Union, restore Russia as a great power globally, build new alliances, and push back against the United States. And whether Russia is partially responsible or not, Putin can point to outcomes that fit his objectives—NATO divided, the European Union in disarray, and America's politics fragmented, nasty, and polarized.

OIL AND GAS HAVE BEEN CRITICAL TO RUSSIA'S REBOUND and to the nation's economy. They also provide a way for Russia to project power other than with military might. As Putin put it, "Oil is no doubt one of the most important elements in world politics, in the world economy." He was once asked if Russia is an energy superpower. "I would prefer to move away from the terminology of the past," he replied. "'Superpower' was the word used during the Cold War." He added, "I have never referred to Russia as an energy superpower. But we do have greater possibilities than almost any other country in the world. This is an obvious fact."[1]

The "obvious fact" is evident in the sheer scale and abundance of Russia's energy resources. It is one of the Big Three of world oil production. It is the second-largest producer of natural gas (after the United States) and is still the world's largest gas exporter. The earnings from oil and gas exports provide the financial foundation for the Russian state and Russian power—in normal times, 40 to 50 percent of the government's budget, 55 to 60 percent of export earnings, and an estimated 30 percent of GDP. Much more than anything else, these resources make Russia a major player in the world economy.

This geological endowment gives Russia global presence. It undergirds its economic relationship with Europe and growing bonds with China. Yet this reliance is also subject to much debate. Former finance minister and deputy prime minister Alexei Kudrin and others have argued that Russia is overly dependent on oil and natural gas, hindering the development of a more balanced and dynamic economy.

THESE DEBATES AND DILEMMAS ARE NOT NEW. FOR A CENTURY and a half, "Russia"—whether the Russian Empire, the Soviet Union, or, since 1991, the Russian Federation—has been a major player in world energy while at the same time heavily reliant on oil and then also on gas.

Russia's oil industry emerged in the nineteenth century both in what today is Azerbaijan, on the western side of the Caspian Sea around the city of Baku and northwest of there and, to a lesser degree, in Kazakhstan, on the eastern side of the Caspian. A British visitor to Baku in the 1880s marveled at what had become the new "oil breadbasket of Europe." By 1898, Russia overtook the United States to become the world's biggest petroleum producer. But after the revolution of 1905—the "dress rehearsal," as Bolshevik leader Vladimir Lenin called it—Russia's once-buoyant oil industry languished and declined.[2]

In the civil war that followed their seizure of power in 1917, the Bolsheviks faced what they called a "fuel famine," a daunting threat to their revolution. "The fuel crisis must be overcome at any cost," said Lenin. "We desperately need oil," and anyone who stood in the way "we will slaughter." To help solve their fuel famine, the Bolsheviks nation-alized the oil industry.[3]

By the early 1920s, the Communists were in control of the entire country. Over the next several years, the oil industry recovered and once again became a player in the global market. In the mid-1930s, Jo-seph Stalin, Lenin's successor, launched purges that engulfed the en-tire country. The oil industry was not spared; the secret police claimed to have "discovered" a "counterrevolutionary, wrecking and spying or-ganization" throughout the industry. Many of its leaders and workers were either sent to the gulag prison camps or summarily executed. The industry ceased being a major factor outside the Soviet Union.[4]

Only in the late 1950s, well after the end of the Second World War, did the Soviet Union return as an oil exporter to the global market. This was made possible by new production in the Volga-Urals region and then the discovery of vast new supplies in West Siberia. But Rus-sia was exporting into a global market that was already oversaturated with rising amounts of Middle East oil.

In response to the surging volumes of Soviet oil, the international companies reduced prices in 1959 and again in 1960. Enraged at the re-sulting cut in their own revenues, oil-exporting countries, led by Saudi Arabia and Venezuela, came together to form a new organization, the Organization of Petroleum Exporting Countries—OPEC.

By the beginning of the 1970s, the Soviet Union's centrally planned economy was failing. It could not produce the goods that people wanted, and what it did produce was shoddy, except for specific sectors, mainly defense. The oil crisis of the 1970s came just in time. The dramatic increases in petroleum prices delivered a massive surge in revenues that rescued the stagnant economy and helped fund a big Soviet military buildup. But this new lease on life would prove only temporary.

In 1985, Mikhail Gorbachev emerged as the new leader of the Soviet Union. Young and energetic, he was determined to reform the economy. But fate was against him. The next year, oil prices collapsed, delivering a terrible blow to the Soviet economy and marking the start of what Yegor Gaidar, former finance minister and acting prime minister, called "the timeline of the collapse of the Soviet Union."[5]

Oil revenues could no longer mask the failures of the centrally-planned economy. "We were planning to create a commission," remembered Gorbachev, "to solve the problem of women's pantyhose. Imagine a country that flies into space, launches Sputniks, creates such a defense system, and it can't solve the problem of women's pantyhose. There's no toothpaste, no soap powder, not the basic necessities of life. It was incredible and humiliating to work in such a government."

It got worse. Oil production began to fall rapidly. In 1989, the chairman of the Council of Ministries bemoaned, "If there is no oil, there will be no national economy." While several factors were converging to push the Soviet Union to its demise, the tanking of oil prices severed the financial lifeline that had kept the economy afloat.[6]

The Russian republic was by far the largest of the "republics"—that is, states—that comprised the USSR, the Union of Soviet Socialist Republics. All the republics had their own parliaments and government agencies, but they had no real power. But now, no longer a rubber-stamp tool of the Soviet government, the Russian republic asserted a new authority. It took control of the Soviet oil and gas assets within its territory and of the petroleum revenues that had gone to the all-union Soviet government. Boris Yeltsin, the president of the Russian republic, and not Mikhail Gorbachev, was now in charge of the oil money.

In December 1991, Yeltsin and the speakers of the Ukrainian and

Belarusian parliaments met in a forest, in a hunting lodge. Over the course of a night, facilitated by large amounts of bison-grass vodka and Soviet-style champagne, they came to a stunning agreement: Invoking the status of their three "republics" as "founding states of the USSR" in 1922, they declared that "the USSR as a subject of international law and a geopolitical reality ceases in existence."

On December 25, 1991, Gorbachev delivered on television what one of his aides called the "death notice" for the Soviet Union—it was dissolving itself. The formerly powerless constituent "republics" would now become independent nations. The Russian Federation, as it became known, would be the main legatee of the Soviet Union. That, among other things, required the transfer of the all-important codes that controlled the disposition and use of the vast arsenal of nuclear weapons. But the enmity between Gorbachev and Yeltsin was such that they could not agree on who would go to whose office in the Kremlin to do the transfer. Finally, two sets of military officers—one representing the Soviet Union, and the other the Russian Federation—met in a Kremlin hallway for the handover of the codes, an epochal event marked only by a quick exchange of salutes.[7]

The disintegration of the Soviet Union fractured the integrated edifice of the oil industry. The original base on the western side of the Caspian Sea was now in the newly independent nation of Azerbaijan. The oil on the eastern side of the Caspian now belonged to newly independent Kazakhstan. The breakdown of the Soviet economic structures left the giant oil industry of West Siberia fragmented and in disarray. During the "spontaneous privatization" of the 1990s, a decade that became known as the "wild '90s," the oil industry in Russia itself was up for grabs. New oil companies began to emerge, gathering up assets.

As the decade proceeded, the economy rebounded, the foundations of a market economy were taking shape, and optimism came with it. There was talk of a *chudo*—a Russian economic miracle—invoking the

"economic miracles" in Western Europe and Japan after World War II.[8]
But then in August 1998, the Asian financial crisis engulfed Russia. The
ruble collapsed, as did oil prices, drying up government revenues. The
new economy stopped working and people weren't paid. The credibility of the Yeltsin presidency was shattered. Yeltsin himself was a
spent force.

IN 1976, THE *LENINGRAD EVENING NEWS* REPORTED THAT A
previously unknown local "judoist" had won a judo competition and had
"for the first time joined the ranks of champions." It predicted that people would hear more about him in the future. This was Vladimir Putin,
age twenty-three. He went on to join the KGB, which sent him as
an operative to East Germany. In 1990, after that country's collapse,
he hurriedly burned secret KGB files and drove home with a valuable
memento—a washing machine that sat atop his car. His home city was
now no longer known as Leningrad but once again as St. Petersburg.
He went to work for the reformist mayor and became deputy mayor.[9]

In 1996, after the mayor lost a reelection bid, Putin found himself
unemployed. He pursued a degree at the local geological institute.
He also went to Moscow looking for a job. The result, beginning in the
state property office, would be a meteoric rise up the ranks of government, until finally, in 2000, he became Yeltsin's designated successor as
president of Russia. Once in the Kremlin, Putin's objectives were to reimpose order, stabilize the economy, renew the authority of the state,
and restore Russia as a major player in the world. Over the two decades
since, he has proved himself, as predicted in 1976, a champion "judoist," but on a global scale, capitalizing on other countries' weaknesses
or mistakes, seizing on opportunities and openings. Energy would be
central to that entire agenda. He understood the power that came from
Russia's oil and gas. Western interlocutors would be consistently

surprised by his detailed knowledge of the energy industry and energy markets and the fluency with which he discussed the intricacies—as much like a CEO, they would say, as a head of state.

Under Putin, the government reasserted its control over the energy industry. Mikhail Khodorkovsky, the head of Yukos, one of the biggest of the new oil companies, and among the most powerful of the new oligarchs, challenged Putin directly, and ended up spending ten years in a prison camp. Yukos assets were absorbed by the state company Rosneft. The company's CEO is Igor Sechin, formerly deputy prime minister, who had worked with Putin in the mayor's office in St. Petersburg in the early 1990s. In 2013, Rosneft took over a major company, TNK-BP, in a $55 billion deal that made it a larger oil producer than ExxonMobil. In 2016, it acquired another company, Bashneft. Today Rosneft produces 40 percent of Russia's total oil. The government owns just over half of the company and has the controlling share.

Similarly, the Russian government holds the controlling stake in the giant gas company Gazprom, headed by Alexey Miller, who was also with Putin in the mayor's office in St. Petersburg in the early 1990s. In 2005, Gazprom acquired Sibneft from another oligarch, renaming it Gazprom Neft (meaning Gazprom Oil). Its CEO, Alexander Dyukov, had headed the St. Petersburg port. Today, only a handful of private oil companies remain. LUKOIL is the largest. Its CEO, Vagit Alekperov, started his career working offshore in the Caspian and then in West Siberia before coming to Moscow as a deputy energy minister in the late 1980s, where he developed the idea of starting a Western-style oil company in Russia.

Both Mikhail Gorbachev and Boris Yeltsin had bad luck when it came to oil, with price collapses that sent the economy spiraling downward. By contrast, Vladimir Putin had very good luck, for petroleum prices recovered as he came to power in 2000 and continued to rise during the BRIC era. Output, which had fallen by almost half with the collapse of the Soviet Union, rebounded. This was made possible by new investment. That included investment by Western companies, which also brought Western technology and practices—all of which had been shut out during Soviet times—and was blended with traditional Soviet capabilities. By the end of 2018, Russian output reached 11.4 million barrels per day, as high as at the peak for Russia in Soviet days.

The value of Russia's oil exports increased eightfold between 2000 and 2012, from $36 billion to $284 billion a year. The annual value of gas exports over the same period increased from $17 billion to $67 billion. As oil and gas revenues mounted, Russia went from economic weakness to strength, paying off international debts, raising wages and living standards, increasing pensions, saving money in "stabilization" funds, spending more on defense, and financing its restoration as a great power.

Russia was a major beneficiary of the commodity supercycle in the BRIC era and the strong demand from emerging market countries that defined it. More than anything, that meant China and its fevered economic growth. A deputy prime minister, in his office on Old Square in Moscow, was once asked about what was going to happen in Russia's economy. He gestured toward the window and pointed to the east.

"Tell me what's going to happen in China," he replied.[10]

Chapter 10

———

CRISES OVER GAS

Russia's supply of natural gas to Europe—about 35 percent of Europe's total gas consumption—is at the center of a geopolitical clash. It comes with a basic question: Is this reliance and multibillion-dollar gas business an instrument of Russian power and influence, or is it a part of a mutually beneficial, geographically determined trading relationship? And if both, where does the balance lie?

Nowhere is this tension more evident than in the fissured and violent relationship between Russia and Ukraine. The consequences reverberate on energy markets, in relations with Europe, and on the relationship between Russia and the United States, with impacts ranging from the U.S. military budget and the domestic strife over the 2016 U.S. presidential election, to the rising hostility between the world's two major nuclear powers, to the impeachment trial of Donald Trump in 2020. Indeed, if one were to point to the single most important reason for the new antagonism between Russia and the West—and the new cold war—it is Ukraine and the bitter and unsettled questions resulting from the breakup of the Soviet Union and the way they have played out.

The linguistic root of "Ukraine" means "edge" or "border land." The territory that became known as Ukraine is mostly an extended plain with few natural borders. Ukraine and Russia both assert a common origin in Kyivan Rus. This medieval kingdom was established by Viking warriors who intermixed with local Slavic tribes in what became known as the "Rus lands," which were ruled from Kyiv (the capital of Ukraine today). Despite their shared lineage in Kyivan Rus, modern Ukraine and Russia clash bitterly over claims of common identity, as Russians portray it, versus separate identities, as Ukrainians assert.

Kyivan Rus disappeared from history when the Golden Horde, the Mongols, sacked Kyiv in 1240. The first maps of Ukraine, including its borders, were drawn around 1640, when it was part of the Grand Duchy that combined Lithuania and Poland. One map was labeled "General Description of the Empty Plains (in Common Parlance Ukraine)."

Fourteen years later, in 1654, a leader of the Cossacks in what today is Ukraine swore allegiance to the tsar of the eastern Slavic principality of Muscovy, which was "regathering" the Russian lands. Historians, politicians, and nationalists continue to argue today whether that allegiance to "Muscovy" was conditional, retaining autonomy, or absolute submission and incorporation into the regathered lands.[1]

Nationalist identity and fervor were developing in Russia's Ukraine before World War I. This was at the same time that full-scale industrialization got underway in the Donbas region in southeastern Ukraine, drawing in Russian speakers from elsewhere in the empire. In 1918, shortly after the Bolshevik Revolution, an independent Ukraine was declared, but it disappeared in the chaos of the Russian Civil War. After the Bolshevik victory, Ukraine became a founding republic of the Soviet Union.

With the breakup of the Soviet Union at the end of 1991, Ukraine was for the first time—aside from that fleeting moment at the end of World War I—no longer an idea, a borderland, a province of an empire. Now, for the first time, it was a sovereign nation.

At independence, Ukraine was "born nuclear," for it inherited nineteen hundred nuclear warheads from the Soviet Union, making it the world's third-largest nuclear state. In 1994, in what is known as the Budapest Memorandum, it gave up those weapons and transferred them to Russia. In exchange, Russia, Britain, and the United States solemnly promised to "respect" the "existing borders of Ukraine."[2]

ONE INSTITUTION MANAGED TO SAIL THROUGH THE MAEL-strom of the post-Soviet collapse intact, though somewhat battered—the ministry of natural gas. It, however, changed its name—to Gazprom. It gained control of the big export pipelines—and the revenues that came from exports—and thus inherited the Soviet-era relationship with the major Western European energy companies.

Gazprom became the largest gas company in the world. It provided gas to keep Russia's domestic economy going, even if bills went unpaid; it maintained its reputation as a reliable supplier to Western Europe. And it delivered desperately needed revenues to the national treasury. Amid the chaos of the collapse, Gazprom represented not only continuity with the past but also Russia's future economic integration with the West. Gazprom insisted it operated as a commercial organization. But for some outside Russia, it was not just a gas company. It was also the palpable ghost of the Soviet-American Cold War, the embodiment of resurgent Russian power, and the instrument for Russia to gain leverage over Western Europe and thus drive a wedge between Europe and the United States. At the heart of the gas issue was Ukraine.[3]

In which direction would Ukraine look for its future? This fundamental question has inflamed relations between Ukraine and Russia ever since the breakup of the Soviet Union. Would it continue to look east and remain under Moscow's sway? Or west, toward Europe and the European Union and, worse from Moscow's point of view, toward NATO and the United States?

Natural gas, and the pipelines that carry it, had bound the two countries together—but now would set them against each other. Gas imported from Russia was Ukraine's major energy source and critical to its own economy and fueling its heavy industry. Moreover, the tariffs—that is, the fees—that Ukraine earned on the transmission of Russian gas to Europe through its pipelines and territory were a major source of government revenue.

But this was not a one-way street. Assuring access to Europe was critical for Gazprom, for the European market was its major source of revenues. That meant Russia also depended on Ukraine; as late as 2005, 80 percent of its gas exports to Europe passed through Ukraine's pipelines. That, of course, had not mattered when Ukraine and Russia were both parts of the same country and were connected by what was called the "Brotherhood" gas pipeline. But now the Soviet Union was gone, and it mattered a lot. Ukraine and Russia were no longer brothers.

The breakup of the Soviet Union in 1991 quickly led to acrimony between the two countries over the price of Russian gas and the tariffs charged by Ukraine for passage through its pipelines. Yet the disputes were largely containable until the 2004 contested Ukrainian presidential election in which the "two Viktors" were pitted against each other. The initial winner in what was widely seen as a rigged election was Viktor Yanukovych, the sitting prime minister and a onetime boxer. Yanukovych's native language was Russian, and he was Moscow's candidate. His opponent was the other Viktor—Viktor Yushchenko, a former prime minister and head of the central bank and a native Ukrainian speaker, for whom Europe was the great calling.

The stolen election ignited massive protests, which converged on Kyiv's Maidan Square in what became known as the "Orange Revolution"—after the colors of Yushchenko's campaign. In a court-imposed runoff, Yushchenko won. The result was a shock to Moscow. Ukraine now had a president who wanted to look west. As if to drive home that point, his wife was an American of Ukrainian descent, a graduate of Georgetown University who had worked in the Reagan administration.

For the Kremlin, the outcome was a powerful warning of the existential threat from "color revolutions." Ukraine's Orange Revolution had been preceded by the "Rose Revolution" in newly independent Georgia, which had swept anti-Russian reformers to power. These color revolutions, as Moscow saw it, were supported by Western nongovernmental organizations and, Moscow suspected, by Western "security services" that were aiming to dislodge Russia from its "privileged sphere" in the rest of the former Soviet Union. Even worse, the Orange Revolution could potentially bring NATO—which the Baltic nations had already joined—to Ukraine's borders with Russia. That, said Putin, would be a "direct threat" to Russian security. And there was the further risk—that the contagion of color revolutions might spread to Red Square in Moscow.[4]

Yushchenko's victory triggered truculent negotiations over natural gas prices. Ukraine was paying only one-third, or even less, of what Western Europeans were paying. Why, said the Russians, should they continue to subsidize Ukraine with cheaper gas, to the tune of billions of dollars a year, when Ukraine was already billions of dollars in arrears on its gas bill, and was now led by a president who wanted to pivot away from Russia? Beyond revenues, Moscow had another objective—to gain control over the all-important Ukrainian gas pipeline system, on which it depended for sending its gas to Europe. This, however, was not on the table. As Yushchenko put it, those pipes, laid down in Soviet times, were the "crown jewels" of independent Ukraine.[5]

On January 1, 2006, with no resolution in sight, Gazprom cut off gas earmarked for Ukraine. But Ukrainians siphoned off gas meant for delivery to Europe, which cut gas supplies to other European countries—and led to a crisis in Russia's relations with Europe.

Russia maintained that the cutoff was not about politics, but simply about economics and "market pricing"—the Ukrainians should no longer get such a steep discount. But for Europe and the United States, the issue was Moscow's demonstration of raw energy power. Russia, said U.S. secretary of state Condoleezza Rice, was using natural gas and oil as a "politically motivated" weapon. "The game just can't be played that way," she declared.[6]

After a few days, the two sides managed to come to a new agreement. But Russia did not get control over the Ukrainian pipeline system. Kyiv would not give up the crown jewels. Left in place was a murky and mysterious company called RosUkrEnergo that played a central role in the gas business.

Three years later, on New Year's Eve, December 31, 2008, Vladimir Putin sought to inject a little humor into his holiday remarks on Russian television. "It's a sure sign New Year is coming," he said. "The gas negotiations are heating up." Hours later, on January 1, 2009, Russia once again cut off gas directed for Ukraine's own consumption. Putin declared that the gas destined for Europe was again being siphoned off and stolen by Ukrainians. He ordered a halt of all shipments of gas into Ukraine. That meant that now no gas was being put into the system for Europe. It took more than two weeks and what Putin called "difficult" negotiations for Russia and Ukraine to come to a new deal.[7]

Yet, remarkably, despite exceptionally cold weather, this second gas crisis resulted in no shortages, except in parts of the Balkans. The Ukrainians had built up ample storage. The Europeans also drew gas from storage.

These crises put a new emphasis on energy security for both Russia and Europe. But the concept of "energy security" meant strikingly different things to them.

Chapter 11

CLASH OVER
ENERGY SECURITY

U ntil recently, we assumed that the energy security regime that had come into existence in Europe was an optimal one," reflected then–Russian president Dmitry Medvedev soon after the gas crisis. "It turns out that it wasn't."[1]

In 2011, Russia's new concept was made clear. The setting was Lubmin, a seaside resort on the northeastern coast of Germany that advertises its beaches as "a paradise for families." Gathering there were Medvedev, German chancellor Angela Merkel, the French and Dutch prime ministers, and the EU's energy commissioner. Also attending was Gerhard Schroeder, Merkel's predecessor as chancellor of Germany and now the chairman of a new pipeline company, Nord Stream.

They were obviously not there for the seaside amenities but rather to turn the valve on a $10 billion pipeline called Nord Stream—actually twin 750-mile natural gas pipelines that ran under the Baltic Sea directly from Russia to Germany. Nord Stream was Russia's solution for its own version of energy security, reducing its dependence on Ukrainian transit by building new pipelines that went around that country.

"Its construction meets our long-term goals," said Medvedev at the dedication. And he generously added, "Of course this is our contribution to European energy security." Chancellor Merkel voiced her approval of the project. Europe and Russia, she said, would "remain linked" in a "safe and resilient partnership" for decades to come. The European Union designated the pipeline "a priority energy project" that would contribute to European energy security.

The gas that arrived at Lubmin had been injected into the pipeline two months earlier, at the port of Vyborg, northwest of St. Petersburg. While pushing the button to start the flow, Putin had been more explicit about Russia's view of energy security. Nord Stream, he said, would end Ukraine's "temptation to benefit" from its "exclusive position." He predicted a "more civilized relationship" between Russia and Ukraine would result. As events turned out, that would hardly be the case.[2]

Russia already had built other pipelines to circumvent Ukraine: Yamal-Europe, through Poland; and Blue Stream, which runs from Russia under the Black Sea to Turkey. But Nord Stream loomed much larger.

FOR EUROPE, ENERGY SECURITY MEANS SOMETHING QUITE different—greater flexibility and diversity of supply. For years, the EU had been searching for a common energy policy. But that was very hard to come by with twenty-eight different countries, with different interests, different endowments, different needs—and different attitudes toward Russia. West Europeans generally welcomed Russian gas imports. Eastern and Central European countries, much more dependent on Russian gas, saw their reliance as a source of vulnerability, reminding them of their former thralldom to Moscow when they were satellites of the Soviet Union. They would point to cases in which the Soviet Union and then Russia, they said, used cutoffs and manipulation of supplies to apply political pressure.

As European energy policy evolved, it had two main objectives. The first, dealing with natural gas, was to build resilience and greater energy security into the natural gas system and push for the formation of a single gas market for the entire continent. Companies increased connections among pipelines to make it easier to move gas from one part of Europe to another. They reengineered pipeline systems so that the direction of gas flows could be reversed, if needed. Investment in LNG terminals and storage was promoted. "Destination clauses," which limited the ability to shift gas supplies from one buyer to another, were eliminated. This package of policies and initiatives would end up reshaping the entire European gas system.

European policy also aimed at doing away with the traditional rigid contracts that ran twenty years or longer and in which price was indexed to oil. The European gas system had been built over many decades on the foundation of these long-term contracts with long-term predictability and long-term relationships. Instead, Brussels now intended to promote competition and transparency. It wanted "markets," a world of buyers and sellers, not "relationships." It was not necessarily against long-term contracts, but it wanted market-related pricing— that is, based on the short-term prices that emerged at the "trading hubs"—the places primarily in the UK and the Netherlands, where pipelines, LNG terminals, and gas trading converged. The EU also wanted contracts to be transparent to prevent what it defined as "anti-competitive" behavior, and it prohibited Gazprom from owning the pipes through which its gas moved across Europe.[3]

The second major thrust of the EU was around climate, aiming at decarbonization and efficiency and making a rapid march to renewable energy. At the forefront was Germany. Under the rubric of its *Energiewende*, or "energy turn," Germany provided extensive subsidies for wind and solar development. Although not the intent, it also ended up indirectly providing large subsidies to Chinese solar companies, which became the main low-cost suppliers of solar panels to the world.

By 2019, 33 percent of Germany's electricity came from renewables. But it had not come cheaply. Germany's Federal Court of Auditors criticized government ministries for exerting "no oversight over the financial impact of the *Energiewende*," for not asking "how much should the *Energiewende* cost the state," and for failing to take into account "reliability and affordability."[4]

In March 2011, a giant tsunami, set off by a massive earthquake, inundated a nuclear power station at Fukushima in Japan, resulting in the worst nuclear power accident since the Chernobyl reactor in Ukraine blew up in 1986. In the immediate aftermath of the Fukushima disaster, the German government decided to shutter its biggest source of non-CO_2 electricity—its large fleet of nuclear reactors. To help meet the gaps in electric generation, it temporarily increased its consumption of coal.

For the entire EU, natural gas comprises about 25 percent of energy consumption. That means that Russian gas, at about 35 percent of total gas consumption, provides 9 percent of Europe's overall energy. After Russia, the next largest source of gas is "indigenous" or "domestic supplies," largely from the Groningen field in the Netherlands and the British sector of the North Sea. Norway, though not a member of the EU, is highly integrated with it economically, and supplies 24 percent of the EU's gas; about 9 percent comes from North Africa, mainly Algeria.

The debate about the political risks of importing energy from the Soviet Union and now Russia has been going on for a long time. The surge of Soviet oil exports to Europe in the late 1950s and early 1960s generated great alarm in the United States. A U.S. senator thundered that the Soviets wanted to "drown us in a sea of oil" in quest of "world conquest." Headlines captured the transatlantic dissension: "Soviet Oil Feeds Dispute in the West" and "Oil a New Soviet Weapon." Washington adamantly opposed what was called the "Soviet oil offensive." For the Europeans, it was more a matter of business. The Soviets were planning a new oil pipeline to Eastern Europe, and West Germany planned to sell the special large-diameter pipe that would be required. But the United States succeeded in blocking the sale. It did not take the Soviets too long, however, to master the technology and build their own large-diameter pipes. The embargo had delayed the pipeline for a grand total of one year.[5]

In the early 1980s, in the first years of the Reagan administration, dissension between the United States and Europe again erupted over Soviet energy exports—this time not about oil, but about natural gas. Western European companies, with their governments' support, were working on a big deal to build a new pipeline to import natural gas from western Siberia. The Reagan administration, which was stepping up defense spending, did not want the Soviets earning money that would fund their own military buildup. Washington also feared that dependence on Russian gas, especially in Germany, could help Moscow generate fissures in NATO and provide a major pressure point if East-West tensions worsened. This was the time "to dig in our heels," President Reagan said, and "just lean on the Soviets until they go broke."[6]

When Germany and the other Europeans showed no signs of stepping back from the deal, the Reagan administration slapped an embargo on exports from Europe that used American technology and know-how that was required for the proposed pipeline. The sanctions, although aimed at the Soviet Union, infuriated the Europeans, who were outraged not only over being potentially shut off from the gas but also because of the loss of manufacturing jobs resulting from the embargo on selling technology and equipment. Finally, an understanding was worked out between Washington and the Europeans that capped Soviet gas imports into Western Europe. The pipeline project went ahead, but so did another new pipeline that would bring natural gas from Norway to continental Europe.

The sanctions, meanwhile, had proved to be what Secretary of State George Shultz called "a wasting asset." They had the impact that sanctions often do—they motivated the Soviets to seek to develop their own technological capabilities to substitute homemade equipment for what was embargoed.[7]

TWO DECADES LATER, NORD STREAM, UNDER THE BALTIC SEA, revived the controversy over pipelines. There was much criticism in Central European policy circles and the media over the political influence that, it was argued, Russia would gain from the deal. The Western Europeans, and particularly the Germans, saw matters differently, as part of a larger complementary relationship involving markets, trade, and investment, a relationship that was made inevitable by geography. Moreover, while they may have depended on the Russians for gas, the Russians depended on them for markets and for revenues. Nord Stream went ahead unhindered, culminating in that inauguration ceremony at Lubmin in 2011.

In the face of all the agitation and criticism that came with Nord Stream, Moscow had a message for the Europeans. At the St. Petersburg International Economic Forum, Gazprom CEO Alexey Miller told a room largely full of Europeans, "Get over your fear of Russia, or run out of gas."

Chapter 12

UKRAINE AND NEW SANCTIONS

On June 23, 2013, whatever remained of the post–Cold War comity began to unravel. On that day, Edward Snowden, a disgruntled contractor for the U.S. National Security Agency, boarded a flight in Hong Kong for Moscow. He was not carrying a valid visa, but the Russians let him in. But he did carry something of enormous value—the "keys to the kingdom" of U.S. intelligence—vast amounts of files he had stolen from the NSA. From that would flow a breakdown and then a crisis centered on Ukraine and its borders that would splinter East and West and jump-start a new cold war.

The theft was a shock to the U.S. government, and the results devastating. A minor amount of what Snowden had taken related to communication between intelligence targets and U.S. "persons"—the reason he gave for the theft. The great bulk, however, had to do with worldwide intelligence gathering, most of which was focused on terrorism and how to counter such threats as IEDs, improvised explosive devices, which were killing and maiming U.S. military personnel.

Snowden's presence in Moscow was facilitated by the Russian government. The source of that information was none other than Vladimir Putin. "I will tell you something I have never said before," Putin told a press briefing on September 3, 2013. Snowden "first went to Hong Kong and got in touch with our diplomatic representatives," who conveyed to Putin that an American "agent of special services" was seeking to come to Russia. Putin recounted that he had said the agent would be "welcome" in Russia provided that "he stops any kind of activity that could damage Russian-U.S. relations." Moscow granted Snowden asylum. Putin did add that he would have preferred not to have had to deal with the Snowden problem. "It's like shearing a pig. There's a lot of squealing and little fleece."[1]

Others, however, thought otherwise—that the Russians must have gotten quite a lot of fleece from Snowden. For, they reasoned, empathy for a "whistleblower" would have hardly been reason enough for Moscow to bear the political cost of offering shelter to Snowden.

The cost quickly became apparent. President Obama was scheduled to meet Putin in September 2013, in Moscow, for their first summit in four years. U.S.-Russia relations had begun to deteriorate after the 2003 invasion of Iraq and had worsened with the 2008 Russian-Georgian War and the 2011 Arab Spring. While the Obama administration had sought to launch a "reset" in the overall relationship, personal relations between the two presidents had been brittle from their first meeting in 2009, when Obama appeared to have been seated in a kiddie chair, while Putin lectured him about the "errors" the United States had made in its dealings with Russia. In August 2013, after Snowden's defection, Obama had reciprocated, saying Putin's "got that kind of slouch, looking like the bored kid in the back of the classroom."

Putin believed that the United States was intent on frustrating his overriding objective—the restoration of Russia as a great global power with a "privileged sphere" in the "post-Soviet space." Yet however fraught the relationship between Moscow and Washington, a summit would have been a path for restoring at least a modicum of a working relationship. But once Moscow had granted asylum to Snowden, who was regarded as the perpetrator of the worst intelligence theft in U.S. history, there was no way that Obama could meet Putin. The summit was canceled. To add an extra sting, Obama would later dismiss Russia as nothing more than "a regional power."[2]

Meanwhile, Ukraine remained caught between East and West. Ostensibly, it was about trade. Moscow was promoting what was called the "Eurasian Economic Union," which would tie together the new countries of the former Soviet Union under Moscow's leadership, inside a common tariff system and a unified economic space. But Ukraine was at the same time discussing with the European Union an "association" agreement for greater economic integration. There was a total incompatibility, indeed a fatal contradiction between the two sets of negotiations, because it was impossible to be inside two mutually exclusive tariff systems at the same time. In other words, if Ukraine completed its negotiations with the European Union, it could not be part of Putin's Eurasian Economic Union.

Moreover, Ukrainian engagement with the European Union would have major geopolitical impact, drawing Ukraine away from Russia. The discussions between Kyiv and the European Union proceeded on a rather technocratic basis, without much attention to the geopolitical issues. For the West, "Ukraine" was only one of many interests that jostled for attention in Brussels and Washington.

But for Russia, Ukraine was, as Obama would later observe, a "core" interest. In Moscow's narrative, it was part of Russia, going back to Kyivan Rus and the Cossack pledge of allegiance to the tsar of Muscovy in 1654. As Putin once summed up this view, "Ukraine is not even a country. What is Ukraine? Part of its territories is Eastern Europe, but the greater part is a gift from us." Later, citing the words of a White Russian commander in the Russian Civil War, he said, "Big Russia and

Little Russia—Ukraine. . . . No one should be allowed to interfere in relations between us; they have always been the business of Russia itself."[3]

Ukraine's economy was a mess, and corruption endemic. The king of corruption was none other than the president, Viktor Yanukovych. Defeated in 2005, the former boxer had climbed back into the political ring and, in a comeback match, was elected president in 2010.

In 2013, Yanukovych was about to sign the association agreement with the European Union when the Russians suddenly realized that it would shut Ukraine out of the Eurasian Economic Union. Moscow raised the ante—and the pressure. It was "either/or." Yanukovych backed out of the EU agreement, his exit lubricated by a $15 billion loan from Moscow.

Ukrainians were enraged. In late 2013, half a million flooded into Maidan Square in Kyiv to protest the abandonment of the European Union agreement and against the rampant corruption and Russian influence. In the freezing December weather, U.S. assistant secretary of state Victoria Nuland passed into the crowds, handing out cookies. Meanwhile, Moscow denounced the demonstrators as "fascists and neo-Nazis."

In February 2014, police opened fire on the demonstrators, killing a hundred of them. Civil war seemed imminent. Three European foreign ministers hurriedly flew in and worked out a deal with Yanukovych and opposition politicians to hasten presidential elections. But the government was disintegrating. Yanukovych's own security detail vanished. Yanukovych abruptly fled to Russia. The United States and the European Union immediately announced their support for the new interim government. One of its first acts was to ban Russian as an "official" language, a position that it had shared with the Ukrainian language. Russian was the prime tongue of many, particularly in eastern Ukraine and Crimea. This unfortunate mistake was quickly rectified, but it had a lasting impact. "The Europeans prevented them from doing that," said Putin. "But the signal had already been sent."[4]

———

WHILE ALL OF THIS WAS UNFOLDING, THE 2014 WINTER OLYM-
pics was taking place in the snow-covered mountains above Sochi, in
the south of Russia—a great celebration of Russia's return from the
abyss of the Soviet collapse. The chief celebrant was Vladimir Putin.
The opening ceremony featured a sweeping musical tribute to Russian
history. Many heads of state were in attendance, including Xi Jinping.
But not Barack Obama, not with Edward Snowden a guest of the Kremlin and not in light of new Russian legislation on homosexuality that
the Obama administration had condemned. Representing the United
States instead was Janet Napolitano, a former member of Obama's cabinet and now chancellor of the University of California.

At some point, amid all the glory and glitter of the Olympics, the
Russian government—presumably Putin and an inner circle—made a decision. Shortly afterward, perhaps in accord with a preexisting contingency plan, "little green men"—paramilitary forces—appeared in Crimea,
the large peninsula that juts out from Ukraine into the Black Sea. These
paramilitary forces were there, it was declared, to protect "oppressed"
Russians living in Crimea. Russia took control of the peninsula.

Over the centuries, Crimea, with its balmy, semitropical weather
in the summertime, had been the favored vacation spot for tsars and
nobles, and then communist leaders, and also for millions of ordinary
Soviet citizens. In 1954, the Soviet leader Nikita Khrushchev had the-
atrically "given" Crimea to the Ukrainian Soviet Socialist Republic—
ostensibly to celebrate the three hundredth anniversary of the Cossacks'
swearing fealty to Muscovy in 1654 and thus, according to that narrative, merging Ukraine with Russia. But Khrushchev was also seeking
to ensure the support of the Ukrainian Communist Party in his battle
for power in the aftermath of Stalin's death a year earlier.[5]

Of course, the gift of Crimea did not matter in Soviet times. It did matter a lot, however, when Ukraine and Russia became separate countries, and not just for reasons of nostalgia and holidaymaking. The Crimean city of Sevastopol was the only warm-water port for Russia's navy, which was there on a lease from Ukraine.

By the middle of March 2014, a Moscow-organized referendum in Crimea supposedly had 96 percent of the people voting to join Russia. The next day, Putin announced the "reunification" of Crimea with Russia. The United States and the European Union, taken by surprise, declared that Russia had overturned the accepted boundaries of Europe and imposed sanctions.

The Ukrainians protested bitterly at the annexation. The Russians had been party to the Budapest Memorandum in 1994, which guaranteed Ukraine's territorial integrity in exchange for its giving up its nuclear weapons. But Moscow insisted that the Budapest Memorandum had been invalidated by what it described as a "coup d'état," allegedly engineered by the West, that had overturned what it asserted was the "legitimate" government of Ukraine.

Then separatists, paramilitary forces, and Russian soldiers "on vacation" started military operations in the Donbas, in southeastern Ukraine, the country's great industrial heartland, still highly integrated with the Russian economy, most notably its defense industry. The pro-Russian separatists seized several cities. The insurgency turned into a war, with the support and involvement of the Russian military.

On July 16, 2014, the United States ratcheted up sanctions on Russia's financial, defense, and energy sectors. It was not clear that the Europeans, who would be more directly impacted economically, would go along. But then the next day, July 17, a shocked world learned that separatists, apparently believing that they were aiming at a Ukrainian troop plane and using a Russian ground-to-air missile, had shot down a Malaysian airliner over eastern Ukraine. All 298 passengers aboard perished, two-thirds of them Dutch. The Europeans joined the new sanctions. In lieu of force, the sanctions became, in the words of Obama treasury secretary Jacob Lew, "the centerpiece of the international response to Russia's aggressive actions in Ukraine."[6]

The war has dragged on ever since, with at least fourteen thousand deaths, widening the split between Russia and Ukraine and between Russia and the West.

One set of sanctions was aimed at specific individuals and organizations that were judged to be either close to Putin or active in Crimea and the Donbas. A second set restricted Russia's access to the global financial system and its ability to raise money in international markets, and at the same time choked off foreign investment into Russia. It made

Main gas pipeline through Ukraine

Sources: IHS Markit; Council of Foreign Relations

0   100 mi
0   100 km

Ukraine

Turmoil in Kyiv, the annexation of Crimea, and a war in southeastern Ukraine have led to a new cold war.

international banks very leery of doing business with Russia, for fear of running afoul of sanctions or inadvertently falling short on some compliance rule and becoming subject to multibillion-dollar fines and public shaming. Such financial sanctions depend on the centrality of the United States in the global financial system and the world economy's dependence on the global dollar payments system that flows through New York—and the danger of being shut out of that system.

Yet there is a risk that the commanding position of the United States—derived from its capital markets and the dollar—could be eroded over time by the overreliance on financial sanctions, because nations will find alternatives. Two years after the United States imposed financial sanctions on Russia, Obama treasury secretary Lew himself warned, "The more we condition the use of the dollar and our financial system on adherence to U.S. foreign policy, the more risk of migration to other currencies and other financial systems in the medium term grows. Such outcomes would not be in the best interests of the United States."[7]

The third set of sanctions was aimed at constraining Russia's energy might. Care was taken to construct sanctions that would not hinder Russia's current oil output, for fear of driving up the price of oil at a time when it was already high. Instead, they were aimed at the new growth areas that were deemed to require Western technology and partners. Western participation in the Arctic offshore was banned. Russia's vast Arctic shelf is little explored but is thought potentially to contain very large oil and gas resources. The U.S. Geological Survey concluded that "the extensive Arctic continental shelf may constitute the largest unexplored prospective area for petroleum remaining on earth." But for Moscow, more is at stake. Russia's advance into the Arctic is aimed at confirming its primacy in a region opening up to commerce and political competition and that Moscow ranks as of great strategic importance. This was made more than evident a few years earlier when two Russian mini-subs planted a titanium version of Russia's flag in the seabed fourteen thousand feet below the North Pole. In response, the foreign minister of Canada, another Arctic power, snapped, "This isn't the fifteenth century. You can't go around the world and just plant flags and say 'We're claiming this territory.'" But that is what Russia did.[8]

Also targeted was shale oil and Russia's immense nonconventional resources, including the huge Bazhenov formation, under the West Siberian basin. Whatever the potential, for a long time there was no technology with which to successfully produce from that complex geology.

But the shale revolution in the United States provided a possible solution for the Bazhenov—horizontal wells and multistage hydraulic fracturing. It was not just Russians who came up with that idea. In 2013, the U.S. Energy Information Administration estimated that Russia's "unproved technically recoverable" shale oil resources were potentially greater than those of the United States.

Western partners, with their know-how and experience, could help a great deal. Finding and developing the "sweet spots" is a matter of patience, capabilities, data, and trial and error. As a Russian petroleum engineer in Siberia put it, "We need to gradually, bit by bit, find the keys." Thus the Russian companies enlisted Western partners and technology.

But with the new sanctions, the Western companies had to drop out. As the Russian engineer observed, Western companies were "afraid to touch the Bazhenov as if it were a fire."[9]

So the Russian companies are on their own, advancing by themselves, improving their capabilities. Eventually they will be able to substitute Russian-made equipment for that which they cannot buy from the West, bearing out Secretary of State George Shultz's already cited dictum from the 1980s Soviet gas controversy—sanctions can be a wasting asset. Still, the sanctions probably have set back the Bazhenov development by half a decade or more. But in an era of oversupply and extensive conventional opportunity in Russia, such a delay is not such a bad thing from Russia's own point of view.

Chapter 13

OIL AND THE STATE

The U.S.–EU sanctions, as well as those of other countries, including Japan and Norway, had been imposed at a time of high oil prices and expectations of a continuing tight market. But then in late 2014 the oil price collapsed, delivering a new shock to a Russian economy and national budget so heavily dependent on oil. A severe crisis seemed inevitable. And indeed, the initial impact was great—capital flight, drying up of both foreign and domestic investment, loss of access to international capital markets, plummeting spending by consumers, and a declining GDP.

But the shock was cushioned by the policies of the Russian central bank. It closed insolvent banks, including those owned by powerful figures, and allowed the ruble to float. The currency lost more than half its value against the dollar. But this flexibility helped to steady the economy. Expenditures by the Russian government are largely in rubles. Thus a fall of 50 percent in dollar revenues from oil would, roughly speaking, still convert into the same amount of rubles within Russia as prior to the collapse.

The devaluation was a great boost to the Russian oil industry. It received dollars from its exports, but most of its expenditures on workers and equipment were in devalued rubles, and so the collapsing oil price had very little effect on industry activities within Russia. Indeed, between 2014 and 2016, Russian oil output increased.

Imported goods became much more expensive for Russian consumers, who were paid in what were now devalued rubles; and they cut way back on such purchases. At the same time, owing to the fall in the ruble, domestically produced Russian goods were now much more competitive not only domestically but internationally. This applied to both manufactures and agriculture, in the latter case also aided by far-reaching reforms in the farming sector. Russia became the largest exporter of wheat in the world—quite a turnaround from the 1970s, when the Soviet Union spent a good part of its oil earnings buying wheat from the United States. Moreover, in retaliation for the sanctions, the Russian government imposed sanctions of its own—banning food imports from Europe. This proved to be a great boon for Russian farmers.

But other parts of the economy were much threatened. The closing of international capital markets put Russian financial institutions and companies that had borrowed in dollars or euros in the precarious position of not being able to meet their debt payments. The Kremlin stepped in with an "anti-crisis" program that provided subsidies and funding. To do so, it drew down its sovereign wealth funds.

Those funds had been built up over several years by Alexei Kudrin, finance minister from 2000 to 2011. He had been seared by the 1998 crisis when the Russian economy went into free fall and the government ran out of money. Kudrin had long been criticized for socking away some of Russia's large oil earnings into sovereign wealth funds and paying off its foreign debt, instead of spending the money right away. But now the wisdom of the "rainy day" funds was being proved. A visitor remarked to Kudrin that people must be thanking him for his prescience—and insistence. He managed a small smile. "Not enough," he said. He was still being attacked.[1]

Yet when it was all added up, the Russian economy proved more resilient to the sanctions and oil price collapse than had been expected. By 2017, the economy had crawled back into positive economic growth, and by 2019 it was growing at 1.6 percent. Yet the crisis had demonstrated once again the risks of being so reliant on oil. Hopes for economic reform were derailed, in part, by the plethora of sanctions and disengagement from the global economy—and by the domestic vested interests that would be challenged by reform. The new isolation instead made companies more dependent on the state and expanded the role of the government in the national economy.

The hopes for reinvigorating economic growth now largely rested with a series of "national projects," aimed at a wide array, including infrastructure, health, and education. These are government initiatives, with a considerable commitment to spending. The Russian economy was returning to state control. Reform would once again have to wait.[2]

Chapter 14

———

PUSHBACK

In late 2015, four years after Nord Stream started operating, survey-ors began to map out a second pipeline route under the Baltic Sea from Russia to Germany. Opposition to Nord Stream 2, as it was known, was much stronger than against the original Nord Stream. Part of the reason was what had happened in between—specifically, Ukraine. Criticism came from parts of Europe, notably Poland and the Baltic countries, as well as the European Union itself—a turnaround from its supportive position on the original Nord Stream. Donald Tusk, presi-dent of the EU Council, was already warning, "Excessive dependence on Russian energy makes Europe weak." Maroš Šefčovič, EU vice pres-ident, put Nord Steam 2 high on the list of "hybrid threats."

That was not, however, how other Europeans, including German chancellor Angela Merkel, saw Nord Stream 2. It was a commercial project, she said, and up to the companies involved—Gazprom and its European partners. In March 2017, the first pipe arrived at a logistics hub in Germany. But the proponents of Nord Stream 2 had not fac-tored in what was unfolding in Washington.[1]

Donald Trump came into the presidency determined to set a new course on Russia. During the election campaign, he had praised Putin as a strong leader, exuberantly declaring that Putin had called him a "genius." (Putin had used the word *yarkii*, which translates as "bright," as in "eye-catching" or "shiny.")

Trump, however, was running in his own lane. "Russia" had become a bitterly divisive subject in Washington. A joint task force from the CIA, the National Security Agency, and the FBI concluded "the Russian government pursued a multifaceted influence campaign" in the 2016 election that included "aggressive use of cyber capabilities" and that "President Putin directed and influenced the campaign to erode the faith and confidence of the American people in the presidential election process" and "demean Secretary Clinton" and "advantage Mr. Trump."

Putin certainly did not hide his distaste for Clinton. It was mutual. She had said of Putin that as a former KGB agent, "by definition he doesn't have a soul." After the outcome of 2011 parliamentary elections in Russia sparked demonstrations, she charged the Kremlin with electoral fraud. In response, Putin accused her of paying anti-Kremlin demonstrators in Moscow. In Washington, after Trump's victory in the 2016 election, "Russia" and Russian intervention in that election was a dominating theme, and the authoritarian nature of the government and corruption a constant refrain.[2]

In an effort to "do something," the U.S. Congress passed myriad new sanctions, targeting individuals said to be close to Putin, and companies and financial institutions. Some aimed to further constrain Russian energy projects and limit Western participation in them.

Normally, U.S. sanctions legislation permits presidential discretion so that they can adjusted as a policy tool in response to changes on the part of the sanctioned country. But, controversial in terms of presidential authority, some of these new sanctions were written permanently into law, eliminating that flexibility. This prevents current or future administrations from using them as leverage in negotiations or to affect behavior. History demonstrates that sanctions written into law, without presidential discretion, are not easily removed. The Jackson-Vanik amendment, passed in 1974 to support Jewish emigration from the Soviet Union, remained on the books for thirty-eight years, although by then the Soviet Union had long ceased to exist and Russia's compliance with the law had been certified many years before. These new congressional prohibitions were testament to the general hostility toward Russia and, among Democrats, continuing rancor over Russian interference in the 2016 election—and to deep distrust of the Trump administration and of Trump himself. That point was noted by Rosneft CEO Igor Sechin, who said, "Sometimes it seems to me that sanctions are imposed on him, not us."[3]

Trump signed major sanctions legislation in August 2017 though he called it "seriously flawed" because it encroached on presidential authority. But he found a novel source for his authority beyond the Constitution. As he said, "I built a truly great company worth many billions of dollars. That is a big part of the reason I was elected. As president, I can make better deals with foreign countries than Congress."[4]

Nord Stream 2 was targeted in several pieces of proposed legislation. There was an assumption among some in Washington that if the pipeline was not built, it would reduce Russian gas going into Europe. But that was not correct. The gas would simply flow through other pipelines, including those in Ukraine and Turkey.[5]

While the eastern members of the European Union wanted new sanctions aimed at stopping Nord Stream 2, the rest of the continent reacted differently. "Europe's energy supply is a matter for Europe, not the United States of America," said Germany's foreign minister and Austria's chancellor in a joint statement. "Instruments for political sanctions should not be tied to economic interests." It was difficult for Europeans to see the connection between Russian meddling in U.S. elections and a natural gas pipeline in Europe. There had to be another

explanation. The head of one of Europe's largest energy companies suggested that the sanctions were a way for the United States "to try to
favor its own gas"—that is, exports of LNG from the United States.
Germany's foreign minister said the same thing. This reading was not
without reason, for the 2017 legislation called for "the export of United
States energy resources to create American jobs."[6]

Wolfgang Ischinger, the chairman of the Munich Security Conference and the former German ambassador to the United States, observed that Americans would be highly riled were Brussels to pass
legislation to block an oil pipeline from Canada to the United States.
Ischinger also pointed to a fundamental lesson of sanctions—they are
more likely to succeed when they are multilateral. Unilateral sanctions
create rancor among allies. The biggest beneficiary of conflict between
the United States and the EU over sanctions would be Russia, which
would warmly welcome a more divided West.

The outlines of a deal over Nord Stream 2, at least in Europe, were
becoming evident. The pipeline was not "just an economic project,"
said Chancellor Merkel in the spring of 2018. "Of course, political factors must also be taken into account." What she had in mind specifically was that a certain amount of gas would be guaranteed to flow
through the Ukrainian system. For its part, Gazprom signaled that it
would maintain some level of exports through Ukraine. But in Washington, thirty-nine senators called on the administration to stop Nord
Stream 2 because it would make "Europe more susceptible to Moscow's coercion and malign influence."

It turned out that the bluntest critic of Nord Stream 2 was Donald
Trump. Across a breakfast table from NATO's secretary-general, he
declared, "Germany is totally controlled by Russia because they're get-
ting 60 to 70 percent of their energy from Russia in a new pipeline.
You tell me if that is appropriate. I think it's not." He wasn't finished.
"Germany is a captive of Russia," he added.

At this, Chancellor Merkel took personal umbrage, having grown

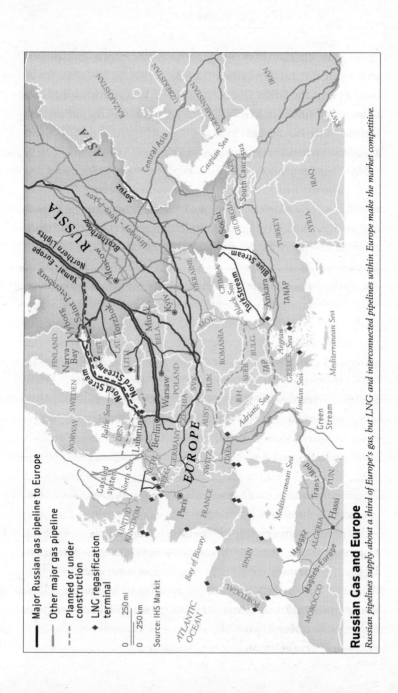

Russian Gas and Europe

Russian pipelines supply about a third of Europe's gas, but LNG and interconnected pipelines within Europe make the market competitive.

Major Russian gas pipeline to Europe

Other major gas pipeline

Planned or under construction

LNG regasification terminal

Source: IHS Markit

0 ____ 250 mi
0 ____ 250 km

up in communist East Germany under the omniscient eye of the Stasi secret police. "I have experienced myself how a part of Germany was controlled by the Soviet Union," she shot back. "I am very happy that today we are united in freedom" and "can make our independent policies and make independent decisions."

Not deterred, Trump renewed the attack via Twitter: "What good is NATO if Germany is paying Russia billions of dollars for gas and energy?" Meeting with the European Commission's president, he promised that the United States would sell "massive" amounts of LNG to Europe. Yet around that time, in the shallow waters off Lubmin, Germany, pipes were being dropped into place for the first eighteen miles of the 840-mile underwater pipeline.[7]

A year and a half later, in December 2019, the $11 billion pipeline was only weeks away from completion. On December 9, Putin was in Paris to meet with German chancellor Angela Merkel, French president Emmanuel Macron, and Volodymyr Zelensky, the new president of Ukraine. Zelensky had become famous in Ukraine for his popular television comedy show "Servant of the People," in which he played a schoolteacher who accidentally becomes Ukraine's president. Now he actually was president, elected with 73 percent of the votes in April 2019. Zelensky described this, his first meeting with Putin, as "a tie for now."

A week later, on December 17, the U.S. Senate passed a multibillion-dollar defense bill. Sanctions on Nord Stream 2 were tacked onto it.

Three days later, on December 20, to the surprise of many, word got out that Russia and Ukraine had concluded a settlement of what had seemed to be their endlessly acrimonious natural gas battle. It was more than a "tie"; it was the deal that Ukraine could only have hoped for: Russia guaranteed five years of volumes of natural gas to Europe through Ukraine, which would assure a level of transit revenues. Even more surprising, Russia agreed to pay Ukraine a $3 billion arbitration claim that Ukraine had won against Gazprom. That was roughly equivalent to one year of transit revenues to Kyiv.

Some few hours after the Russians and the Ukrainians had finally settled their long-running conflict, Donald Trump, at Andrews Air Force Base in Maryland on his way to Florida, signed the defense bill, imposing sanctions on the Nord Stream 2 pipeline.

The sanctions targeted just one company—a Swiss firm that owned the pipe-laying barge, at the time the only one of its kind in the world equipped for this project. The barge stopped work almost at once; the company had no choice but to comply. Both Germany and the European Union expressed outrage at what they saw as extraterritoriality and America's illegally intervening in Europe's domestic affairs. "This is a very important project," said Chancellor Merkel. "It has been legitimated by the new Europe law. We need to carry it through." For its part, Gazprom responded that it would finish laying the pipe itself. But not quickly. It had already bought a pipe-laying barge as a precaution, but it would take many months to properly outfit the ship for the job. There were new legal challenges and threats of new sanctions. And there, in the meantime, at the bottom of the Baltic Sea, Nord Stream 2 lay in a state of suspended animation, almost but not quite finished.[8]

WHATEVER THE PIPELINE ROUTE, EUROPE WOULD NEED ADditional natural gas imports to make up for declining domestic supplies. The Groningen gas field in northern Holland, discovered in 1959, was the biggest domestic source of gas within Europe, and the foundation on which the original European gas system had been built. It still ranks among the top ten gas fields in the world. But its days are numbered. Owing to its particular geology, production over many years has led to subsidence, sinking of the topsoil, which has triggered tremors and earthquakes, causing cracks and damage in houses. The Dutch government has imposed severe restrictions on production and a likely total shutdown by 2022.

This would not affect new discoveries of the Netherlands' offshore, but it would mean that Europe would lose what had been its largest domestic source of gas. Europe will need additional imports. Some will come from Azerbaijan through a new pipeline system that reaches into Italy. Some might come from Israel and Cyprus. Some would come from LNG. But Russian gas, whatever the route, would also benefit from the declining domestic production.

Yet the concern about Russia's potential leverage from gas exports does not fully recognize how much both the European and world gas markets have changed. The gas market in Europe has become a real market of buyers and sellers, rather than a system based on inflexible long-term contracts. And LNG has become a truly global industry—one that, declared the European Union, "can give a real boost to Europe's diversity of supply and hence greatly improve energy security."[9]

There already were a number of terminals in Western Europe to receive the LNG, regasify it, and inject it into the European pipeline system. But not in Eastern Europe. The first country to remedy this was Lithuania, which was completely dependent on Russian gas and was paying higher prices than other countries. It opened its first LNG receiving terminal in 2014. At the ceremony, the country's president described the facility as "a guarantee not only of our energy but also of our economic independence." She added that Russia would "no longer be able to exert political pressure" by manipulating gas prices. Just to be sure that no one missed the point, the terminal was christened "Independence." Lithuania's energy minister tried to be a little more diplomatic. "The Russians are very good people, but it is difficult to negotiate with them," he said. "We built a small LNG terminal to have a stronger position in negotiations with them. And it worked. Gazprom reduced its price." A year later, Poland, also until then totally dependent on Russian gas, opened a much larger LNG import terminal.[10]

Europe now has more than thirty receiving terminals for LNG, which can be ramped up on short notice. They are also part of an increasingly dense global network. Worldwide, over forty countries now import LNG, compared to just eleven in 2000. Exporting countries have increased from twelve to twenty. Overall global LNG demand in 2019 was almost four times larger than in 2000, and liquefaction capacity is expected to increase by another 30 percent over the next half decade. Methane molecules from a growing number of countries now jostle and compete with one another for customers across the globe.

European buyers today have choices in the new competitive global market. They will put together portfolios of pipeline gas and LNG that fit their needs and economics and their risk calculations. So long as warfare continues in Ukraine, politics will fuel rancorous discussion of European gas. But with Europe now part of a global market, political risk is draining out of Europe's gas supply.

Ukraine is no longer directly dependent on Russian gas, but rather imports gas that may or may not be Russian molecules through Slovakia, Hungary, and Poland. Moreover, domestic production supplies about two-thirds of total demand, and the share could go higher, as Ukraine may possess the largest natural gas resources of any country in Europe. Some 80 percent of current production comes from the state-owned gas company. The second-largest among the private firms is a company called Burisma. It represents just 5 percent of domestic production, but it has gained oversized fame outside the energy sector. It is the company that Donald Trump wanted investigated in Ukraine, owing to the fact that Hunter Biden, son of former vice president Joe Biden, sat on its board. It was that famous "quid pro quo"—linking the provision of U.S. aid to an investigation of the company and the Biden connections—that became the basis for the impeachment of Trump in 2019.[11]

A SURPRISING NEW COMPETITOR TO GAZPROM HAS ENTERED the LNG business—within Russia itself. Since 2009, Russia has been exporting LNG from its Far East island of Sakhalin, north of Japan. Yes, vast volumes of natural gas had been discovered in the Yamal and Gydan peninsulas, above the Arctic Circle. The reserves in the southern part of the Yamal Peninsula are or can be connected by pipeline for export. But it was assumed that the huge reserves in the frozen northern part of Yamal would never be developed because of the remote location and high cost, thus leaving that gas permanently "stranded."

But not quite everyone agreed. Leonid Mikhelson, the CEO of the independent Russian company Novatek, was determined to develop LNG export capacity in the north of the Yamal Peninsula. The main inhabitants of this barely populated region are several thousand Nenets, partly nomadic people who move with their reindeer herds, which they supplement by hunting polar bears. In the language of the Nenets, "Yamal" means "end of the land," and that is what the remote northern part of the peninsula literally is—a harsh, vast, bleak, and treeless land that juts out into the forbidding ice pack of the Arctic Ocean and is underlaid by permafrost. The region is so far north that it is completely cloaked in darkness in the winter and bathed in perpetual polar sunlight during the summer.

The region in which Yamal LNG intended to operate is three hundred miles from the North Pole and often unreachable by land and cut off by winds, fog, and blinding snow that forces helicopters to turn back midflight. Temperatures fall below minus 43 degrees Fahrenheit in winter. To bring the LNG industry to that ultra-cold region, all sorts of special equipment and construction techniques had to be developed. There was also virtually no infrastructure. At a cost of $30 billion, an entirely new port city of Sabetta had to be constructed. And then there is the vast expanse of ocean ice. In addition to nuclear-powered icebreakers, fifteen new ice-class tankers had to be designed and built, at a cost of $320 million each, to move through the Arctic waters. One advantage of the project, however, is that the extreme cold makes the gas refrigeration process easier than in the hot climate of the Persian

Gulf, enabling the units to produce more gas than their official "name plate" maximum capacity.

Already very challenging, the project became even more so in 2014, when Ukraine-related sanctions cut off Novatek's access to Western finance. In order to survive, the $27 billion LNG project needed a new injection of money, and quickly. The Chinese came through with a $12 billion loan and also became partners in the project, along with the French supermajor Total, which had joined earlier. The Russians had historically been reluctant to allow large-scale Chinese ownership of upstream assets. But now there was no choice.

The decision to go ahead with the project in 2013 had been met with much skepticism and doubt in both Russia and the international LNG industry. But by December 2017, the first cargo of LNG was ready to leave the new port of Sabetta. At the dedication ceremony, on a freezing cold day, a parka-clad Putin said that "good people" and "good professionals" had "warned me . . . 'Do not do this.'" He pointedly shifted his gaze across the four hundred people gathered for the ceremony. "The reasons they cited were very serious." Despite the extreme skepticism, he added, "Those who started this project took the risk." He continued, "This is not just an important event in our country's energy sector," but also for "developing the Arctic and Northern Sea Route." He added, "All this is interconnected and secures Russia's future."[12]

That first cargo was sold on the spot market. It ended up in a British terminal, where it was bought by another company. From there, it became part of a cargo that was rushed to Boston to keep freezing New Englanders warm during unexpectedly cold weather. The arrival of Russian molecules in Boston harbor created consternation and outrage. One U.S. senator declaimed that the landing of the molecules, even though no longer Russian-owned and intermixed with molecules from other countries, "undermined broader foreign policy goals regarding Russia." But the local utility had no choice but to buy whatever cargo was available, as the deep freeze threatened to leave the region short of heat. "During the cold snap," said the utility, "LNG was absolutely vital in meeting customer needs."[13]

Massachusetts sits near the vast volumes of inexpensive gas in the Marcellus shale, which would have enabled it to avoid Russian molecules. But environmental activists and regional politicians have unwaveringly blocked construction of a new pipeline from Pennsylvania.

In August 2018, Yamal LNG dispatched its first cargo to China, going east along the Arctic coast, through the ice of the Northern Sea Route. Yamal LNG had come in on time and on budget. The *Financial Times* observed another noteworthy aspect of the project. "No other business venture," it said, "better illustrates Russia's resilience in the face of international sanctions."[14]

The path of the tanker also delineated what has become known as the NSR—the Northern Sea Route. This fulfills a major Russian objective, the opening up of a transit route between Europe and Asia through the Arctic Ocean. It has been facilitated by the retreat of the Arctic ice, although with more variability than sometimes recognized. For instance, in September 2014, the ice extent was 50 percent greater than it had been in September 2012. The route cuts the distance between Shanghai and Rotterdam by about 30 percent, and in the process avoids both the narrow Malacca Strait and the Suez Canal. This opening has been welcomed by Japan, South Korea, and especially by China, which, describing itself as a "Near Arctic State," applies its own distinctive name to the route—the Polar Silk Road.[15]

Overall, however, the Northern Sea Route is still supplementary, owing to the challenges of ice and weather. There's a fee, for instance, for icebreaking. But it is crucial for Yamal LNG, which targets Asia and can deliver LNG to China in just twenty days.

But Europe is also a market. This holds out the possibility that Russian pipeline gas, whether it comes through Ukraine or Nord Stream 2, will face a new competitor in Europe—Russian LNG. LNG projects for Russia's Arctic gas make clear that Russia will become the fourth major pillar for LNG supply in the 2020s, along with the United States, Qatar, and Australia. These Arctic projects will give Russia the same advantage that Qatar achieved earlier this century—the flexibility, as Putin put it, to go either "eastward" or "westward." And Yamal LNG, said Putin, "is one more confirmation of the status of Russia as one of the world's leading energy powers."[16]

The development of Arctic LNG also points to a major geopolitical shift with worldwide impact—the *povorot na vostok*, Russia's "pivot to the east."

Chapter 15

___

PIVOTING TO THE EAST

n May 2014, Vladimir Putin, accompanied by a large entourage of government ministers and businessmen, swept into Shanghai, welcomed on a state visit by President Xi Jinping. Russia's "pivot to the east" had taken on a new urgency. For it was just two months after Russia had annexed Crimea, and now the war was beginning in southeastern Ukraine. The European Union and the United States were responding with their initial package of sanctions, and relations were deteriorating fast. Chancellor Merkel, who a few years earlier had spoken about the "safe and reliant partnership" with Russia, now decried it as a violator of "basic principles" and international law. She added an extra sting; Putin, she said, was "living in his own world."

What Shanghai would demonstrate was that, in Putin's world, the topography of Asia loomed ever larger. This meeting would make clear how much more intertwined energy and strategy had become.

Shanghai in May 2014 was Putin and Xi's seventh meeting in fourteen months. But this time was different. As a leading Russian commentator put it, "The previous rosy dreams of integration with the

West" were over in the face of "the West's attempts to organize its international isolation." China provided an alternative. "We share the same priorities both on a global and regional scale," said Putin. Russia and China were united in opposition to what they described as "unipolarity" and the U.S. "hegemonic" international system, and to the promotion of democracy and regime change, abetted by activists and NGOs. Instead, they championed multipolarity and, more than anything else, "absolute sovereignty"—especially their own.[1]

Top of the agenda was a massive natural gas deal. The negotiations had been grinding on for a decade, but getting a deal done was now a priority. China was determined to use more natural gas to fuel its growing economy and ease its stifling pollution. Russia needed to diversify away from its dependence on European customers and anchor its future in a market with a huge appetite for oil and gas and with a country that was more compatible in both strategy and economics, including the "Beijing consensus" of state-directed capitalism.

All this reinforced the pivot. A Russian Foreign Policy Concept paper declared that Russia needed to adapt to a situation in which "global power" is shifting "to the Asia-Pacific region" and to move quickly to become "an integral part of this fast-developing geopolitical zone." It was also pretty clear that the pivot was mainly toward one country—China. Once asked if he was putting "too many eggs in the China basket," Putin replied, "We have enough eggs, but there are not that many baskets where these eggs can be placed."[2]

Between 2006 and 2013, Chinese gas consumption had tripled. Yet despite the decade of negotiations, the "big deal" on gas was mainly stuck on one question—price. Moscow wanted prices commensurate with what it charged Europeans and indexed to oil (which was still high), while Beijing wanted lower prices in line with domestic energy prices and competitive with coal.

The negotiations in Shanghai were difficult and dragged on. But neither side could afford to leave without a deal. The final negotiations went on until four in the morning. Later that day came the announcement of the big deal—valued at $400 billion over thirty years. The contract would make China the second-largest market for Russian gas, after Germany. The Chinese would also provide the financing for a massive new $45 billion, thirteen-hundred-mile "Power of Siberia" gas pipeline. "This will be the biggest construction project in the world for the next four years, without exaggeration," Putin said after the signing. But, he ruefully added, "Our Chinese friends drive a hard bargain as negotiators." But the deal did send the message that in the future, Russia would not have to depend so heavily on selling gas to Europe.[3]

These deals and their setting broadcast a geopolitical message—about the burgeoning strategic relationship between China and Russia. It was a long time coming. China and the Soviet Union had been bitter rivals over which was the leader of the communist world and vanguard of the global revolutionary movement. Mao Zedong was rather blunt about how he felt about the Soviets. "They were renegades and scabs," he said, "slaves and accomplices of imperialism, false friends, and double-dealers." The rivalry erupted in 1969 into a shooting war on their border in the Far East.[4]

In the 1990s, after the collapse of the Soviet Union, the Russians became apprehensive about Chinese expansion into the sparsely populated Russian Far East. But once installed as president, Putin put a new emphasis on China and Asia. As estrangement from the West grew, he found a willing partner in Beijing. Xi Jinping's first stop on his first foreign trip as president in 2013 was Moscow. China became Russia's largest trading partner. The respective roles were very clear. China provided manufactures, consumer goods, and finance; Russia, oil, gas, coal, and other commodities—and geopolitical alignment.

Putin described China as "our strategic key partner." Xi fully reciprocated, calling the two countries each other's "most trustworthy strategic partners." This partnership manifested itself in many ways. Russian naval units participated in Chinese naval exercises in the South

China Sea, and Putin declared that "a non-regional power"—that is, the United States—should stay out of the South China Sea dispute. Chinese troops marched through Red Square with Russian soldiers on the seventieth anniversary of the victory in World War II in Europe, although China had not fought in Europe. Russia sold China advanced Su-35 fighter jets and S-400 missile defense systems. It had previously refused to sell such weaponry for fear that the Chinese would reverse engineer and copy the weapons. But after sanctions and with the deepening relationship with China, that concern was shunted aside. The expansion of military ties with China, said Russia's defense minister, was "an absolute priority."[5]

This was the kind of relationship that Putin valued—great power to great power. "It's always about global leadership—not arguments about some second-rate regional issue," he once said. "The competition is among the world powers. That's the law. The question is what are the rules by which this competition is developing." On what the rules should be, Moscow and Beijing pretty much agreed.[6]

The pivot is very evident in energy terms. It has been facilitated by the construction of the $25 billion, 2,800-mile ESPO (Eastern Siberia–Pacific Ocean) oil pipeline. In 2005, just 5 percent of Russia's oil exports went to China. It rose to almost 30 percent, and Russia eclipsed Saudi Arabia as China's number one supplier. A financial backbone to the oil trade is provided by the $80 billion of prepayments that China has made to Rosneft for oil supplies to be delivered over the next twenty-five years.

There are clearly issues that could impede this strategic partnership. Russia continues to fear large numbers of Chinese pouring across the once-contested border to fill the empty space in the underpopulated and economically stagnant Russian Far East. Some estimate that two to five million ethnic Chinese have migrated legally and illegally into the Russian Far East. Others put the figures much lower.

Whatever the number, Putin once warned that if the economic decline in the region was not reversed, the Russian population in the Far East would end up speaking Chinese.[7]

In China, some refer to part of the Russian Far East as "Outer Manchuria," insofar as it was transferred from China to Russia under two of the "unequal treaties" of the nineteenth century. But at this time, the frontier is firmly entrenched, with no disagreement over the borders.

What Xi called the "new level of cooperation" between the two countries was on display in what became known as "pancake diplomacy" at an economic forum in Vladivostok in September 2018. Putin and his chief guest, Xi Jinping, took time out from the conference proceedings to don blue aprons, cook the Russian pancakes known as blinis, cover them with caviar, and toast each other with vodka.[8]

At that same time, a massive war game was taking place across the Russian Far East—the largest since Soviet exercises in 1981. The exercise included Chinese troops, along with Chinese jets, in the weeklong simulation of a major, multifront war.

Almost a year later, in the summer of 2019, the Russians and Chinese began joint air patrols in the Pacific. The new arrangement was announced to the world when a joint patrol flew into South Korea's air defense identification zone, which ended up with South Korea scrambling its own jets.

All of this was a clear message not only of preparation for possible conflict in the future, but also of geopolitical alignment today.

Chapter 16

━━ ━━ ━━

THE HEARTLAND

This new comity between Russia and China is all the more remarkable in light of the Chinese advance into Central Asia. In the 1990s, before Ukraine took center stage, Central Asia and the Caucasus had been the regions in which Russia vigorously sought to reassert its primacy in the post-Soviet space. The "near abroad," as Moscow called it, became the focus of a geopolitical clash—not with China, but rather the United States. Central Asia—encompassing what are now Kazakhstan, Kyrgyzstan, Tajikistan, Turkmenistan, and Uzbekistan, plus Azerbaijan—was the very center of the Eurasian landmass that one of the fathers of modern geopolitics, Halford Mackinder, in a famous address to the Royal Geographical Society in 1904, had identified as the "geopolitical pivot of the world"—the "heartland."[1]

In the years following the collapse of the Soviet Union, Moscow was determined to ensure that these Central Asian countries remained in Russia's privileged sphere, despite their new independence. The United States, some Russians insisted, had engineered the collapse of the Soviet Union to build a band of independent states that would keep Russia weak—and get its own hands on Caspian oil. For the United States and Europe these new nations were independent countries that should be permitted to develop their own identities and economies. That was what life was meant to be like in the globalized world after the Cold War. Moreover, if these countries were weak and unstable, they would fall back into Russian hands or become prey to neighboring Iran. For some of the countries, oil and gas were essential for their independence. "We used oil for our major goal," said Ilham Aliyev, who would become the president of Azerbaijan, and that goal was "to become a real country." To do that, they needed to lay down a new infrastructure map, in which pipelines flowed not north into the Russian system, but from east to west, to the Black Sea, ensuring independence.[2]

There was an additional reason for the Western interest. The Gulf War in 1991 had liberated Kuwait and yet had also engendered a new sense of insecurity about the heavy reliance on the Middle East. "The Caspian region will hopefully save us from total dependence on Middle East oil," said then–U.S. energy secretary Bill Richardson. Washington did not want to see oil development in the region stymied by Russian opposition and competition—or reassertion of control. Vice President Al Gore became a champion of pipelines for Caspian oil. "The security of the world's oil and natural gas resources continues to be the highest interest of the United States and its allies," he explained. "Today, the boundaries of the region over which U.S. interests are concentrated expanded to the Caucasus, Kazakhstan, and Siberia."[3]

Thus ensued the great Caspian pipeline derby, a struggle that went on for a decade and pitted Russia against the United States and Britain, but also involved other countries. The Russians were opposed to flows that went west. But after much invective, threats, and political machinations, the key westward pipelines went ahead. One begins in Azerbaijan, just south of Baku, heads west for more than a thousand miles, crosses fifteen hundred rivers, high mountains, and sensitive terrains, and goes south across Turkey to the port of Ceyhan on the

Mediterranean. The Caspian Pipeline, which takes Kazakh oil to world markets, goes from Kazakhstan through southern Russia to the Black Sea. From there tankers pick up the oil and carry it through the Bosporus and into the Mediterranean and onto world markets. That pipeline is Kazakhstan's essential connection to the world.

By the first decade of the twenty-first century, both pipelines were operational, connecting Azerbaijan and Kazakhstan to global markets. Azerbaijan's output has tripled since the breakup of the Soviet Union and is now around 800,000 barrels per day. But the real oil powerhouse is Kazakhstan, whose output has grown from 570,000 barrels per day to 2 million. Those countries together are now producing more oil than the combined Norwegian and British output from the North Sea. The buildup of production from the long-delayed huge Kashagan field and expansion of Tengiz in Kazakhstan will further push up output. Oil and gas development are indeed underwriting their independence.

But pipelines would eventually also run in the other direction, not only from east to west, but also west to east—that is, from Central Asia to China.

And no one understood this emerging new geopolitics better than Nursultan Nazarbayev. He had been the leader of Kazakhstan since becoming first secretary of its Communist Party in 1989. With independence in 1991, he became president. Buoyed by oil revenues, Kazakhstan's economy had by 2019 grown almost eightfold since 2000. Nazarbayev has been an adroit balancer among the great powers—Russia and China and the United States. He also sought balance at home. A majority of the country is ethnic Kazakh, but 25 percent are ethnically Russian and Ukrainian, largely in the north of the country; and he worked to maintain the domestic balance, including establishing in the middle of the country the new capital city.

The connections between China and Kazakhstan are very important to both. Kazakhstan is rich in the natural resources that China needs. Far larger in territory than all the other Central Asian countries combined and with a long border with China, it is a major trade corridor for China. For Kazakhstan, the China market and Chinese investment will do much to determine its future prosperity.

In 2019, Nazarbayev did something unprecedented in the "post-Soviet space." Aged seventy-eight, he suddenly announced his resignation. He did not want his "legacy" to be the turmoil of an uncertain transition. Moreover, he would remain the "father of the nation" in a role somewhat reminiscent of Lee Kuan Yew's in Singapore after he stepped down as prime minister. Nazarbayev's successor as president is Kassym-Jomart Tokayev, who had been both prime minister and foreign minister and who speaks Chinese.[4]

The Central Asian countries welcome the trade and investment from China, recognizing that Beijing has no interest in changing their autocratic political systems, does not criticize them and their elections, and does not support human rights activists. Government officials in those countries might still conduct business in the Russian language with which they had grown up, but Chinese economic influence has become significant. Yet the rapid growth of the Chinese presence—and dependence on China—also causes anxiety in the Central Asian countries. At the popular level, there is suspicion and resentment of Chinese influence and such matters as Chinese acquiring farmland. At the government level, officials worry about how to balance between Moscow and Beijing. Russia's continuing engagement helps prevent these countries from becoming too tightly embraced by China. One Central Asian leader, asked about that challenge, did not reply in words. Instead, he silently responded by smiling and locking his arms in a very tight hug.

The Chinese are careful to convey that they are not aiming to supplant Russia's "privileged" position in the region. Still, Moscow, even while deepening its own ties with China, eyes the Chinese expansion with some wariness. Whatever the rhetoric, Chinese investment in energy and infrastructure in the region does weaken the underpinnings of Russia's privileged sphere and supplants Russian influence. But for now, Russia sees itself as the prime beneficiary of investment from and trade with China—and, beyond economics, from the strategic relationship, which has become a major fact of international life.

THE RELATIONSHIP BETWEEN THE RUSSIAN AND CHINESE presidents has taken on a very personal character. Xi was Putin's guest at the St. Petersburg International Economic Forum in June 2019. Putin apologized for keeping Xi up late, to what was 4 a.m. China time. "But," he explained, "we talked about everything." Xi added, "We never have enough time." Two weeks later at a conference in Tajikistan, Putin surprised the Chinese leader with a present for his birthday—a large box of his favorite Russian ice cream. An appreciative Xi beamed that Putin was his "best friend."[5]

However deep the level of friendship between the two leaders, the respective commercial interests between the two countries are disproportionate. For Russia, China's market is crucial. China now accounts for 11 percent of Russia's total exports, and in terms of energy it will only grow. For China, Russia is an important and reliable supplier of energy imports, but accounts for only 2 percent of China's total exports. Russia is also a key part of Beijing's strategy of energy diversification. Russian oil and gas lessen dependence on the Middle East and on waterborne transport, which the Chinese worry could be interdicted by the U.S. Navy.

Yet, economically, the United States is much more important to China than is Russia. In 2018, before trade wars and the coronavirus, China exported $35 billion worth of goods to Russia, compared to $410 billion to the United States. When sanctions were placed on financial dealings with Russia owing to Crimea and Ukraine, the major

Chinese banks observed them. It was not worth losing access to the dollar system and international capital markets for a few deals in Russia. That business was left to a handful of special banks.

The pivot was very much in evidence on December 2, 2019. Five and a half years after the signing of the mega natural gas deal in Shanghai, the massive 1,865-mile Power of Siberia natural gas pipeline was ready to begin flowing gas. Vladimir Putin, in Sochi, and Xi Jinping, in Beijing, were joined by a complicated video hookup to control rooms on each side of the Russian-Chinese border to mark the inauguration of the pipeline. From their respective control rooms, Alexey Miller, CEO of Gazprom, and Wang Yilin, chairman of China National Petroleum Company, asked their presidents for approval, which each expeditiously gave. With that, the technicians opened the valves just in time to begin to supply China for the winter. "This is a genuinely historical event," said Putin as the gas began to flow, "not only for the global energy market, but above all for us, Russia and China."

But there was still a burning question about Russia's future. Who would succeed Putin when his presidential term ended in 2024? In the spring of 2020 the answer became clear. Putin would succeed Putin. A new constitutional revision would allow him to serve as president until 2036—more or less for life. That way, he explained to the Russian parliament, he could continue to be "the guarantor of the country's security, domestic stability, and evolutionary development"—essential, he went on, because domestic and foreign enemies were "waiting for us to make a mistake or slip up." This would also enable him to continue to guide Russia's energy development and fortify the "turn" to the East. And that last would mean, more than anything else, further strengthening the relationship with China, where his friend Xi Jinping had already been elected president for life.

But then the coronavirus began to spread to Russia. People fell ill along the Russian-Chinese border in the Far East. The border was closed. The funds for the long-term major "national projects" had to be redirected to the immediate needs of battling the pandemic. In Moscow, people were instructed to stay home.

April 22, 2020, had been set as the date for a "people's referendum" to endorse the constitutional change. But the coronavirus forced its postponement until late June. May 9 was to be the great celebration of Victory Day, the seventy-fifth anniversary of the defeat of Nazi Germany and the enormous costs of World War II. It was slated to be an extravaganza that also celebrated resurgent Russia and its might and military prowess—and the president who had presided over that resurgence. But it too had to be postponed.[6]

THE PREVIOUS DECEMBER HAD BEEN MARKED BOTH BY THE inauguration ceremony for the Power of Siberia pipeline and Trump's signing into law the sanctions on the Nord Stream 2 pipeline. The juxtaposition highlighted the changing maps of geopolitics and energy—both East and West. The opening of the valve on the Power of Siberia pipeline demonstrated the fundamental role of energy in the strategic partnership between Russia and China. It is not all there is to the partnership, of course. Moscow and Beijing are conjoined by their emphasis on "absolute sovereignty," their rejection of the "universal" values and norms propounded by the West, their reliance on state-dominated economies, and their opposition to what they call the would-be "hegemonic" position and "unilateralism" of the United States. But energy is a very important part of this new geopolitical nexus. A relationship that was once based on Marx and Lenin is now grounded in oil and gas.

CHINA'S
MAP

Chapter 17

———

THE "G2"

The piling up of all the "Gs" can get a little confusing. There is the G7, the annual meeting of the world's leading industrial nations, which for a time was also the G8, and then, after Russia was expelled, became the G7 again. Then there is the G20, "the major economies"—the G7 and the European Union plus big "emerging" markets, including China, India, Brazil, and Saudi Arabia. Once thought by some as destined to become the "board of directors" of the global economy, the G20 remains more of a discussion and coordination group.

Then, to confuse further, there is the "G2." Except there is no G2, at least not officially. Yet it is very real, in the sense that it is the most decisive grouping of all. It has more say over the future of the world economy—and indeed, over the rest of this century—than any other grouping. The G2 comprises just two countries —the United States and China—which together represent about 40 percent of the world's GDP and 50 percent of its military spending. The G2 is not an alliance or a forum for decision-making. Rather, it underlines the importance

of the relationship between these two countries—and their new rivalry—
and its impact on the entire world.

Not so long ago, the United States and China were thought to be
ever more bound together by their interdependence—integrated supply chains (iPhones designed in the United States but manufactured in
China), total trade in 2018 (before the trade war) of $738 billion, $116
billion of U.S. investment in China, and Chinese investment in the
United States totaling $60 billion—plus more than 360,000 Chinese
university students in the United States, contributing $13 billion to the
U.S. economy.[1]

This interdependence was propelled by China's joining the World
Trade Organization (WTO) in 2001—what Bill Clinton would call
"one of the most important foreign policy developments" of his presidency. The aim was to bring a rapidly rising China inside a market system for global trade that would, as Clinton put it, mean that China's
actions "would be subject to rules embraced and judgments passed by
135 nations." It would open the Chinese market to U.S. business and
support global economic growth. Interdependence and "engagement"
would promote a convergence of interests that would reduce the risks
of conflict. This set of ideas added up to what we can call the "WTO
consensus." Despite the criticism at the time, there did not seem to be
any obvious alternative in the face of China's growing economic
might.[2]

But the "WTO consensus" has broken down, and the G2 is fractured. Engagement is giving way to estrangement—trade wars and conflict over economic and security issues, talk of "decoupling" of the two
economies, an arms race, and what is coming to be seen as a battle for
economic models and indeed for primacy in the rest of this century.
All this seemed to point to the development of a new cold war, albeit
of a different kind than that between the United States and the Soviet
Union. The disruption and terrible human and economic costs of the
novel coronavirus epidemic in 2020 led to a real, if temporary,

decoupling, as air travel was was canceled and trade constrained; re-criminations mounted, and hostility reached a new level.[3]

Does all this mean that China and the United States are headed for what Harvard professor Graham Allison called the "Thucydides Trap"? Named for the ancient Athenian military historian, the concept depicts the risk of war arising from the collision between a "dominant" power and a "rising" power. The many examples begin with the war in the fifth century BC between "dominant" Athens and "rising" Sparta, which Thucydides chronicled. It went on for thirty years and left both city-states devastated. The other case studies include the naval race and economic competition between Britain and Germany that culminated in the First World War. At the end, both victors and vanquished were much worse off, and the carnage laid the ground for the Second World War. None of these historic cases, of course, involved large arsenals of nuclear weapons—nor, of course, cyber war.

While the validity of the Thucydides Trap is debated, it has become part of the vocabulary. "There is no such thing as the Thucydides Trap in the world," said Xi Jinping on a visit to Seattle. Yet he warned, "Should major countries time and again make the mistakes of strategic miscalculation, they might create such traps for themselves."[4]

From the collapse of Soviet communism in 1991 until the 2008 global financial crisis, the U.S. design for global economic management had been generally accepted. But the 2008 disaster blew up in the heart of the American economy or, as the Chinese saw it, "in the core of the capitalist world." The "Chinese model" of a state- (and party-) managed economy offered an alternative. Moreover, China was the engine that first pulled the world economy out of the crisis in 2009 and back to recovery. China no longer felt the need to look to the United States for either guidance or role models. In the Chinese view, the financial crisis was "a watershed in the history of U.S.-China relations" that was "compelling the United States to treat China as a coequal." In the aftermath of the financial crisis, a Chinese strategist published a book titled *The China Dream: Great Power Thinking and Strategic Pos-ture in the Post-America Era*. It argued that conflict with the United States was inevitable. It was a bestseller in China.[5]

This shift would be reinforced by the changing balance in the world economy. China has become what Britain had been during the Industrial Revolution—the manufacturing "workshop of the world." A few examples: China today is the world's largest producer of steel (al-most 50 percent), aluminum, and computers—as well as the rare earths necessary for electric vehicles and wind turbines. In one three-year period, 2011–13, China consumed more cement than the United States did in the entire twentieth century. It has financial heft. SAFE—the State Administration for Foreign Exchange—holds foreign reserves totaling $3 trillion—about one-third of it U.S. government debt.[6]

It is also rapidly becoming a country of consumers, as Beijing seeks to shift the economy from export-driven to consumer-driven. In 2000, 1.9 million cars were sold in China, 17.3 million in the United States. By 2019, the number was 25 million in China and 17 million in the United States. The weight of China in the world economy was made clear by the novel coronavirus. When the SARS epidemic began in 2002, China accounted for only 4 percent of world GDP. When the coronavirus hit in 2020, it was 16 percent, meaning that the economic impact would reverberate around the world even before the coronavi-rus shut down much of the rest of the world.[7]

When GDP is measured by exchange rates, the U.S. economy is still larger than China's. By the other major measure of GDP—purchasing power parity—China is already the largest economy in the world. By that measure, it overtook the United States in 2014. (Just to note, Ger-many's economy overtook Britain's in 1910, four years before the out-break of the First World War.) But one reality check is in order for China's future growth—demographics, the consequence of the one-child policy and social changes. "No country has ever gone gray at a faster rate," demographer Nicholas Eberstadt has observed. Over the next two decades, the number of older people will increase dramatically while the number of people of working age who drive economic growth will decline just as dramatically. Other challenges include the scale of domestic debt and the structure of the economy.[8]

When it comes to oil, the difference between the two countries is stark. China imports 75 percent of its petroleum, which Beijing sees as a major vulnerability and is one of the drivers of its strategic policy. The United States used to share such concerns when its import levels were high. But owing to shale, no longer.

The rivalry between the two countries is certainly evident in military capabilities. "Our military has always fought with great spirit," Xi said on a visit to a division famed for fighting U.S. forces during the Korean War. "In the past we had more spirit than steel. Now we have plenty of equipment." It also has the money. Over the last two decades, China's military expenditures have grown sixfold. In the latest comparative numbers, it is $240 billion, compared to America's $634 billion. The third and fourth spenders are far behind—Saudi Arabia and Russia—each at around $65 billion. China's military has, in the words of a RAND Corporation assessment, "transformed itself from a large but antiquated force into a capable modern military." It has "narrowed the gap" with the United States. Crucially, it has the "advantage of proximity in most plausible conflict scenarios, and geographical advantage will likely neutralize many U.S. military strengths." It has also focused on developing a "wide variety of missiles, air defense, and electronic capabilities" that could neutralize U.S. capabilities from ships to satellites.[9]

The United States is hardly standing still. In response to the emerging capabilities of what is now called a "great power/peer level" competitor, the U.S. military is making a major shift in focus, strategy, and weapons. The U.S. Marines, for instance, are going through a transformation, in the words of its new force design, away from two decades of fighting on land against "violent extremists in the Middle East."

Instead, it is to become an agile naval expeditionary force able to move with great speed and in dispersed fashion from island to island in the Pacific in order to neutralize a Chinese navy that is capable of attacking traditional U.S. military assets.[10]

China's preponderance in Asia continues to increase. In early 2017, just days after becoming president, Donald Trump yanked the United States out of the Trans-Pacific Partnership, which would have encompassed twelve nations that border the Pacific—though specifically excluding China—in a new bloc representing 40 percent of world trade. It would have asserted U.S. commitment to Asia and given other Asian nations a counterforce to the powerful magnetic field of the Chinese economy. For those nations, it was as much political as economic. Trump's action was seen in Asia as marking a retreat from the region—and an opportunity for China to fill the vacuum. Indeed, one Chinese called the U.S. withdrawal "a huge gift" to Beijing. No longer hampered by a counterweight, China proceeded with negotiations for its own trade agreements with Asian countries—excluding the United States.[11]

Xi demonstrated China's new great power status when he hosted twenty-nine leaders of other countries at a Beijing forum. He made clear that China, unlike the United States, would not lecture them about human rights nor support democracy activists. "We have no intention," said Xi, "to interfere in other countries' internal affairs, export our own social system, . . . or impose our own will." His message was warmly welcomed by the leaders drawn to Beijing by the prospect of China's economic largesse.[12]

The rivalry of the G2 is most evident in two arenas. One—the South China Sea—involves, literally, geographic maps. The other—what is known as "Belt and Road"—represents an effort to redraw the map of the global economy. In both, energy is deeply intertwined.

There are other danger points, beginning with the fundamental issue of all—Taiwan. That Taiwan is not an independent country, and that it not move toward independence, is the oft-repeated "core interest" of China. Beijing is explicit that it would use military force and, if required, go to war to prevent any such move. Another danger point are tiny uninhabited but strategically located islands northeast of Taiwan that both China and Japan claim. North Korea and its nuclear weapons and missile program are a focus of great concern. Yet the South China Sea constitutes what has been described as "the greatest point of tension" directly between the United States and China. Or, as Admiral James Stavridis, former NATO supreme commander, put it: "The South China Sea is the most dangerous potential confrontation between the United States and China."[13]

Chapter 18

"DANGEROUS GROUND"

I n April 1933, French naval captain Georges Meesemaecker set sail from Saigon, in France's colony of Vietnam. The South China Sea into which he headed was the province of fishermen. For world trade, it was a backwater. The captain's mission was to extend the sovereignty of the fraying French empire in Indochina to its remote outer limits. This he would do by establishing "possession" of a group of "land features" in the South China Sea known as the Spratly Islands, named for a British sea captain who in 1843 had sighted one of them.[1]

As islands go, the Spratlys are anything but impressive—"quirks of geology, flyspecks on the map, barely protruding above water." They, combined with a perilous multitude of hidden rocks, reefs, and shoals hidden below the waves, create innumerable dangers for sailors. For two centuries, the area has been marked on nautical maps as "Dangerous Ground." Even today, U.S. government sailing instructions advise that "avoidance of Dangerous Ground is the mariner's only guarantee of safety." Flyspecks though they may be, the Spratlys occupy crucial territory on the map; they sprawl across about 160,000 square miles—an area the size of Michigan, Iowa, and Illinois combined—several hundred miles from Vietnam and slightly closer to the Philippines. They are more or less in the center of what today is the most important waterway in the world—the South China Sea.

The flotilla entrusted in 1933 with its grand imperial mission comprised just three ships—a gunboat, a fishing trawler, and a hydrographic boat. The mission had been ordered to "prevent a foreign power from claiming sovereignty." The French were worried about the Chinese, but more so an expansionist Japan, which was trying, the French feared, to "penetrate Europe-dominated waters." The French declared that their claim was watertight, based on the assertion of sovereignty made by the Vietnamese kingdom of Annam going back a century earlier.[2]

Aided by good weather, Meesemaecker's expedition sailed to nine of the tiny islands, on each of which the same ritual was repeated. A proclamation of sovereignty was signed by the captains of the ships and then placed in a bottle that, in turn, was inserted into a boundary marker that was placed in the ground. And then on each of the islands a French flag was hoisted. At that, a trumpet was sounded. It was a lonely scene, bounded by the silence of the empty sea.[3]

Today there is a growing struggle over sovereignty in the South China Sea—over who controls the Spratlys, as well as another island group closer to China and Vietnam called the Paracels, and other tiny "land features" that barely jut out from the waves—and indeed the sea itself. It is a battle over a host of critical matters—oil and gas resources, both known and purported; a substantial part of the world's fishing resources; control of the world's most important sea lanes and, potentially, the trade that goes through it. It is also about national identity, a shifting strategic balance, and the changing relationships of China both with its neighbors and with the United States. And, specifically, it is about whether naval warships pass freely or not through its waters. With all that comes the risk of armed clashes, whether intended or accidental.

The South China Sea, described as the "world's most critical waterway," stretches from the Indian Ocean to Asia and the Pacific Ocean. It is bordered by Indonesia, Malaysia, Brunei, the Philippines, Vietnam, China, and Taiwan. Singapore is just beyond its limits. Through its waters pass $3.5 trillion of world trade—two-thirds of China's maritime trade, and over 40 percent of Japan's and 30 percent of total world trade. The flows include fifteen million barrels of oil a day—almost as much as goes through the Strait of Hormuz—as well as a third of the world's LNG. Eighty percent of China's oil imports pass through it. Its waters are crucial for food security. Ten percent of the world's entire fish catch comes from it, 40 percent of its tuna. It provides much of the seafood consumed both in China, the world's largest consumer of fish, and Southeast Asia. It has even been suggested that "the value and importance of the South China Sea's fish stocks" make "fish a strategic commodity." Conflicts over fisheries also inflame public opinion in the countries that border the sea.

Those waters are also fraught with risk. "A single irresponsible action or instigation of conflict," warned Vietnam's prime minister, "could well lead to the interruption of these huge trade flows, with unforeseeable consequences not only to regional economies but also to the entire world."[4]

By comparison, in 1933, Meesemaecker's mission was far from the center of global contention. Its objective was much simpler—to try to consolidate France's position in the region. But news traveled slowly in those days. It was only later in the year that word of the mission finally reached the Chinese Ministry of Foreign Affairs, which was outraged. But after a few weeks of reflection, the Chinese Military Council warned, "We need to cool down the game with the French." The reason? "Our navy is weak and these nine islands are not useful for us now."[5]

Others, however, were hardly prepared to cool down, for the territorial integrity of China was a rallying cry for nationalists outraged by what became known as the "Incident of the 9 Islets." They were already mobilized by the "humiliation" of the "unequal treaties" with foreign countries, beginning, after the First Opium War, with the Treaty of Nanking in 1842. It required a defeated China to lease Hong Kong to Britain and grant "extraterritoriality," which meant that British citizens would be subject to British law, not Chinese. A whole series of subsequent "unequal treaties" over the nineteenth century gave European nations, including Russia, as well as Japan, preferential commercial and extraterritorial legal rights in Chinese coastal cities, along with political control within defined concession areas. All this undermined China's sovereignty and heralded its weakness. A climax of the "humiliation" came in 1919, when the Versailles treaty awarded Germany's Shandung concession to Japan. This ignited student demonstrations in Beijing on May 4, 1919—what became known as the "May 4 Movement"—a landmark for modern Chinese nationalism.

The Republic of China, founded in 1912, was supposed to modernize the country and regain sovereignty. But by the beginning of the 1930s, China had degenerated into a fragmented country. Chiang Kaishek, leader of the Nationalists and heir to the Republic of China, was fighting both warlords and Communists. In 1931, the Japanese seized control of Manchuria, where a substantial part of China's industry was located, and breached the Great Wall.

In the face of this turmoil and continuing dissolution, any further diminution of Chinese authority, no matter how distant, stoked outrage and alarm. But without the naval power to counter the French advance, "the Chinese government," as one historian has written, turned to warriors of another kind—"its mapmakers."[6]

Various maps were promulgated between 1933 and 1935 that asserted Chinese sovereignty into the South China Sea, reaching almost a thousand miles from the Chinese mainland, and with Chinese names for the various islands (in the words of a recent government document), "reviewed and approved." A singular cartographic combatant led the charge—Bai Meichu, one of China's most influential and respected geographers. His work was inspired not only by longitudes and latitudes but also by nationalist passion. "Loving the nation is the top priority in learning geography," he said, "while building the nation is what learning geography is for." To drive home the point and educate the country, he had already in 1930 produced a "Chinese National Humiliation Map." One of his aims, he said, was "to help the common people to be patriotic."

In 1936, he drew a map for his *New China Construction Atlas*. It included a U-shaped line—some would call it a "cow tongue"—that snaked down the coastlines along the South China Sea almost to the Dutch East Indies, now Indonesia. Everything within that line, he asserted, belonged to China. As he put it in an annotation, the South China Sea was "the living place of Chinese fishermen. The sovereignty, of course, belonged to China."[7]

Almost nine decades later, Bai Meichu's map is at the heart of today's struggle over the South China Sea.

After World War II, in 1947 and 1948, the Nationalist government, drawing directly on Bai Meichu's 1936 map, promulgated a new map showing China's control over the Spratlys as well as the Paracel Islands, and going all the way down to the James Shoal, off the coast of what today is Malaysia. This delineation, said an official statement at the time, was followed by Chinese "government departments, schools, and publishers before the anti-Japanese war, and it was also recorded on file in the ministry of interior. Accordingly, it should remain unchanged."[8]

China at that time was engulfed in the climactic struggle between Chiang Kai-shek's Nationalists and the Communists, led by Mao Zedong. The Communists prevailed. Late in the afternoon on October 1, 1949, from atop the Gate of Heavenly Peace in Beijing, Mao proclaimed the communist People's Republic of China. "The Chinese people have stood up," he declared. China, as Mao had put it, would "no longer be a nation subject to insult and humiliation." As for Chiang Kai-shek, he escaped to Taiwan with his Nationalist followers, along with many others who were fleeing the communist advance. In later decades, Taiwan became prosperous and a major source of investment in the People's Republic, although shadowed by the risk of forcible reunification.[9]

On the mainland, the new communist government adopted Bai Meichu's map, outlining its claims for the South China Sea. Today, the Chinese map—and its view of history—continues to be known as the 9-Dash Map. In 2013, a prominent Chinese geographer declared that Bai Meichu's map "is now deeply engraved in the hearts and minds of the Chinese people." Generations of Chinese grade-school students have been taught that the southernmost part of China is Zeng Ansha, otherwise known as James Shoal, an underwater reef about fifty miles off the coast of Malaysia. On a flight into China today, flip open the in-flight magazine on Air China and you will find the long cow tongue of the 9-Dash Line imprinted in a dark line on the map of the South China Sea, reaching all the way down to Malaysia and Indonesia.

Chapter 19

THE THREE QUESTIONS

The clash over the South China Sea is about islands and "territorial waters." It revolves around three issues:

First, who owns the tiny "land features" that protrude out of the waters of the South China Sea? The importance of this question is that jurisdiction over the waters flows from "land features." This question of sovereignty is primarily an issue between China and the nations of Southeast Asia. China says its claim to sovereignty is rooted in history, which it succinctly set out in a December 2014 paper: "China has indisputable sovereignty over the South Sea Islands and the adjacent waters. Chinese activities in the South China Sea date back to over two thousand years ago. China was the first country to discover, name, explore, and exploit the resources of the South China Sea and the first to continuously exercise sovereign power over them. From the 1930s to the 1940s, Japan illegally seized some parts of the South China Sea islands during its aggression against China. At the end of the Second World War, the Chinese government resumed exercise of sovereignty over the South China Sea islands."[1]

Other nations challenge the Chinese claim to sovereignty, arguing that for many centuries Southeast Asian and Arab merchants dominated trade in the region. Moreover, other countries and legal scholars argue that "historic rights" is too vague and ambiguous a claim to be the basis of sovereignty. Vietnam, the Philippines, Malaysia, Brunei, and Taiwan all dispute various Chinese claims and assert their own claims out into the South China Sea.

None of these contested islands amount to much. All of them together add up in terms of territory to no more than three times the size of Central Park in New York City. To make matters even more complicated, there is disagreement about whether some of the "land features" in the sea even count as islands at all in international law or are only "rocks which cannot sustain human habitation or economic life of their own." And "artificial islands, installations, and structures do not possess the status of islands," and thus do not have legal rights to the waters around them. Yet they can become facts.[2]

And that is what is happening in the South China Sea, where China is dredging millions of tons of rocks and sand to construct artificial islands atop a series of coral reefs. These new islands become "facts on the water"—and military bases that provide anchorage and missile batteries as well as runways that can handle China's strategic bombers. While both Malaysia and Taiwan have engaged in territorial reclamation in the South China Sea, the areas involved are very small. Nothing matches the speed and scale on which China is advancing its projects—to date some thirty-two hundred reclaimed acres. These islands would enable China to maintain continuing air patrols over the South China Sea. That, in turn, could permit it at some point to implement an air defensive identification zone over the region—which would inevitably be challenged.[3]

The second question is whether the South China Sea itself—that is, the water—constitutes international waters, high seas, or is part of the national territory of China. That is a matter of concern to the countries in the region, those nations whose trade passes through those waters, and the commercial shipping companies and navies of the world. Does the 9-Dash Map assert that 90 percent of the entire South China Sea itself is territorial waters of China? The original 1947–48 map describes itself as "The Map of the Chinese Islands in the South China Sea." More recently, a spokesperson for the Chinese Ministry of Foreign Affairs declared, "China enjoys indisputable sovereignty over the South China Sea" as well as the islands, and Chinese legal specialists claim that it has "authority over the South China Sea."

Many nations make their marine claims on the basis of the 1982 United Nations Convention on the Law of the Sea, which had involved negotiations with more than 150 countries over fourteen years. China's assertion that the 9-Dash Line is a national boundary, which encompasses the sea itself, rests not on this convention, but rather, Beijing says, on the "historic claim" with "a foundation in international law, including the customary law of discovery, occupation, and historic title." Other nations reply that the international law "does not recognize history as the basis for a maritime jurisdiction." To affirm its position, in this view, China would have had to exercise effective sovereignty over the South China Sea for a long and sustained period of time in a "well-known" way. Some would say that China will be able to assert this claim by establishing and maintaining an entrenched position in the years ahead.[4]

The third question is about the EEZ—Exclusive Economic Zone. The concept of the EEZ was established by the Convention on the Law of the Sea. The EEZ is different from territorial waters, which generally extend twelve miles from shore. The EEZ reaches out two hundred miles from shore. For most countries the EEZ entails only "economic" rights—to the fish in the waters and the oil and gas and minerals below in the seabed. The Chinese contend that the EEZ also gives it control over who passes through those waters. That directly pits the United States and China against each other. For the key question, very much at the heart of the dispute, is, as the international lawyer Robert Beckman puts it, not "freedom of navigation" in itself, but "freedom of military activities in open ocean."[5]

The United States has repeatedly said that it is not taking sides in the disputes on the South China Sea but only wants to see them settled peacefully in accord with the UN Convention. The Chinese call this position hypocritical because, owing to opposition in the U.S. Senate, the United States never actually signed the convention. But on the question of "freedom to navigate," the United States takes a firm stand. For it regards freedom of the seas and open navigation as fundamental to the law of the seas—including "freedom of military activities in open seas." It is on that basis that the U.S. Navy operates around the world.

China, however, asserts that foreign navies must have its permission to sail in China's EEZ—whether the claimed EEZ in the South China Sea or off the coast of China directly. According to the U.S. position, which is generally accepted by most other nations, the U.S. Navy can, for example, operate just beyond the twelve-mile limit off Shanghai without asking "by your leave" of China. Beijing rejects that premise, although, ironically, it would afford China the right to do exactly the same just beyond twelve miles off San Diego.

In 2012, 123 miles off the coast of the Philippines, China took control of the Scarborough Shoal (named for an unfortunate British tea clipper shipwrecked there in 1784) and closed it off to Philippine ships.

The Philippines hardly had any military of its own with which to respond. So it used the only weapon available to it—the UN Convention on the Law of the Sea. It brought a case challenging the 9-Dash Line to an international tribunal established in The Hague. Vietnam associated itself with the claim. In 2016, the tribunal delivered its verdict—wholly in the Philippines' favor. It rejected the legal and historic claims of the 9-Dash Line. China denounced the decision and made clear that it would not recognize it, let alone abide by it or the Philippines' "illegal claims." The decision changed nothing.[6]

At this point, there seems to be no definitive means to resolve the differences among the different countries. The gulf is too wide. For instance, Vietnam, relying on the same argument formerly used by the French colonial administration, says its rights go back to 1816, when the Vietnamese kingdom claimed sovereignty over the Spratlys. The Chinese reply that this claim is invalid because at the time, Vietnam was a "tributary" state to the Chinese empire, not an independent country, and thus was in no position to assert sovereignty over anything. The struggle over the South China Sea has turned into what Robert Beckman, whose specialty is the law of the sea in Southeast Asia, calls "the game of maps." He adds, "The maps tell you everything."

The 9-Dash Map may be the starting point and the defining document for the game. But there are other maps. Hanging on the wall in the chambers of Judge Antonio Carpio in the Supreme Court of the Philippines on Padre Faura Street in Manila is a map published by a Jesuit in 1734. It identifies the Scarborough Shoal with the Tagalog word *Panacot* for "threat" or "danger." That matters because Tagalog is the main native language of the Philippines.[7]

In 2014, Vietnam mounted its own map exhibit to support its claims in what it calls the East Sea. That same year, another player made its entry—rather, reentry—into the "game of maps": Taiwan. After all, the 9-Dash Line had started with the Nationalists. Taiwan decided to put some of its original maps on public display. They were consistent with the People's Republic's 9-Dash Line. But, said Taiwan's president, Ma Ying-jeou, the maps demonstrated that what the Nationalist government was claiming in 1947 with its dashed map were the islands in the South China Sea, but not all the waters.

President Ma added a plea—that the interested nations solve their differences peacefully. For otherwise, he warned, two trains were on track to crash into each other. Or, to use a more appropriate metaphor, two navies were on a collision course.[8]

Chapter 20

———

"COUNT ON THE WISDOM OF FOLLOWING GENERATIONS"

n previous decades, flare-ups over the South China Sea receded as the nations in the region focused on economic growth. For no country was that truer than for China, with its 10 percent or more annual economic growth and what it would call its "peaceful rise." And peace was the absolute necessity to allow such growth and China's expanding role in the global economy.

Deng Xiaoping knew, from all his life experience, the costs of war and upheaval. Paramount leader for two decades, Deng masterminded China's move toward the market and its integration with the world economy. Once one of Mao's most important lieutenants, he had been purged twice. That gave him ample time, first during years of exile, as a laborer in a tractor factory, and then under house arrest in Mao's later years, to reflect on what had gone wrong with the revolution. He witnessed the huge, tragic cost and massive waste of Mao's grand schemes and personally suffered from Mao's rule and the Cultural Revolution. His son, thrown out a window by fanatic young Red Guards during the

Cultural Revolution, was left permanently paralyzed in a wheelchair. By nature, Deng was a pragmatic problem solver. While working as a communist propagandist during his student days in France, he had also run a Chinese restaurant. Later in life, he said that he had never actually read Karl Marx's *Das Kapital* in full because he didn't have the time.[1]

Although Deng is said to have once told the Vietnamese that the South China Sea islands "have belonged to China since ancient times," he also laid out a strategy for dealing with the issue: "This generation is not wise enough to settle such a difficult issue. It would be an idea to count on the wisdom of following generations to settle it." (He had conveyed the same idea on the East China Sea.) Better, was his message, to let people get on with their lives—and expand their economies and raise their incomes. In 2002, China and ASEAN—the ten-member Association of Southeast Asian Nations—had worked out the outlines for a Code of Conduct that pointed to a pragmatic resolution.[2]

Deng's approach, what has been called "ambiguity on all fronts," was more or less working—until 2009. In March of that year, Chinese naval and air forces and fishing boats intercepted a U.S. Navy ship, USS *Impeccable*, which was tracking Chinese naval operations about seventy-five miles off the island of Hainan. Here was the clash over the concept of EEZ. This was well outside China's twelve miles of territorial waters, though within its exclusive economic zone. To the United States, these were open waters; to China, they were under Chinese suzerainty. Chinese naval vessels ordered the *Impeccable* to leave. At one point the *Impeccable* executed a direct stop to avoid a collision.

Two months later, in May 2009, in response to a dispute over the sea with Vietnam and Malaysia, China delivered a "Note Verbale" to the United Nations declaring its "indisputable sovereignty over the islands in the South China Sea and the adjacent waters." Of special significance in the note was a short parenthetical phrase that was anything but parenthetical in importance—"see attached map." That was the

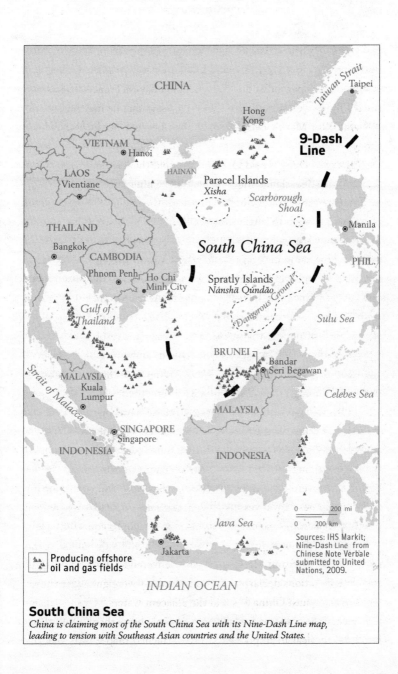

South China Sea
China is claiming most of the South China Sea with its Nine-Dash Line map, leading to tension with Southeast Asian countries and the United States.

9-Dash Line map, based on Bai Meichu's 1936 map—officially introduced as part of a legal document and a key moment in the "game of maps."

In June 2009, almost exactly a month later, a Chinese nuclear submarine collided with a sonar array being towed by the USS *John S. McCain* (named for the senator's father). The American ship was participating in naval exercises with six Southeast Asian countries. A new era of contention had begun for the South China Sea.[3]

In July 2010, the ASEAN nations were holding their seventeenth Regional Forum, this time in Hanoi, hosted by Vietnam. A few months earlier, some Chinese officials and strategists had begun to describe the South China Sea as a "core interest." Those were highly significant words, for they were the same formulation applied to the ultra-important questions of Tibet and Taiwan, and they could be interpreted as fighting words. "Core interest" had generated pushback from the United States and set off alarms for Southeast Asian countries. Prior to the meeting, several of them, including Vietnam, had urged the United States to say something strong on the subject at the Hanoi meeting. And that is exactly what Secretary of State Hillary Clinton intended to do.

In her speech, Clinton said that while the United States did not take sides on the specific questions of sovereignty, the impasse should be solved on a multilateral basis, preserving "unfettered access" to the South China Sea. "The United States, like every nation," she said, "has a national interest in freedom of navigation, open access to Asia's maritime commons, and respect for international law in the South China Sea." "National interest" had been "a carefully chosen phrase," as Clinton later said, as a riposte to the "core interest" emanating from Beijing.

As Clinton finished her speech, Chinese foreign minister Yang Jiechi was, as Clinton observed, "livid." He asked for an hour's adjournment and then returned. He was still furious.

Clinton's remarks, he declared in English, were "virtually an attack on China." Looking right at Clinton, he warned against "outside interference." Nobody was "threatening the region's peace and stability." The United States, he seemed to be saying, was conspiring against China. Then, fixing his eyes on Singapore's foreign minister, he intoned, "China is a big country and other countries are small countries, and that is just a fact."

Yang then turned his gaze directly to Vietnam's foreign minister, who was chair of the meeting, and said to him, "We socialists must stand together."

Vietnam's foreign minister said nothing. The room was quiet. After further reflection, the foreign minister came up with a diplomatic response. "It's time for lunch," he said. "Let's go to lunch."

But what was left on the table was the jarring juxtaposition of "interests"—China's "core interests" versus the United States' "national interests."[4]

Chapter 21

THE ROLE OF HISTORY

The collision over the South China Sea is a complex matter. For as Singapore's eminent diplomat Tommy Koh, who led the UN Law of the Seas negotiations, observed, "The South China Sea is about law, power, and resources, and about history." Indeed, it is a confrontation vexed by history.[1]

A symbol of China's historic claims is Admiral Zheng He, otherwise known as the "Three-Jeweled Eunuch." He was the paramount admiral among several eunuchs who commanded China's huge fleets in the fifteenth century. The ships carried a wide range of Chinese goods and the most advanced ordnance of the day—guns, cannonballs, and rockets. The biggest boats were the massive "treasure ships," as much as ten times bigger than the ship that, almost a century later, would carry Christopher Columbus to the New World.

Zheng He's first voyage, in 1405, put to sea with an armada of 317 ships, of which sixty were the treasure ships. He commanded six more voyages, each taking two or three years, some of which went to the Arabian Peninsula, and even on to the east coast of Africa. All along the way they traded with the locals. They also projected the power of China—in Zheng He's words, "making manifest the transforming power of imperial virtue." Zheng He's fleets brought back not only a wide variety of treasures and products, but also rulers and ambassadors to China to pay homage and tribute in person to the emperor.

In 1433, on a voyage homeward, Zheng He died. The great navy did not long survive him. The Confucian bureaucrats, the eunuchs' great rivals, argued that the fleets were wasting money needed to resist the encroaching Mongols. They wanted to destroy the navy because they saw it as the eunuchs' power base. Foreign travel was forbidden. Eventually, China's huge oceangoing fleet was burned. The Three-Jeweled Eunuch and his voyages—and his legacy—were expunged from history.[2]

But in modern times, Zheng He has been resuscitated as the great symbol of China's trading relationship with Southeast and South Asia, and as "the most towering maritime figure in Chinese history." The admiral was popularized with a popular miniseries on Chinese television. On the six hundredth anniversary of Zheng He's first voyage, a $50 million museum dedicated to him opened in Nanjing. It is overseen by a literally towering statue of the admiral, testament to Zheng He's lasting significance for China's historic claims to the South China Sea.[3]

While economic requirements and broad ambitions help shape China's maritime strategy, Beijing also has a very specific reason—the island of Taiwan. In 1996, fearing that a leading candidate in Taiwan's presidential election might move Taiwan toward a declaration of "independence," Beijing launched missile tests and live fire in waters near Taiwan, effectively blockading the island's western ports. President Clinton responded by dispatching two aircraft carrier groups into the Taiwan Strait, between Taiwan and the mainland, ostensibly to avoid "bad weather." The crisis subsided, but the lesson for Beijing was clear—the importance of sea power. The result was a new determination to build a powerful navy to ensure that such a moment would never again occur.[4]

On a clear Sunday morning in late August 2014, Chinese naval personnel gathered in the port of Weihai. They were there not to mark a victory, the usual reason for such a gathering, but rather a defeat—China's loss to the Japanese in the First Sino-Japanese War of 1894–95, which included the destruction of the Chinese fleet at Weihai. As a result, Japan gained control over both Korea and Taiwan, and Weihai itself passed under British control—altogether a particularly humiliating chapter in the "Century of Humiliation."

That morning, white chrysanthemums and red roses were scattered over the waters to mourn the Chinese losses in the battle for Weihai. The most prominent speaker was Admiral Wu Shengli, at the time the Commander of the People's Liberation Army Navy.

"The rise of great nations is also the rise of great maritime powers," Wu said. "History reminds us that a country will not prosper without maritime power." The Century of Humiliation, he said, was the result of insufficient naval strength, which the defeat a century before had demonstrated. "Having the ceremony in the old battle waters is a way to remember the tragic part of history," he said. But today "the sea is no obstacle; the history of national humiliation is gone, never to return."[5]

In looking for analogies for what might be unleashed by the U.S.–China naval competition, one is drawn back to that twentieth-century example of the Thucydides Trap—the pre–World War I Anglo-German naval race—and the vast tragic consequences that flowed from it. For the naval race was a defining geostrategic competition in the years before the First World War. It obsessively preoccupied both London and Berlin. It fueled suspicions about intentions and programs that fed upon themselves, convincing people that war was increasingly inevitable.

So worrying is this analogy that in his book *On China* Henry Kissinger ends with an epilogue titled "Does History Repeat Itself?" entirely devoted to this question. Yet he does conclude—with some uneasiness—that "historical analogies are by nature inexact."[6]

Chapter 22

OIL AND WATER?

mports" for China means everything from iron ore from Brazil to soybeans from Iowa to components from Vietnam that will be assembled into products in China and sold into the world. But inevitably, imports mean energy. For energy has been the foundation of the country's extraordinary economic growth. Energy fuels the global manufacturing platform that is the "workshop of the world." Rising incomes in China mean more building, more infrastructure, more cars, more air travel, and ever-more energy use. In 2009, China overtook the United States to become the world's largest energy consumer; today it alone represents almost 25 percent of world energy consumption. Chinese demand will continue to grow, but the rate will slow as services and domestic consumption become a larger share of the economy.

Despite the growth of nuclear power and renewables, 85 percent of China's energy today comes from fossil fuels. The structure of China's energy economy differs from North America's and Europe's; it is heavily reliant on coal—almost 60 percent of its total energy, compared to just 11 percent in the United States. Oil is the leading source of energy today in the United States—37 percent. In China, it is half that—20 percent. Natural gas is just 6 percent of total demand, though growing fast, compared to 32 percent in the United States. China's coal imports may fluctuate, but basically coal is a secure, domestic source. Not so for oil, and not so for natural gas; and that makes the geopolitics of energy a priority for Beijing.

The modern Chinese oil industry is less than sixty years old. For Mao's communist government, the cutoff of oil supplies during the Korean War in the early 1950s and then the later rift with its "big brother," the Soviet Union, made "self-reliance" in oil an absolute priority. That became a possibility in 1959, with the discovery in Manchuria of a giant oil field named Daqing—which means "Great Celebration." By the 1980s, the domestic petroleum industry was meeting the nation's needs and also producing a surplus of oil that was exported, principally to Japan.[1]

But economic growth sparked domestic demand for oil that eventually grew faster than domestic production. In 1993, China crossed an historic line—one that would shape its perspectives to the present day. It became a net importer of oil. Since then it has become increasingly dependent on imports.

In the 2000s, alarms went off in Beijing about fear of peak oil supply, the world's running out of oil, and the economic growth machine faltering. Apprehension grew about the possibility of bruising competition for constricted supplies, especially with the United States. Similar fears were also current in Washington in those years.

In the decade and a half following its entry into the World Trade Organization in 2001, China's oil consumption increased two and a half times over. It is currently the eighth-largest oil producer in the world, at 3.8 million barrels per day. But its demand has surged far ahead of domestic supply. It has become the world's largest importer of oil: by the beginning of 2020, 75 percent of total demand.

The concern was not only the volumes of oil, but also whence they came. China is the biggest customer for oil flowing out of the Persian Gulf and through the Strait of Hormuz. The South China Sea is the superhighway for petroleum imports. Most of them, whether from the Middle East or Africa, pass through the narrow Malacca Strait that loops up into the South China Sea.

That is what led then-president Hu Jintao to warn in late 2003 about what he called the "Malacca Dilemma"—the risks that come from dependence on that strait. "Certain powers," he warned, could disrupt China's oil supply line. His warning came several months after the U.S. invasion of Iraq. Beijing had trouble believing that the invasion could be about something so abstract as "democracy." It had to be about something concrete, and that had to mean oil. If the United States was worried enough about access to oil to invade Iraq, then China certainly should worry as well. Thus control over the South China Sea took on a new priority, both because of its importance as a transit route and because of the potential resources under its seabed.[2]

About half of the world's oil tanker shipments pass through the South China Sea, not only to China, but also to Japan and South Korea. For Japan and South Korea, the possible risk of disruption would come from actions by China. For China, however, there is only one "certain power"—the United States and, in particular, the U.S. Navy. Chinese strategists focus on a potential crisis scenario: Taiwan threatens independence, and China responds with force. The United States, in turn, responds by cutting the South China Sea oil line to China. What would happen next—escalation in unpredictable ways? Here, then, is part of Beijing's strategic rationale for the 9-Dash Map and its assertion of sovereignty over those waters.

And now that sea lane is not only the highway for oil, but also for natural gas. Traditionally natural gas has been a small player in the Chinese energy mix, but it has taken on new importance.

In the winter of December 2017, the days in Beijing were unexpectedly clear, crystal clear, and bright blue, though very cold, with a

wind that cut sharply across one's face. The characteristic noxious winter pollution, which would blot out both the sun and the mountains to
the north in a haze of smog, burning the eyes and causing pain in
breathing, had disappeared. Air pollution is a health problem and a
drag on China's GDP. It is also a potent social and political issue, and
the government has promised to meet the public's demand for clean
air. What that meant was the physical demolition of old coal-burning
equipment in northeast China, from which the winter winds would
carry pollution down to Beijing. But the winter was colder than expected, leaving the region short of energy, and hospitals and facilities
without heat. This became a national crisis. The only quick recourse
was to import more natural gas, and as that happened, global prices for
LNG spiked. The result that winter was not only a lack of heat in the
shivering cold, but also a severe shortage of natural gas.[3]

Beijing has moved to ensure that that does not happen again. China
has domestic gas production and is developing shale gas. It imports
pipeline natural gas from Central Asia and now, of course, from Russia through the Power of Siberia pipeline. Currently, half of China's gas
imports are in the form of LNG. China will soon be the largest importer of LNG. But the LNG still has to get there, and that means a
good part of it passes through the South China Sea.

But what about the possibility that oil and gas resources are
waiting to be discovered deep beneath the South China Sea?

On May 2, 2014, a flotilla of high-powered tugs towed a huge Chinese drill ship into waters 120 miles east of Vietnam and 180 miles
south of China's Hainan Island. Known as *HD-981*, it was a technological marvel in scale and capabilities—forty stories tall, built in 2011 at
a cost of a billion dollars, capable of drilling ten thousand feet under
the seabed. The drilling rig was accompanied by as many as eighty other
Chinese ships, as well as aircraft. The Vietnamese, surprised, denounced the arrival of *HD-981* and scrambled over thirty ships to
challenge *HD-981*'s right to be in those waters, which Vietnam also

claimed. China snapped back that the waters were part of its "inherent territory."

The waters around the drilling rig turned into a deadly serious dodge-'em game, as dozens of Chinese and Vietnamese ships rammed one another, with the loss of at least one Vietnamese ship. The Chinese used high-powered water cannons to ward off the Vietnamese ships, which Beijing described as "the most gentle measures we can take when trying to keep the other side out." Protests erupted in Vietnam itself, with street demonstrations, the torching of Chinese (and Taiwanese) factories, and the deaths of a number of Chinese nationals. As many as seven thousand Chinese nationals had to be quickly evacuated.

One night in mid-July, two and a half months after its arrival, *HD-981* abruptly sailed away. Now the Chinese internet erupted with nationalist fervor, criticizing the government for "bowing" to U.S. pressure. Officials dismissed the charge, saying that *HD-981* had completed its work and was hastening to beat the typhoon season. The Chinese and Vietnamese were sufficiently shaken that, at the end of 2014, both countries pledged to avoid "megaphone diplomacy," which, they said with some understatement, can "trigger volatility in public opinion."

The *HD-981* standoff was one of the most visible of a series of confrontations over oil and gas exploration and fishing rights that have continued in the South China Sea. In July 2019, the Russian company Rosneft, in partnership with Vietnam's state oil company, was drilling in what Vietnam describes as its EEZ, when several Chinese ships abruptly appeared, trailing back and forth in the same waters. The Chinese insisted that the disputed waters were part of their EEZ. Vietnam's president, Nguyen Phu Trong, while calling for restraint, declared that Vietnam would "never compromise" on its sovereignty. China's defense minister responded: "We will not allow even an inch of territory that our ancestors have left to us to be taken away." In October, the Chinese ships sailed away. But they did so only after Rosneft and its Vietnam partner had wrapped up their drilling.[4]

But will the prospects for the South China Sea live up to expectations? Current production from the South China Sea itself is about 900,000 barrels per day, which is less than 1 percent of world consumption in 2019. But what about the future? The highest estimate comes from a Chinese source that estimated 125 billion barrels of reserves still to be discovered—roughly on the same scale as Iraq's and Kuwait's. Other projections are much lower. The U.S. Energy Information Administration estimates about twelve billion barrels of undiscovered oil, which corresponds with estimates from international oil companies.

It is also thought that the most likely discoveries will be closer to the shores, and only one-fifth would fall into contested waters. The more central and controversial area of the South China Sea is estimated to be less prospective, owing to the thinness of the sediments and insufficient pressure. Moreover, the considerable sea depths and lack of infrastructure would make the development of any resources expensive. In any event, it is thought that the bulk of any discoveries is likely to be natural gas, not oil.

Of course, as a famous American wildcatter said at the beginning of the twentieth century, "Only Dr. Drill knows for sure." Continuing exploration would provide more clarity about the actual resource. At this point, based upon what is known, new discoveries would be meaningful for the countries around the South China Sea and for the companies involved, but would not be significant in the global oil balance. Natural gas discoveries could be larger, but would be higher-cost and would have to compete with growing flows of LNG and pipeline gas.

For Beijing, as Singapore prime minister Lee Hsien Loong put it, the South China Sea, as well as the Malacca Strait, "must be kept open to protect China's energy security." What really counts for China's energy security is not the unproven resources that may lie deep under the seabed, far beneath the sea lanes, but rather the sea lanes themselves and what traverses them.[5]

Chapter 23

— — — ▬ —

CHINA'S NEW TREASURE SHIPS

T he South China Sea dispute is also about trade itself.

How China's extraordinary economic surge came about is the result of many things. But it would not have happened without a revolution that was born in the U.S. port of Newark, New Jersey—a revolution in shipping that would change the map of global trade and prove transformative for the world economy—and for China.

The revolution was instigated by someone largely unknown in China and indeed the rest of the world—an entrepreneur from a small North Carolina town once known as Shoe Heel. Yet Malcom McLean, otherwise known as "Idea-a-Minute" McLean, is one of the most consequential figures in the history of transportation.

Starting off with a tiny trucking company that he built into a major enterprise, McLean went on to unleash the container revolution in world shipping that is the foundation of today's global economy. There is nothing romantic or flashy about a container; it is a steel box that might be twenty or forty feet long and eight and a half or nine and a half feet high. As one author has written, "it has no engine, no wheels, no sails; it does not fascinate those captivated by ships and trains and planes, or by sailors or pilots." It is just a metal box. But it has taken distance out of the economic equation by driving down transport costs to a fraction of what they had been, by speeding loading and unloading of ships, and by facilitating the easy movement of containers among ships and trucks and trains. By so doing, it turned manufacturing into a global rather than a local or regional business. Without "this humdrum innovation," the business theorist Peter Drucker wrote, "the tremendous expansion of world trade—the fastest growth in any major economic activity ever recorded—could not possibly have taken place."[1]

McLean had never been on a ship. He just wanted to lower costs and save a few dollars moving his trucks from the East Coast down to Texas. The idea came to him when he had to wait in his truck while longshoremen "laboriously moved cargo one piece at a time." As his frustration grew, he wondered, Why not just lift the whole truck body in one go and move it over longer distances by ship?

On April 26, 1956, cranes at the port of Newark, New Jersey, lifted up fifty-eight truck bodies, minus their wheels and cabins, and put them on a surplus World War II tanker bound for Texas. "We are convinced that we have found a way to combine the economy of water transportation with the speed and flexibility of overland shipment," McLean announced. This was the beginning. By the early 1960s, containers were becoming a real business, with McLean and his company in the lead. No longer did shipments have to be broken down into boxes and crates and sacks and hoisted around by hordes of longshoremen, taking up many days in port and adding substantially to costs. Instead, packed into containers, they could be lifted by cranes, with the operator high above in a cabin, and moved between shore and ship. The world of longshoremen that had been depicted just a couple of years earlier in the 1954 film *On the Waterfront* was on its way out.

In 1965, the first regularly scheduled container ships began to operate between the United States and Europe. But McLean had his eye on Asia, beginning with supplying U.S. troops in Vietnam. That launched containerization into the Pacific Ocean.

His next step was to detour ships on their way back from Vietnam, now empty of cargo, to Japan to pick up containers filled with inexpensive goods destined for U.S. customers. Manufacturers in the Asian "tigers"—South Korea, Taiwan, Hong Kong, and Singapore—followed suit. It was the spread of this innovation, and the networks and system that implemented it, that integrated East Asia into the world economy.[2]

In 1980, the year Deng began his reforms, McLean initiated the first container service to China. Two years later, the state-owned China Ocean Shipping Company launched container service to the West Coast of the United States. China's rapid economic growth would not have been possible without the container fleets to carry its goods to global markets with very little additional cost. This applied both to goods manufactured in China and the supply chains for which it was the hub.

Seven of the world's top ten container ports too are Chinese—Shanghai the largest—and China normally accounts for over 40 percent of the world's container shipments. Trade in turn is almost 40 percent of China's GDP, making that flow of exports outward, and the flow of oil and other commodities inward, essential for economic growth and indeed the foundations of political and social stability. Containerization became the backbone of global commerce. Yet the full extent of both China's and the world's dependence on containerization—and the supply chains that rely on it—really only became apparent when the coronavirus temporarily shut down much of the world trade in 2020.

McLean died in 2001. On the day of his funeral, the shipping industry honored him; worldwide, container ships sounded their whistles to recognize the man who had done so much to knit together the global economy. In his obituary, the *Journal of Commerce* called containerization "the most significant development in shipping since the shift from sail to steam power."[3]

For China, these container ships are China's twenty-first-century "treasure ships," the true descendants of Zheng He's fifteenth-century great fleet. They are the vessels that have carried China's economy into its current position in the global economy and world trade.

Chapter 24

THE TEST OF PRUDENCE

The South China Sea may be about oil and natural gas and flows of trade. Yet, as the International Institute for Strategic Studies puts it, "maritime disputes in the South China Sea are at their heart about power politics." From a strategic point of view, it is one of the most contentious issues between China and the United States. Chinese spokesmen frequently describe America as seeking to "contain" and "encircle" a "rising" China. For the United States, it is about freedom of seas and relations with the countries of Southeast Asia. Since World War II, the United States has been the guarantor of the open sea lanes that underpin global stability and the $90 trillion world economy. And it is on those open sea lanes that China has flourished.[1]

The relationship of the G2—between the "rising power" and the "dominant power"—has become more complex as the balance shifts between the two countries—and as the economies of the rest of Asia become increasingly integrated with China's. As one Singaporean diplomat put it, "China will loom larger and larger in all of our economic lives. China is creating new realities, building up a web of economic interests—trade, investment, infrastructure—extending out from south-west Asia to Southeast Asia, binding it into one economic sphere."[2]

XI JINPING IS OF A NEW GENERATION—THE FIRST CHINESE leader born after World War II. His father, a veteran of the revolution, had risen to vice premier, before being purged and imprisoned. In the words of a book prepared by China's Foreign Language Press, after Xi's father was "wronged and disgraced, Xi experienced tough times. During the Cultural Revolution he suffered public humiliation and hunger, ex-perienced homelessness, and was even held in custody on one occasion." He was then sent off as an "educated youth" to the countryside, where he labored for seven years. For a time, he lived in a cave and slept on "earth beds," and then in a work camp. "Life was very hard," he has said. It also hardened him. After the Cultural Revolution, he went to uni-versity and then rose up through government and party positions, be-coming a member of the Politburo in 2007 and emerging as the future leader. For Xi, the Communist Party is paramount and central, the de-fining organizing principle for China's rise, and its discipline and con-trol are essential.[3]

In November 2012, several days after becoming party leader, Xi led the new Politburo on a very short field trip—just across Tiananmen Square to the National Museum of China. Passing by the myriad trea-sures of China's five-thousand-year history, he halted at an exhibit called "The Road to Rejuvenation," which depicted the suffering and humiliation that China had suffered at the hands of the imperialists and its path to revival under the Communist Party.

"Now everyone is discussing the China dream," he said in front of the exhibit. "I believe that realizing the great revival of the Chinese people is the greatest dream of the Chinese nation in modern times."[4]

In 2013, Xi also became president. In the years since, he has launched what has been called China's "third revolution," following on those of Mao and Deng. This aims at a new era of modernization that would lift the economy up the value chain and create "a moderately well-off society." He has reasserted the primacy of the Communist Party and the state's dominating role in the economy, initiated a massive anticorruption campaign, promoted a more assertive great power role for China on the world stage, elevated the navy and the air force, and reined in the internet. In 2018, the National People's Congress made him president for life, breaking with the post-Deng tradition of term limits for presidents. The congress also elevated "Xi Jinping Thought" to the level of "Mao Zedong Thought" and "Deng Xiaoping Thought."

In turn, Xi declared that China now "stood tall and firm." He evoked a "mighty east wind" that would carry China forward. And, with a message that he said was aimed at those "who are accustomed to threatening others," and clearly referring to the South China Sea, he declared, "It is absolutely impossible to separate a single inch of territory of our great country." That determination would be backed up, he said, by a continued military buildup aimed at a "world-class armed forces" and what he called a "modern combat system with distinctive Chinese characteristics."[5]

Incidents and near collisions continue in the South China Sea between Chinese ships and American navy vessels making "freedom of navigation patrols," including in October 2019, when a Chinese destroyer came within forty-five yards of an American destroyer, forcing it to "jam on the brakes." Other nations—Japan, Australia, and Europeans—are conducting similar patrols through those waters. The dangerous game of maps also occurs in the airspace above the South China Sea. In one instance, two Chinese jets came within fifty feet of an American aircraft. In another, the radio crackled on a U.S. surveillance plane flying over the Spratly Islands (which the Chinese call the Nansha Islands) with six different warnings.

"U.S. military aircraft," said the Chinese military. "China has sovereignty over the Nansha Islands and adjacent waters. Leave imme-diately and keep off to avoid any misunderstanding."

"This is a sovereign immune United States naval aircraft conducting lawful military activities beyond the national airspace of any coastal state," replied an American airman, reading from a carefully worded card. "In exercising the rights guaranteed by international law, I am op-erating with due regard for the rights of and duties of all states."

Shortly after, the Chinese news outlet *Global Times* issued "a thumbs-up" to Chinese airmen for defending Chinese territory.[6]

In response to the growing competition, the United States intro-duced its own new map for the entire region—what Secretary of State Mike Pompeo called the "map of the Indo-Pacific," which firmly places India in the region as a major counterbalance to China. "Make no mis-take," said Pompeo, "the Indo-Pacific, which stretches from the United States west coast to the west coast of India, is a region of great impor-tance to American foreign policy" and "a big part of America's interna-tional economic future." The United States, he added, "will oppose any country" that seeks "domination in the Indo-Pacific."

Backing this up, the 2019 U.S. defense budget identified what it called China's "military and coercive activities" in the South China Sea as one of the Pentagon's top priorities and reasons for increased mili-tary spending.[7]

THE CONTEST OVER THE SOUTH CHINA SEA AND THE U.S.-China rivalry create dilemmas for the countries that comprise the As-sociation of Southeast Asian Nations—ASEAN. It was founded in 1974 as the Vietnam War was coming to an end. "It was a very dark time," remembered Singaporean ambassador Tommy Koh. With the United States about to withdraw from Vietnam, fear was widespread that the entire region would fall to communist revolution. But events did not work out that way at all, and today communist Vietnam is integrating into the global market economy. Unlike the European Union, the ten countries that comprise ASEAN have widely different political systems. But they are becoming an increasingly connected economic community of six hundred million people.[8]*

The countries face continuing challenges in balancing and then rebalancing. "Southeast Asia is integrated in security terms with the United States but in economic terms with China," said Chan Heng Chee, former Singaporean ambassador to the United States. Moreover, the Chinese message, repeated many times over to the other countries with coastlines along the South China Sea, is to say, "We are a geographical reality. The U.S. alliance is a geopolitical concept."[9]

The ASEAN counties are increasingly integrated with China in terms of trade. In 2005, U.S. trade with ASEAN was 50 percent higher than was China's. Today it is reversed: ASEAN's trade with China is 50 percent higher than with the United States. Yet at the same time, the ASEAN countries are also seeking to bolster security relations with the United States as "the only realistic counterweight" in order to assure their own independence of action in the region.[10]

China's arms buildup has led other countries in the region—ASEAN, as well as Japan and Australia—to spend increasing sums of money on weapons. As former Australian prime minister Kevin Rudd once observed, "It's as plain as daylight that there is a significant military and naval buildup across the Asia-Pacific region, that's a reality." Some of the ASEAN countries have a different worry—about, as a Singaporean strategist put it, a "U.S. that is distracted with an inability to focus on geostrategic issues." There is another kind of risk. As one observer asked, Are the nations of Southeast Asia and China and the United States to be hostage to the risk of "aggressive behavior" by

---

* The ten ASEAN member countries are Brunei, Cambodia, Indonesia, Laos, Malaysia, Myanmar, the Philippines, Singapore, Thailand, and Vietnam.

individual commanders "that spreads out of control" and "that could have dire consequences"?[11]

China has been a great beneficiary of the open world economy and the free flow of commerce on the world's oceans that the United States has championed, and its growth objectives are far more likely to be achieved in a stable, peaceful world than in one riven by confrontation and disrupted by conflict.

Some measures could reduce the risks in the South China Sea. The ASEAN countries and China are negotiating a code of conduct to reduce tensions in the region. But one of the sticking points is the Chinese proposal that would give it a veto over other countries conducting military exercises with the United States.

A much stronger military-to-military dialogue and greater transparency about programs could help to mitigate the growing "strategic mistrust" and the uncertainty about the critical but often murky question of intentions. A step was taken in that direction with two confidence-building measures. In 2014, the United States and China agreed to alert each other to major military exercises in the region and adopted "rules of behavior" for managing naval and air force encounters. Tempering the passions of populist nationalism in the countries bordering the sea would give governments more flexibility to resolve differences. Clarification on the treatment of "land features" and their "rights" in the adjacent waters of these "features" would be important. Another issue, one of the most vexing, is an understanding of the legal status of the exclusive economic zones.

Or perhaps the best that can be hoped for—as a play on the "MAD" (mutually assured destruction) of the U.S.-Soviet nuclear standoff in the Cold War years—may be "MAA," "mutually assured ambiguity." But seeking to address issues in a multilateral framework, with a critical role for ASEAN, would help modulate the conviction that the South China Sea is fundamentally a standoff between China and the United States.[12]

In terms of energy, the clash of nations can be eased with the recognition that the offshore waters of the South China Sea are unlikely to be another Persian Gulf in terms of supply and that the most important contribution to energy security is the secure passage of tankers through its waters.

For Japan and South Korea, which also depend mightily on the South China Sea as the highway for much of their exports and imports, its control by China would be regarded as a potentially major threat. As former Japanese foreign minister Yoriko Kawaguchi put it with great simplicity, "Sea lanes are important to Japan."

Those two nations would become increasingly alarmed, in the words of a retired Japanese admiral, by "China's unilateral ambition to monopolize the whole South China Sea" and achieve "control of most of the sea lanes of communications" and "the lifelines" of Japan and Korea. Yet both Japan and South Korea are increasingly interconnected with China. Japan's exports to China are about the same as those to the United States—about 20 percent in each case of total exports. Korea is more dependent—27 percent of its exports go to China, as opposed to 12 percent to the United States.[13]

Between the G2, the interdependence is extensive: General Motors sells more cars in China than in the United States. Before the Trump trade war, up to 60 percent of U.S. soybean exports went to China, and Apple sold $40 billion a year of iPhones. China was also expected to become the biggest market for U.S. LNG.

Yet there is no shortage of sharp and divisive economic issues between the two countries—intellectual property rights and theft; the requirement that U.S. firms had to do joint ventures in China; hidden subsidies; cyber intrusion. And, yes, trade wars.

WHEN DONALD TRUMP BECAME PRESIDENT, HE CHANGED THE setting on the table. No longer, in the view of his administration, is

China an economic partner, albeit a challenging one. Now it is an economic adversary as well as a strategic rival. The WTO consensus is out the window. Trump denounced the 2001 WTO agreement as "the greatest job theft in history." Once in the White House, he declared, "Trade wars are good, and easy to win," although not indicating which ones he had in mind. Those of the 1930s had proved neither good nor easy to win, and ended badly for all concerned.[14]

The trade war is part of a striking shift in the overall approach of the United States. The Trump administration's National Security Strategy represented a sharp break from the position of five previous presidents. Those presidents and their administrations had not hesitated to criticize China over a range of major issues. But they all sought to emphasize "engagement" and the positive potential: "cooperative relationship" (Reagan); working together to contribute to "regional stability and the global balance" (George H. W. Bush); "broader engagement" (Clinton); "constructive relationship with a changing China" and "responsible stakeholder" (George W. Bush); "deepen our cooperation" (Obama).[15]

Trump's National Security Strategy throws "engagement" and the WTO consensus out the window. It casts China as a deeply menacing geopolitical rival and at the top among America's "adversaries." It is a "revisionist" power (along with Russia) that is seeking "to shape a world antithetical to our interests and values."[16] The Department of Defense followed up with its National Defense Strategy, which declared that "China is a strategic competitor using predatory economics to intimidate its neighbors while militarizing features in the South China Sea."[17]

The language highlights a fundamental shift away from two decades of a "war on terror" to a new strategic era—rivalry with Russia, and much more so with China. As Secretary of Defense Mark Esper put it, "China number one; Russia number two." For China was now seen as seeking to "displace" the United States from its global position and developing the capabilities to do so. For two decades, in this view, the United States had been both oblivious and distracted and lulled into denial by China's entry into the World Trade Organization in 2001. "If I was a Chinese strategist," said a senior Pentagon official, "the only thing I would regret is that I did not keep the United States asleep for longer."[18]

In an October 2018 speech, Vice President Mike Pence elaborated on the new perspective. China, he said, had "masterminded the wholesale theft of American technology" and has built "an unparalleled surveillance state" and "an Orwellian system." It is seeking "to erode America's military advantages" and "push the United States of America from the Western Pacific." But, Pence added, "We will not be intimated." In what is otherwise a deeply divided and partisan Washington, Pence's remarks reflect what has become a broad consensus across party lines, as would become evident in the 2020 presidential campaign.[19]

China responded with its own white paper, "China's National Defense in the New Era." While more moderate in some parts—it talked about areas of cooperation between China and the United States, including their respective militaries—it too defined the "new era" as one of great power "strategic competition." In the Chinese version, the fault lay with America's "growing hegemonism, power politics, unilateralism." The United States, it said, is pursuing "absolute military superiority" and has "undermined global strategic security." The Asia-Pacific, it said, "has become a focus of major country competition" because "countries from outside the region"—that is, the United States—"illegally enter China's territorial waters and the waters and airspace near China's islands and reefs, undermining China's national security." China, it declared, would resist all threats to its position.

The white paper went on to blame "external forces"—again, the United States—for fueling the Taiwanese independence movement. Just to be clear about what was already abundantly clear, Senior Colonel Wu Qian, in presenting the white paper in Beijing, said that in the event of a move toward Taiwanese independence, the People's Liberation Army was "ready to go to war."[20]

Together, the Chinese white paper and two U.S. papers, one analyst soberly observed, provide "a clear warning of growing strategic rivalry between an existing and emerging superpower"—a rivalry "that will shape the future of both China and the U.S. for decades to come." In the national security establishments of both nations, each side is focusing on the other as the future enemy.[21]

The continuing rancor over the coronavirus and its origins in Wuhan made relations worse. And then, in May 2020, after months of demonstrations in Hong Kong, some violent, Beijing imposed its security laws and security systems on the semiautonomous Hong Kong. China was abrogating the concept of "one country, two systems," which had been the governing principle since the British handover of Hong Kong to China in 1997. Beijing asserted that this was a purely "internal" matter. The international reaction was strong. By coincidence, at almost exactly the same time, the White House issued a new strategy on China. "Beijing," it said, was "seeking to transform the international order to align with" the communist party's "interests and ideology." In response, the administration has adopted "a competitive approach," which includes "a tolerance of greater bilateral friction." The friction would continue to increase.

UNDERLYING THE TRADE WAR IS A DEEPER CONCERN OVER "FUture technologies." A high-tech algorithmic arms race is already under way, led by competition for preeminence in artificial intelligence.

Beijing's "Made in China" 2025 strategy, enunciated in 2017, aims to make China a leader in ten high-tech industries. The Chinese point out that their per capita income is only about one-sixth that of the United States and that it needs to move up the value chain to avoid the stagnation of the "middle income trap." The goal is also to "catch up" with the United States. In his speech, Pence responded that "the Communist Party" of China could end up "controlling 90 percent of the

George Mitchell, left, was determined to extract natural gas from shale rock. People said it was impossible, but he was stubborn. And his company needed new gas to supply Chicago.

In 2016, a tanker passed through the expanded Panama Canal carrying the first cargo of U.S. liquefied natural gas (LNG) to China. Oil and gas exports now loom large in U.S.-China trade battles.

Yuhuang Chemical of China made a multi-billion-dollar investment in St. James Parish, Louisiana, because of inexpensive natural gas. It also bought the old high school, enabling the parish to build a new one.

In 2010, Mark Papa predicted that shale oil would be a North American "game changer." It turned out to be a global game changer.

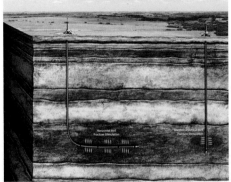

The Permian Basin in West Texas and New Mexico is now the world's second-largest producing area, propelling the United States to become the world's largest oil producer in 2018, ahead of Saudi Arabia and Russia. Horizontal drilling taps multiple zones containing shale oil.

President Donald Trump told India's prime minister, Narendra Modi, that he looked forward to India's buying more U.S. LNG. But, he added, he was "trying to get the price up a little."

With gasoline prices high in 2012, Barack Obama flew to a pipeline junction at Cushing, Oklahoma, to declare that his administration would "cut through" red tape to "get done" the building of the southern segment of the Keystone pipeline, which is meant to start in Canada.

In 2016 and 2017, protestors tried to block the last 1,320 feet of the 1,172-mile Dakota Access pipeline, built to move new Bakken oil out of North Dakota, replacing 740 railcars a day.

"The technocrats deceived us," said Mexico's president, Andrés Manuel López Obrador, left, who restricted new investment in Mexico's energy and sought to restore the "sovereignty" of the state oil company, Pemex.

A stunned Russian family watches in December, 1991, as President Mikhail Gorbachev announces the "death notice" for the Soviet Union.

The "Judoist." Vladimir Putin applies his judo skills not only on the judo mat but also on the world stage, taking advantage of openings and weaknesses to restore Russia as a great power.

German chancellor Angela Merkel and then–Russian president Dmitry Medvedev, center, turn the valve in 2011 to open the first Nord Stream pipeline, bringing Russian natural gas under the Baltic Sea directly to Germany. The European Union called it a "priority energy project".

ST.LOUIS POST-DISPATCH

SUNDAY EDITORIAL SECTION

SUNDAY MORNING, JUNE 4, 1961   NEWS ANALYSIS AND INTERPRETATION   SECTION B PAGES 1—10

Oil a New Soviet Weapon In Economic and Political Offensive Against West

U.S. Moving Toward Policy Revision to Threat of Reds' Increased Production and E —Officials Uncertain of Main Danger of Deals That Open Way to Infiltration.

By RAYMOND P. BRANDT
Chief Washington Correspondent of the Post-Dispatch.

WASHINGTON, .

SHARPLY INCREASED PRODUCTION and steadily risi of Russian oil have confirmed fears in American official a circles that the Kremlin has begun an economic and politica against the West with petroleum as a powerful weapon.

In the words of one official, "Oil is Russia's largest, easiest and most flexible export." Both in production and marketing, the Communists have decided advantages over the competitive, capitalistic producers, who have billions of dollars invested in the overseas trade.

Fidel Castro's seizure of American-owned refineries in Cuba to process Russian crude oil dramatized the need for a new Ameri-

tions to maintain their ref money to foreign government traction of oil. The World Ba the United States has the lar interest, is maintaining this p loans, and the Development L affiliate of the International Administration, has made no

The Export-Import Bank,

For many decades, debate has raged about how much political power Moscow would gain from Europe's importing oil and natural gas from the Soviet Union and now Russia, as in this article from 1961.

A specialized barge lays pipe under the Baltic Sea in 2019 for the controversial Nord Stream 2 pipeline, which is meant to carry additional Russian gas to Germany. U.S. sanctions forced work to stop just short of completion, straining U.S.-German relations.

In 1564, Cossack leader Bohdan Khmelnytsky, seeking a new ally in a war against Poland, swore allegiance to the tsar of Muscovy—competing narratives of which figure in today's struggle between Russia and Ukraine.

After the pro-Russian president of Ukraine fled in the face of huge popular protests in 2014, "little green men"—Russian paramilitary forces—suddenly appeared in Crimea. Russia annexed Crimea, leading to western sanctions and the beginning of a new cold war.

Ukrainian president Volodymyr Zelensky meets Donald Trump in September 2019, two months after a phone call between them triggered the impeachment of Trump by the U.S. House of Representatives and then acquittal by the Senate in 2020.

Volodymyr Zelensky went from playing a schoolteacher-turned-president in the Ukrainian television comedy "Servant of the People" to actually becoming Ukrainian president—and having to deal not only with Russia, but also with the United States.

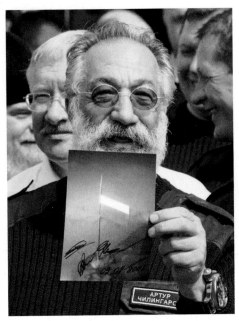

A polar scientist holds a photo of a titanium Russian flag planted 14,000 feet under the North Pole. It symbolizes Russia's claims on the Arctic Ocean, including the ice-filled Northern Sea Route that is essential for shipping Russian LNG to Asia.

At the "end of the land" on a remote Arctic peninsula, Yamal LNG was developed in the face of great skepticism. Russia will become one of the world's major exporters of LNG, capable of cutting through the ice eastward to Asia or westward to Europe.

Pancake diplomacy and Russia's pivot to the east. Russian president Vladimir Putin explains how to make blinis—Russian pancakes—to Chinese president Xi Jinping in 2018. At the same time, the Chinese military joined in Russia's large war games in the Far East.

The huge "Power of Siberia" natural gas pipeline carries Russian natural gas to China. A signature project for Putin—here he signs the pipe—he called it "a genuinely historical event" for Russia and China.

Zheng He, the fifteenth century Chinese admiral known as the "Three-Jeweled Eunuch," commanded giant "treasure ships" as far as Africa. Expunged from history after his death, he now embodies China's historical claims to the South China Sea.

Britain's Royal Navy defeats China in the First Opium War, 1839–1841, giving Britain control over Hong Kong and initiating what China calls the "Century of Humiliation."

In 1936, Chinese geographer Bai Meichu drew a map depicting the South China Sea as Chinese waters. "Loving the nation is the top priority in learning geography," he said. His map, described as "deeply engraved in the hearts and minds of the Chinese people," is the basis of today's 9-Dash Line.

地 理 學 系 教 授

白 眉 初 先 生

Shortly after becoming secretary of China's Communist Party in 2012, Xi Jinping led the Politburo to the National Museum's "Century of Humiliation" exhibit. "Now everyone is discussing the China dream," he said.

A Beijing traffic jam. In 2019, 25 million new cars were sold in China, compared to 17 million in the United States. China's oil consumption has tripled since 2000; it now imports 75 percent of its oil.

China has reclaimed 3,200 acres in the South China Sea, turning them into military bases. The runway on Subi Reef, in the Spratly Islands, is almost two miles long.

A Chinese Coast Guard vessel, above, turns water cannon on a Vietnamese ship in 2014 as the "most gentle measure we can take" in a dispute over oil and gas rights in the South China Sea, after the giant Chinese *HD-981* ship, below, begins exploratory drilling.

"Containerization" was the best idea that Malcom "Idea-a-Minute" McLean ever had—"the most significant development in shipping since the shift from sail to steam power." It knitted together the global economy, enabling China to become the "workshop of the world."

A huge container ship in Shanghai's port. Seven of the ten largest container ports in the world are in China.

Chinese president Xi Jinping addresses leaders from forty countries at a Beijing forum for "Belt and Road," which targets an estimated $1.4 trillion in infrastructure and energy investment in dozens of countries. China has "no intention," said Xi, "to interfere in other countries' internal affairs."

Khorgas was a crossing point through the mountains on the ancient Silk Road. Today, on the border between China and Kazakhstan, it is the giant "land port" for speeding containers filled with goods from China on to Europe.

In *Kung Fu Yoga*, shown at the Silk Road International Film Festival in China, Indian and Chinese archeologists team up to seek ancient treasure. The Indian archeologist says cooperation between the two countries would "be in line with the One Belt One Road policy." "So well said," replies the Chinese archeologist, as his professor, played by Jackie Chan, looks on.

world's most advanced industries," which would enable it "to win the commanding heights of the 21st century economy."[22]

The battle over technology has already been joined over 5G connectivity and, specifically, the Chinese company Huawei. The largest telecommunications company in the world and at the forefront of 5G—fifth-generation cellular network technology—Huawei has become one of China's great national champions, a symbol of its technological and commercial prowess. Washington argues that Huawei's technology provides a back door for Chinese government surveillance and possible manipulation and that the company is covertly connected to the government and the Communist Party. It banned Huawei from U.S. telecommunications networks and is trying to persuade other countries to do the same. Indeed, in a manner reminiscent of the British and German battleships before World War I, fifth-generation cellular—5G—along with Huawei has become in this era the embodiment of the new rivalry.

Amid the rising tensions, a warning about the dangers of "mutual animosity and zero-sum calculations" came from Robert Zoellick, who in 2005, as deputy secretary of state, had originally outlined the concept of China as a "responsible stakeholder." A decade and a half later, Zoellick pointed out that the word "stakeholder" in the title of his speech had been followed by a question mark. Nothing was for sure. But, said Zoellick, even with all the evident tensions, China had in fact engaged in multiple ways within the international system that the United States, Western Europe, and Japan had originally created. But now, he added, the United States was pushing China "into championing its own parallel, separate system with very different rules." And that could be very costly, both economically and politically.[23]

Economic interdependence has been the ballast to the military and strategic rivalry, but that ballast appears to be at risk of being dumped overboard into the South China Sea, along with the more cooperative attitude that went with it. This new instability in the economic relationship adds to the risk that "accident," confrontation, or clash will not be managed and contained.

The obvious fact is that neither the United States nor China is "going away." While tensions are rising, the G2 are hitched together on the same planet. Although there is little possibility of a grand bargain, the application of practical solutions, combined with prudence, can help mitigate the risks. Better that, it would seem, than to have to borrow from mariners' maps and apply the legend "Dangerous Ground" to the new geopolitical maps of the twenty-first century.

Chapter 25

———

BELT AND ROAD BUILDING

Nursultan, formerly Astana, the capital of the Central Asian country of Kazakhstan, is a partly new city. In Soviet times, the old section, beaten down by the harsh winters and the bitter winds that sweep down from Siberia, had yet another name—Tselinograd. It languished as a backwater, until chosen as the new capital of newly independent Kazakhstan.

Today, the city is in its new parts a gleaming, futuristic metropolis with soaring buildings designed by world-famous architects—all made possible by the rising oil revenues that have flowed into Kazakhstan since it became independent in 1991. Of all the Central Asian countries, Kazakhstan has worked the hardest to integrate itself into the global economy. It has established the Astana International Financial Center, governed by British commercial law, to be the financial hub for Central Asia. Kazakhstan also has oil, by far the most of any of the Central Asian republics, and as a result is the wealthiest of them.

In 2013, Chinese president Xi Jinping came to the city to deliver a major speech at Nazarbayev University. As part of the university's mission to connect Kazakhstan with the world, all the classes are taught in English. In his speech, Xi unveiled China's new map for the world economy—"One Belt One Road," in which China, historically known as the "Middle Kingdom," would become the middle of a redrawn world economy. The program would tie China together with all of "Eurasia"—the continents of Europe and Asia seen as one vast entity—through infrastructure, energy, investment, communications, politics, and culture. Subsequently, the realm that Xi identified would be broadened to include the Middle East and Africa. China would be the engine of development, the partner of choice, the lead financier, the promoter, and the grand strategist.

On that day, Xi summoned a figure from the annals of history— Zhang Qian, who in the second century BC had been the Han dynasty's special envoy to Central Asia. After enduring ten years of imprisonment by a nomadic tribe, Zhang was able to get back to China and report to the emperor about the potential value of trade with the heretofore unknown west. Now, more than two thousand years later, President Xi grew poetic, even mystical. "As I stand here and look back to that episode of history, I could almost hear the camel bells echoing in the mountains and see the wisp of smoke rising from the desert."[1]

Zhang Qian's reports marked the beginning of the development of trade routes that ran to the west, first to Central Asia and Persia, and then, at least intermittently, as far as the Roman Empire. This transcontinental trade route had no particular name. Only in 1877 was it dubbed *Die Seidenstrasse*—"the Silk Road"—by Baron Ferdinand von Richthofen, a German geologist and geographer who had been dispatched to China to scout mining opportunities and a possible route for a railroad to Europe. He chose "Silk Road" because one of its trades was driven by the passion for Chinese silk on the part of ancient Romans—at times, apparently, overly passionate, for silk was criticized by a Roman senator for promoting adultery by revealing women's bodies too explicitly.[2]

The "Silk Road" was not a single road, but rather a series of trails and paths around the Taklamakan Desert, leading, often perilously, from one oasis town to another, and then over forbidding mountains. But for thousands of years it served as an extraordinary conveyance for goods—from silk and spices to leather goods and musical instruments— and for culture, peoples, religion, and vocabulary. By this route, the invention of paper made its way westward, first used for wrapping, and only later for writing. And now in Kazakhstan in 2013, Xi described Beijing's grand plan as the "new silk road."[3]

A month after the Astana speech, Xi, in front of the Indonesian parliament, introduced the second part of the new strategy. This time he recalled the spirit of the Three-Jeweled Eunuch, the fifteenth-century admiral Zheng He, whose voyages had touched down in what is today Indonesia before sailing on to the "Western seas"—leaving "many stories of friendly exchanges," said Xi, "many of which are still widely told today." And now, he said, China would work with Southeast Asian countries "to build the Maritime Silk Road of the 21st century."[4]

The Chinese have shortened "One Belt One Road," in English, to the BRI—"Belt and Road Initiative." In Chinese, the nuance is a little different—"Belt and Road Strategy." In truth, it is both an initiative and a strategy. For simplicity, it is often just called Belt and Road, or just BRI. And as it has come to be a concept applied to projects around the world, it has also become an all-encompassing brand.[5]

SINCE THE BREAKUP OF THE SOVIET UNION IN 1991, MAJOR advances have been made in Central Asia in highway systems, rail connections, and air transport. But by far the biggest international investment has gone into the large oil and gas resources and into the pipelines to carry those resources to the world market.

As China's energy needs grew rapidly in this century, access to Central Asian energy moved up the priority list. Turkmenistan, with abundant natural gas resources, became the largest gas exporter to China. In Kazakhstan, Chinese companies represent about 20 percent of the country's oil output. A fourteen-hundred-mile pipeline from the Caspian shore of western Kazakhstan carries oil across the country to China, making Kazakhstan an important source of diversification for Beijing.

Beyond energy, China had a very specific security interest in Central Asia. Its large Xinjiang region in the northwest shares borders with Kazakhstan, Kyrgyzstan, and Tajikistan, as well as with Afghanistan and Pakistan. Strengthening relations with these neighbors would help control the East Turkistan Islamic Movement, a violent jihadist group that was affiliated with and provided jihadists to other Central Asian and Mideast extremist groups. The Xinjiang region includes the Tarim Basin, which is one of China's main domestic sources of oil and gas and where the Uighurs, a Muslim Turkic people, have traditionally been the dominant ethnic group, though with Kazakhs and other groups and now with many Han Chinese as well. In Xinjiang, the Han are now at least 40 percent of the population—up from 6 percent in 1949. Following a series of terrorist attacks that left scores dead, Beijing has clamped down with great severity. It has established large camps for Uighurs and other Muslims, holding perhaps as many as a million people. Chinese officials describe them as "education and training centers." Critics describe them differently—as "mass incarceration" and indoctrination. These camps have become a focus of controversy and stimulated much international protest.

After the U.S. House of Representatives passed a bill calling for sanctions and restrictions on transactions with Chinese companies involved in the region, the Foreign Ministry in Beijing denounced the bill as "gross" interference "in China's internal affairs" and said that China's policy is "about fighting violence, terrorism and separatism" and "advancing deradicalization." This campaign—along with Beijing's new security system for Hong Kong and the continuing rancor over the novel coronavirus—adds further strains to China's relationships with the United States and other Western countries.[6]

Belt and Road is focused on energy, infrastructure, and transportation with an overall potential investment estimated at about $1.4 trillion—a scale never before seen and at least seven times larger, measured in today's dollars, than the Marshall Plan, the U.S. initiative to rebuild Europe after World War II.[7]

It is not just physical goods that China is seeking to export to Eurasia and the world beyond; it is also aspects of China's own economic model. Infrastructure investment has served for decades as an economic growth engine for the country's development, which can be revved up when growth appears to be waning.

An initial impetus for Belt and Road was the expectation, following the global financial crisis, for slower global economic growth for China. Development in Eurasia would stimulate growth, create new markets for Chinese industries suffering from extensive overcapacity, support jobs in China, and create new opportunities for Chinese firms.

THERE IS ALSO THE GEOPOLITICAL ELEMENT. IN 2011, THE Obama administration announced a "pivot" to Asia. For Washington, this reflected war-weariness, a drive to shift away from the Middle East and Afghanistan, with the seemingly endless wars and heavy costs, and instead toward the most dynamic part of the world economy. The pivot was meant to be a message of strategic reassurance to Asian countries, other than China, that the United States was committed to the region, was not ceding it to Chinese predominance, and indeed would increase its engagement. The pivot was subsequently rebranded as a "rebalancing."[8]

In Beijing, the U.S. pivot was derided as "this ill-thought-out policy of rebalancing," and was portrayed as an American strategy to "contain" China, alienate its neighbors from Beijing, and prevent it from assuming its rightful role of predominance in Asia. In 2012, an influential article advised "marching west" to counterbalance America's "rebalance" toward Asia and to blunt competition with the United States.[9]

Going west came to mean "One Belt," otherwise known as the "Eurasian Land Bridge," which enables China to expand its economic clout and its influence and relevance through Central Asia, the Middle East, and Russia and into Europe. These land connections would also offset one of China's great strategic fears—about the U.S. Navy and the "Malacca Dilemma." Going west would, in the words of a Chinese general, provide China with a "strategic hinterland and international space."

The maritime initiative—called the "One Road"—swings down around Southeast Asia and South Asia to Africa and then up past the Arabian Peninsula to the Mediterranean and on to Europe. It is punctuated by what some call the "string of pearls," a series of expanded ports that will promote Chinese trade and provide anchorage for the Chinese navy.

China would bring technology, finance, and capabilities to work at scale, as well as the ability to get things done expeditiously. What it would not bring is what the United States and Europe promote—"democracy" and "freedom," opposition parties and NGOs, and critiques of internal political practices and elections. What the West champions as "universal values" the Chinese dismiss as "western values." With China at the helm, there would be no agenda of "regime change," no support for "color revolutions," no championing of human rights activists. China instead would recognize and respect "absolute sovereignty."[10]

One of China's great tools for achieving its goal of "connectivity" is its ability to mobilize capital on a massive scale. A new, wholly Chinese Silk Road Fund has been created with upwards of $60 billion. China also established a new Asia Infrastructure Investment Bank (AIIB) to finance developments along these corridors. This new institution reflects China's dissatisfaction with what it considers its inadequate say in the "governance" of the World Bank, not commensurate with its standing in the world economy, and with the tendency toward "political correctness" in the World Bank's lending (solar and wind, but no oil and gas extraction, and certainly no coal). The Obama administration wasted no time in opposing the creation of a new Chinese-dominated international financial institution. To Washington's dismay, Britain rushed to be the first European country to sign on, beating out Luxembourg. Other stalwart U.S. allies hastened to join. All this left the United States and Japan as the odd men out. There's no question who leads the AIIB: China subscribed 32 percent of the capital and has 30 percent of the voting rights. Yet at the same time, the biggest recipient of AIIB loans so far has been India.

Belt and Road has not only become an organizing principle for Chinese foreign and economic policy. It has also become a bulwark of political, academic, and popular discourse. The number of academic articles in China dealing with BRI ballooned from 492 in 2014 to 8,400 in 2015. China, the world's largest movie market, established the annual Silk Road International Film Festival to bring movie people together from a host of countries to encourage joint production and highlight films that celebrate BRI values. In a film called *China Salesman*, a Chinese mobile phone salesman outwits European rivals in a hard-fought battle for a contract in North Africa. In *Kung Fu Yoga*, Chinese and Indian archeologists team up to seek an ancient treasure.

"We could increase the cooperation in archeological research between China and India," declaims the Indian character. "It would also be in line with the One Belt One Road policy."

"So well said!" replies the Chinese character.[11]

HOW MANY COUNTRIES ARE CANDIDATES FOR BELT AND ROAD?
Many numbers have been bruited about—up to 131. But Chinese officials go out of their way to say that there is no specific number, that instead, as one official put it, Belt and Road "is not a geographic concept.

It's about development, about strategically important projects that are financially sound." Eastern and Central European countries, some of them members of the European Union, have aligned with it, as have Italy and Greece. A Chinese company now owns a controlling share in Piraeus, the port for Athens. The president of Panama asked Xi Jinping if Panama could be part of the BRI. Certainly, Xi replied. "Connectivity" is basic to Belt and Road, and the Panama Canal provides major "connectivity" to the world economy. China happens to be the second-largest user of the canal; a Chinese company has purchased the largest port facility on the canal, and another has proposed a $4.1 billion high-speed railway project for Panama. China finally clarified the number of eligible members of the Belt and Road; it was open to "all nations." Yet the "core" countries could be narrowed down to a much smaller number that have signed "comprehensive strategic partnerships" with China.[12]

Khorgas, once a crossing point through the mountains on the ancient Silk Road, is today the site on the Chinese-Kazakh border of a giant "land port," a huge railway and transport hub that will speed containers from China on to Europe. Regular train service already delivers containers of Chinese goods to Europe in about half the time it takes for a container ship, though at higher cost. The first regular freight train service to the Middle East reached Tehran in 2016. China is spearheading a $6 billion high-speed railroad project punching through mountains in Laos that will connect China with seven countries and will at some points require tens of thousands of Chinese workers on location. It has invested in a multitude of energy projects. It is promoting the sale of its high-speed trains and its technology for ultra-high-voltage power lines, which China's State Grid company developed to efficiently carry electricity generated in western China over long distances. After the United States pulled out of the airport in the Central Asian republic of Kyrgyzstan, which had been used to supply allied troops in Afghanistan, the Chinese stepped in with a billion-dollar proposal to upgrade it.[13]

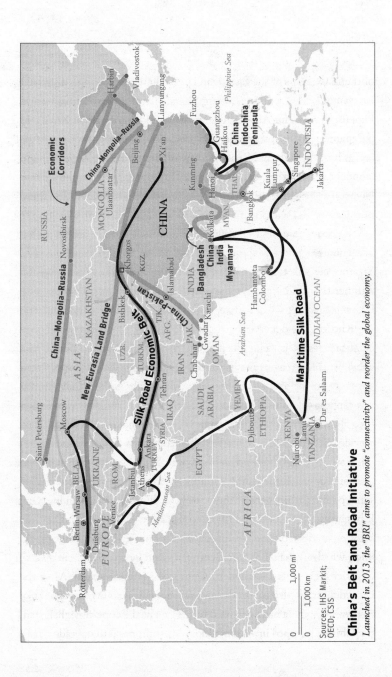

China's Belt and Road Initiative

Launched in 2013, the "BRI" aims to promote "connectivity" and reorder the global economy.

Sources: IHS Markit;
OECD; CSIS

_____

YET GRAND STRATEGY AND ON-THE-GROUND IMPLEMENTATION
are two different things. Chinese financial institutions do not want to
repeat the experience of the "going out" period in the 2000s, when Chinese companies paid top dollar for foreign assets. "Our performance will
evaluate our decisions," observed one official. "Projects have to be not
only strategically important but financially sound." Many deals have
been complicated to negotiate, taken longer than expected, and sometimes been rescinded or fallen apart. In recipient countries, decision
makers argue among themselves about what they are giving up and
whether they are ceding part of their economy to China and what crit-
icisms they will face domestically in the future if they do a project—
but also if they don't.[14]

The biggest project to date, by far, is the $62 billion China-Pakistan
Economic Corridor. Almost 70 percent of it is for electric power–
related investment. The rest is for highways, oil and gas pipelines, and
most notably a major port at the coastal town of Gwadar, located strategically on the way to the Persian Gulf and the Suez Canal. A putative landmark for the Maritime Silk Road, the port could be a convenient
stopping point for the Chinese navy. Gwadar is connected to China not
only by sea. At great cost and with extraordinary difficulty—including
surmounting the sixteen-thousand-foot Khunjerab Pass, the highest
border crossing in the world—a modern expanded highway system has
been built to move goods from China to Gwadar. As the Hong Kong
newspaper the *South China Morning Post* pointedly observed, the Gwadar port-and-road system provides "an alternative shipping route to the
Malacca Strait, which is frequently patrolled by the United States."[15]

The Chinese investment in the power sector is increasing electricity supplies, desperately needed in a power-short country, and thus helping Pakistani manufacturing and export. Yet at the same time, Pakistan's
import bill has shot up owing to the costs of importing of goods from

China for the projects. The country is behind on repaying the Chinese loans, and its indebtedness is rising rapidly. This sent Pakistan back to the International Monetary Fund in 2019 for its twelfth bailout since the late 1980s. And because of the large role of the United States in the IMF, this brings Washington into the business of BRI. "There's no rationale for IMF tax dollars—and . . . American tax dollars that are part of that funding," said U.S. secretary of state Mike Pompeo, "to go to bail out Chinese bondholders—or China itself."[16]

This points to what Christine Lagarde, formerly managing director of the IMF and now head of the European Central Bank, diplomatically called the risk of "problematic increase in debt." Critics call it the "debt trap," which they say could provide political and economic leverage to China. When countries cannot make their debt repayments, Chinese entities can take control. Case study number one is the port of Hambantota, in Sri Lanka, for which China provided $1.1 billion in loans. But the port has had little traffic and there was no chance of its ever earning sufficient revenues to repay the loans. So in exchange for cancellation of the debt, a state-owned Chinese company took a ninety-nine-year lease. But the Sri Lankan government, pressed hard by India, did extract a promise from the Chinese not to use Hambantota, strategically located on the Indian Ocean, for military purposes.[17]

One of the most explicit challenges to the debt question came from Mohammad Mahathir, Malaysia's long-serving prime minister who returned as prime minister in 2018. While on a trip to China, he abruptly announced the cancellation of $23 billion worth of rail and pipeline deals with Beijing. At age ninety-three, Mahathir had no hesitation in speaking his mind, warning his hosts against "a new version of colonialism." Malaysia, he explained, could not afford such debt. "I believe China itself does not want to see Malaysia become a bankrupt country," he said.

But a few days later, Malaysia's foreign ministry hastened to make clear that the country remained completely committed to participating in Belt and Road. After a 30 percent cut in the price of the rail project, Mahathir himself announced his support for Belt and Road, although he took advantage of the 2019 BRI conference in Beijing to pointedly call for "freedom of passage" in the South China Sea.[18]

In response to criticism about debt, and Beijing's own worries about repayment, Xi in 2019 presented a "debt sustainability framework," aimed at preventing borrowers from falling into the debt trap. This was codified by the ministry of finance as a "policy tool."[19] But a year later, the devastating economic impact of the coronavirus crisis on so many countries meant that their debt burdens would become a far bigger problem.

MANY COUNTRIES WANT THE INVESTMENT AND WANT TO BE sure that they are part of this new global economy, and they do not want to be left out of the China-driven "globalization 2.0." At the same time, however, they want to ensure their own independence of action and will seek to balance the growing Chinese presence by engaging with Russia and the United States. The United States remains, after all, the most important economy in the world, and very important in terms of security.

And yet the United States is seen by many countries as stepping back, increasingly unpredictable and no longer reliable, which increases the allure of engagement with China. As one Chinese official observed, "The pullback of the United States is helping us."

Yet Belt and Road's massive mobilization of capital has galvanized action in Washington. When the Trump administration first came in, one agency on the list to be axed was the Overseas Private Investment Corporation, a government agency that lowers the risk of investing abroad for U.S. firms by providing political risk insurance. But now OPIC has been combined with another U.S. agency to become the U.S. International Development Finance Corporation, with global infrastructure investment its major goal. It is slated to have $60 billion of lending capacity, which happens to just about match China's Silk Road Fund.

Russia, for now, is willing to see its Eurasian Economic Union overlap with the BRI. Russia does not have the means to be a significant global funder, and it needs foreign investment itself; and, of course, this fits with its pivot to the east and alignment with China on the issue of "absolute sovereignty." Yet at the same time, Russia continues to see Central Asia as its sphere of influence and will likely look askance at some point if these countries slip into what is a de facto Chinese sphere.

The one country in the region that is most concerned by Belt and Road is the other emerging giant—India. Its economy is about one-third the size of China's. Yet it too is integrating with the world economy and is "rising," with growing regional security interests. It sees Belt and Road as an engine of Chinese domination, potentially leading to what is described as the "encirclement" of India, as it also worries about Chinese naval activity in the region. "Connectivity itself cannot override or undermine the sovereignty of other nations," said Prime Minister Narendra Modi. India and China clash over competing maps. They argue about territorial claims in the "roof of the world," the Himalayas, and indeed have gone to war over those claims.

India has unveiled its own "Act East" policy, motivated in significant part, one scholar has written, by "deep distrust" of Belt and Road, "which it assesses to be a strategic project designed for political or security gains." Act East has four objectives: "securing" the Indian Ocean; deepening relations with Southeast Asia; strengthening strategic partnerships with the United States, Japan, and Australia; and "managing differences" with China. Although India has been cited as part of one of the BRI corridors, and Xi and Modi embraced the spirit of "Chennai Connect" after meeting in that city on the Bay of Bengal, rivalry seems inevitable. The competition is already evident in naval exercises and the growing militarization of the Indian Ocean.[20]

Yet for now, for many nations, China is offering one of the best deals in town. Countries, looking at where the money for infrastructure and energy investment is going to come from and wanting to lock in their place on the new map of the global economy, are concluding that there is advantage to attaching themselves to a rising China and an engaged China, rather than to an America that may increasingly seem to some both inconstant and receding.

MAPS OF THE MIDDLE EAST

Chapter 26

LINES IN THE SAND

I n the summer of 2014, a video floated up on the internet. It featured a militant from what had been until recently a virtually unknown jihadist group, pounding his feet on the sand on the border between Iraq and Syria. Behind him were abandoned shacks, pockmarked with bullet holes, which had until recently guarded the frontier between the two countries. He was, this militant said, stomping his feet to signify ISIS's elimination of the "Sykes-Picot" line between Iraq and Syria.

Sykes-Picot, he said, was finally dead. "We don't recognize the border and never will." His words were underscored by images of border posts being blown up. "We will break all the borders," he declared.

This video came after seven months of shock. In January 2014, this group—the Islamic State in Iraq and Syria (ISIS)—charged out of eastern Syria and stormed across the border into western Iraq. It moved forward ferociously in pickup trucks and captured armor, overcoming whatever Iraqi forces and militias stood in its way, capturing one town after another, committing shocking atrocities all along its route.

In June, ISIS's drive was finally stopped at the gates of Baghdad. That did not, however, prevent ISIS from proclaiming that it was now the new caliphate. Its territory at the maximum comprised one-third of Iraq, and, including its hold in Syria, was more than three times the size of Israel and Lebanon combined—and more than half the size of Great Britain. Altogether, as many as eight million people now found themselves trapped under its rule.

"This is beyond just a terrorist group," said the U.S. secretary of defense. "This is beyond anything we've seen."[1]

ISIS's ambition was to replace borders and nation-states with a caliphate, an empire based not on national sovereignty but on the authority of Islam and the strictures of the seventh century—a caliphate that would take on the world in a global jihad.

ISIS's offensive was but the latest challenge to the map of the Middle East that had been laid down a century ago, a system to which the names Sykes and Picot are so inextricably bound. ISIS ignited a new crisis for a region that had been rocked by turbulence for a century, arising from war and the collapse of an empire, the competition of great powers, Arab nationalism, religious fervor, ideological clash, dynastic ambitions, imperial dreams, American intervention, a Jewish state, and competition for oil. All of this would unfold in a region critical for the world's energy—and thus to the global economy—but also at a time when confrontation between Saudi Arabia and Iran had become central to the region's future.

Yet all of this was coming at a moment when the future of energy has become less clear—and subject to increasing dispute. Could the countries and a region that depend so much on oil continue to rely on it going forward? Would oil hold the same sway in the world economy and geopolitics in the rest of this century as it has in the past?

But first, who were Sykes and Picot?

———

IN LATE DECEMBER 1915, A CAREFUL OBSERVER IN LONDON
might have caught sight of a youngish Englishman slipping unobtru-
sively day after day into the French embassy. But that in which he was
engaged was anything but unobtrusive—drawing a new map for the
Middle East to replace that of the soon-to-be-defunct Ottoman Empire, a map of colonial "spheres" that would in due course become
nation-states with modern definitions of sovereignty.

World War I had catapulted Mark Sykes, a writer of travel books
and Tory member of Parliament, into position as the British government's foremost Middle East expert in London. Sykes had been obsessed with the region since as a child his father had taken him on a
trip across the Ottoman Empire. After Oxford, he had wandered around
it, his travels becoming the basis of a series of books. His latest, just before the war, was *The Caliph's Last Heritage.* To some, given his obsession with the region, he was the "Mad Mullah." The First World War
brought him back to London, where his credentials qualified him as an
expert on Mideast policy, which in turn put him into the critical role
of drawing a new map for the Middle East.

His job was to meet in secret in the French embassy with François
Georges-Picot, a more senior French diplomat who was once described
as seeming "never to have been young." Picot came from a family identified with France's imperial ambitions, and he had been consul-general
in Beirut. Despite their differences, Sykes and Picot were united in their
conviction that something had to be done to replace the five-century-
old Ottoman Empire.[2]

At its maximum, the Ottomans' rule extended over most of the
Middle East and North Africa and into southeastern Europe. There
were regions, but no nations, within the empire. Well before the outbreak of the First World War, the Ottoman Empire was already in decay,
its finances a shambles. Then, at the beginning of the war, it allied with

Germany and Austro-Hungary against Britain, France, and the Russian Empire. Now Britain and France were determined, as Mark Sykes put it, to ensure that the Ottoman Turkish Empire "must cease to be." But what would follow in its place?

For Britain, the significance of the Middle East lay in its strategic position on the routes to India, including the Suez Canal. Moreover, the Ottoman sultan in his role as caliph, the spiritual leader of Muslims and protector of Islam, had called for a jihad—a holy war—against the British, who in turn were alarmed by the potential impact on the Muslim subjects of the British Empire in India and the protectorate of Egypt. French ambitions were more commercial, but were also wrapped up in a collage of religion, history, "mission historique," and the determination to "reap the harvest of seven centuries of French endeavors" going back to the Crusades (although one English official did note, "The Crusaders had been defeated, and the Crusades had failed.").[3]

IN DECEMBER 1915, SYKES WAS SUMMONED TO A MEETING IN Prime Minister Herbert Asquith's office to discuss Britain's plans for the future of the Middle East. Asquith himself had been skeptical of assuming responsibility for Mesopotamia (as what roughly Iraq was then called) and "tackling every kind of tangled administrative question . . . with a hornet's nest of Arab tribes." But now there was general agreement with what Sykes proposed—nothing less than drawing a "line in the sand" for the postwar Middle East, establishing new spheres of influence under direct European control.

Shortly afterward, Sykes began those secret visits to the French embassy to meet with Georges-Picot. By January 3, 1916, they had come to agreement. The "line in the sand" stretched from near Haifa on the Mediterranean coast all the way to Kirkuk, near the Persian frontier. North of the line would be under French protection; south of it, British. There was more to the map than that. In the "blue area," France

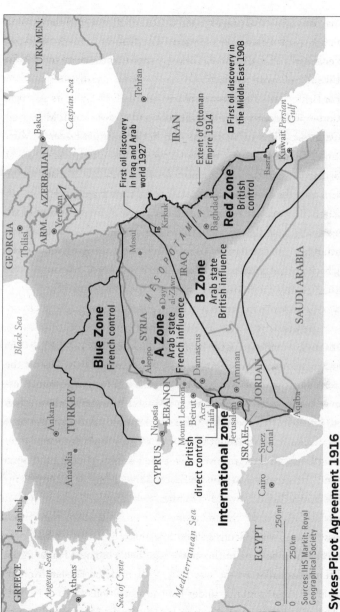

Sykes–Picot Agreement 1916

The British and French drafted this interim map during World War I for the Middle East after the end of the Ottoman Turkish Empire.

Sources: IHS Markit; Royal
Geographical Society

Blue Zone
French control

A Zone
Arab state
French influence

B Zone
Arab state
British influence

Red Zone
British control

International zone
British direct control

First oil discovery
in Iraq and Arab
world 1927

Extent of Ottoman
Empire 1914

First oil discovery in
the Middle East 1908

GREECE
Athens
Aegean Sea
Sea of Crete
Mediterranean Sea
Istanbul
Ankara
TURKEY
Anatolia
Black Sea
GEORGIA
Tbilisi
ARM.
Yerevan
AZERBAIJAN
Baku
Caspian Sea
TURKMEN.
Tehran
IRAN
Mosul
Kirkuk
MESOPOTAMIA
IRAQ
Baghdad
SYRIA
Aleppo
Dayr
al-Zawr
Damascus
Amman
Beirut
Mount Lebanon
Acre
Haifa
Jerusalem
LEBANON
CYPRUS
Nicosia
ISRAEL
JORDAN
Aqaba
Suez
Canal
EGYPT
Cairo
SAUDI ARABIA
Basra
Kuwait
Persian
Gulf

0   250 mi
0   250 km

would exert direct control; in the "red area," Britain would do the same. Over one issue Sykes and Georges-Picot fiercely argued—Palestine and control of the Holy Land. Eventually, they agreed that Britain would get two ports—Haifa and Acre—and a territorial strip for a railroad that would connect to Mesopotamia. The rest of Palestine would be placed under some undetermined form of international administration. The Russian Empire, ally in the war, signed on for its own territorial ambitions in the northern parts of the Ottoman Empire.[4]

Sykes-Picot may loom large in historical memory, but less understood is that the two men were not drawing their map on a blank sheet of paper. Their effort relied heavily on previous maps—those of the Ottoman Turkish Empire, and the vilayets, as provinces were called, that had been established in 1864. The three eastern vilayets—today Iraq—were based on Baghdad, Mosul, and Basra. The vilayets of Aleppo, Damascus, and Dayr al-Zawr comprise modern Syria. Beirut was a vilayet, and Mount Lebanon was a self-governing unit. Jerusalem and the territory around it were an independent district, reporting directly to Istanbul. The Hejaz, western Arabia including Mecca and Medina, was a vilayet; and so was Yemen, at such uncertain times as it was actually under the control of the Ottomans. There was no effort, one scholar has written, "to draw the provinces' boundaries along ethnic lines." It was atop the polyglot Ottoman map that Sykes and Picot drew their own lines, also without consideration for ethnic divisions.[5]

But this was not the only plan for the postwar Middle East. Some British officials in the Mideast promoted the idea of an Arab "revolt," which, it was hoped, would rally the Arab populations against their Turkish rulers. The British found their leader in the charismatic Prince Faisal. He was the son of Hussein, the sharif—that is, guardian—of Mecca and emir of the Hejaz, the western side of the Arabian Peninsula. Hussein had unique authority as a direct descendant of Mohammed. He also had a memorable phone number—Mecca 1.

Prince Faisal had a particular advocate—a young Englishman who had originally gone out to the Middle East to do archeology. This was T. E. Lawrence. He later became Lawrence of Arabia. With the outbreak of the war, he became a British intelligence officer based in Cairo. Lawrence's two brothers were killed on the Western Front in Europe. "They were both younger than I am," he wrote, "and it doesn't seem quite right, somehow, that I should go on living peacefully in Cairo." That, he would hardly be doing. Lawrence's consuming passion was to ignite the Arab revolt. He organized guerrilla attacks on the Turkish railway system, and he ensured that Faisal and Arab horsemen were at the forefront when the Allied forces took Damascus from the Turks.[6]

Mapmaking was further complicated when in November 1917, Foreign Secretary Balfour sent Baron Lionel Rothschild a public letter declaring that Britain would "favour the establishment in Palestine of a national home for the Jewish people and will use their best endeavours to facilitate the achievement of this object, it being clearly understood that nothing shall be done which may prejudice the civil and religious rights of existing non-Jewish communities in Palestine."

But how was this to be implemented? In late May 1918, an unlikely traveler took a tramp steamer from Suez to Aqaba, on the northeastern corner of the Red Sea, and then continued on by car and camel and foot into the desert, finally arriving at a valley where he found Prince Faisal, son of the sharif of Mecca, encamped with his forces. This was Chaim Weizmann, a Russian-born Jew who had earned a Ph.D. in chemistry in Switzerland. He had gained much credit in Britain during World War I for his chemical breakthrough that facilitated the production of explosives. He was also a leader of the Zionist movement and a prime promoter of the Balfour Declaration.

"A very honest man," said Weizmann afterward of Faisal. "Handsome as a picture." At two subsequent meetings, Faisal espoused their affinity as "cousins by blood . . . the two main branches of the Semitic family, Arab and Jew." The two men seemed to agree on a Jewish "home" in Palestine based on the Balfour Declaration and Jewish immigration.

But Faisal added that any agreement was predicated on "the independence of the Arab lands" and uniting "the Arabs eventually into one nation," under the Hashemites, as the family of Sharif Hussein and Prince
Faisal was known. But this was definitely not in the cards in postwar
peacemaking.[7]

Yet another factor would further complicate matters. In 1908,
after seven years of hardship and disappointments, the first Middle
Eastern oil had been discovered in a remote part of Iran. World War I
was proving the criticality of oil, not only to fuel the battleships, but
also for the recent inventions that had become vehicles for war—
trucks and motorcycles and tanks and airplanes. Oil and the internal
combustion engine had quickly redefined mobility and warfare. All
this would make oil the most important strategic commodity in the
world.

In the final months of the war, a secret report on "The Petroleum
Situation in the British Empire" noted that Britain depended upon the
United States for most of its oil. If Britain were to remain the dominant naval power, it could not rely upon a United States whose president, Woodrow Wilson, was so inalterably opposed to the very concept
of empire. Thus, Britain needed "to obtain the undisputed control of
the greatest amount of Petroleum that we can." That meant "Persia and
Mesopotamia." Already in control of Persian oil, the British concluded
that securing "the valuable oil fields in Mesopotamia" and assuring an
independent oil supply before the "next war" was a "first class war aim."[8]

WITH THE COLLAPSE OF FOUR EMPIRES—GERMAN, RUSSIAN,
Austro-Hungarian, and Turkish—resulting from the war, the maps of
Europe and the Middle East would have to be much redrawn. That was
the task at the Versailles Peace Conference, which met outside Paris in
1919.

No one expected to be more centrally involved in negotiating the future of the Middle East than Britain's Middle East expert, the "Mad Mullah." But when Mark Sykes arrived in Paris, he was already desperately sick, "poisoned," as he said, from something he had picked up in Aleppo, dangerously thin, and subsisting largely on canned milk. Still, he was determined to help shape the postwar Middle East. But just a month after the Versailles conference opened, Mark Sykes, further weakened by the Spanish flu pandemic, died in his room at the Hotel Lotti.[9]

Much of the Sykes-Picot Agreement had been confirmed before the conference began. But not quite everything. Sykes-Picot had assigned Mosul, the capital of the vilayet of the same name, with a large Kurdish population, to the French sphere. But at Versailles, after what was described as a "dogfight," the British managed to wrest control and add it to their own sphere. The main reason was the expectation that, as one British official wrote, "The greatest oil-field in the world extends all the way up to and beyond Mosul."

The Versailles Treaty established a League of Nations and provided for "mandates"—a sort of halfway house between colonies and spheres of influence on the one hand, and President Woodrow Wilson's idea of "self-determination" on the other. The "mandatory" powers were to guide their charges to eventual independence.[10]

The mapmaking that began with Sykes and Picot continued with the Treaty of Sèvres in 1920 and concluded, in 1923, with the Treaty of Lausanne. They built on the lines drawn in Sykes-Picot, with some significant adjustments. Turkey was shorn of Arab-speaking lands. France was to have a mandate over both Arab Syria and Lebanon, a new country with a majority Maronite Christian population. "Mesopotamia" was to be a British mandate, and Arabia under British influence. Palestine was also to be a British mandate. Britain was to be responsible for creating a "national home" for the Jewish people in Palestine. The part of the mandate east of the Jordan River became Trans-Jordan, a separate Arab area with its own local government. In Turkey, a brilliant general, Mustafa Kamal, later Ataturk—"Father of the Turks"—overthrew the remnants of the Ottoman regime and replaced it with a secular modernizing republic. He secured Turkish sovereignty over Anatolia, and recovered the hinterlands of Aleppo, in Syria, leaving that city close to the Turkish border.

Thus was the modern map of the Middle East drawn, the result of negotiations that began in 1915 and only ended in 1923. But amid all the different steps, "Sykes-Picot" would be the lasting symbol of how the nation-states of the region had come about. It would also be, for more than a century thereafter, the target and rallying cry for those who would seek to overturn the order that the map had shaped.

British officials were determined to yoke the three eastern distinct vilayets or provinces of the Ottoman Empire into one country under their mandate. At first called "Mesopotamia," it became Iraq. This included the Kurds in Mosul and the north, the Sunnis in Baghdad and the center and the west, and the Shia in Basra and the south. The problem was that these groups did not share a common identity. Basra was oriented toward the Gulf and south to India; Baghdad's connections ran east to Persia; and Mosul looked west, to Turkey and Syria. The divisions went beyond that. The population included Kurds; Assyrian Christians (sometimes called Nestorians) who had fled from Turkey; Jews (the largest single group in Baghdad); Yazidis, clustered around Mount Sinjar; plus Turkomans, Persians, Armenians, Chaldeans, and Sabeans. Arabs, of course, were the overall majority, but they were divided between the much more numerous Shia and the minority Sunni.

The British decided that the solution to this vexing problem of uniting the country was to install a king. They found their candidate in Prince Faisal, the leader of the Arab revolt and son of the sharif of Mecca. He was available and recently unemployed, having just been dismissed by the French as the king of Syria, where he had reigned only briefly. His ascension to the new throne in Baghdad was approved in a plebiscite by an astonishing 96 percent of the vote, even more astonishing in a country that was largely illiterate. He was crowned on August 23, 1921, to the tune of "God Save the King," hastily called up since the country did not yet have its own anthem.[11]

Six years later, in 1927, came another landmark—oil was discovered in northeast Iraq. "No longer would we writhe under the cynical comments of those who offered to wager that they would drink all the oil found in Iraq," said one of the engineers on the project. "From now on we were really on the map." Though oil had been found almost two decades earlier in Iran, this was the first discovery in the largely Arab lands of the Middle East.

The mandate for Iraq ended in 1932, and Iraq became an independent nation, achieving Faisal's great goal. Faisal, however, had little opportunity to savor the sovereignty, for he died just a year later. He left behind a plaintive testament as to the future of his country. "It is my belief that there is no Iraqi people within Iraq," he wrote. "There are only diverse groups with no patriotic sentiments." After his death, coups by military officers became endemic to Iraq's political life. Syria's exit from the French mandate took longer and was more tumultuous; it did not become fully independent until 1946.[12]

Two years later, in May 1948, in accordance with a United Nations resolution calling for two states in Palestine and in the wake of the Holocaust, Israel declared its independence. Five Arab nations immediately launched a war, but it ended with an Israeli victory, which established a Jewish state in the Middle East. Throughout the Middle East, the Arab defeat was a source of lasting anger. It was also transformative for an Egyptian officer who had fought in the war: Gamal Abdel Nasser had grown up an Egyptian nationalist; he was now an Arab nationalist. Four years later, in 1952, he led a coup that sent the corpulent King Farouk packing, along with the British influence that had dominated Egypt.

NASSER WAS THE FIRST POST-WORLD WAR II LEADER WHO
set out to overturn the maps that came with the end of the First World
War and, as he wrote, eliminate the "barbed wires" that "marked the
frontiers separating and isolating countries," and instead unite their
peoples in a single "Arab nation." His nationalization of the Suez Canal
in 1956 and success in resisting a subsequent invasion by Britain,
France, and Israel did make him the "hero" for which, he had declared
a few years earlier, the Middle East was waiting. Promoting his brand
of militant Arab nationalism, Nasser railed against the system that had
emerged after World War I. His targets, amplified by his powerful
Voice of the Arabs radio, were the United States, Britain, Israel, and the
"feudal" and "reactionary" Arab monarchies, most notably Saudi Arabia and Kuwait. The oil of the Middle East, he declared, was one of the
key "sources" of power for the "Arab nation." Toppling the oil-rich monarchies would provide him with that power.[13]

Nasser's new order appeared to be on the way when military officers, pledging "loyalty" to him, seized power in a coup in Syria. This led,
in 1958, to a "merger" of Egypt and Syria into what was supposed to be
a single country, the United Arab Republic. But then in 1961 other officers seized power in Damascus and promptly withdrew Syria from
the new "state." The following year, Nasser sent troops to intervene in
the civil war in Yemen, expecting a quick victory that would expand
his reach. Instead it turned into a long battle against royalist guerrillas
and a proxy war between Egypt and Saudi Arabia. Iran joined with
Saudi Arabia to support the guerrillas in resisting the Egyptian forces,
one result of which was the establishment of an Iran-Arab Friendship
Society, with offices both in Tehran and Riyadh. Nasser would end up
calling Yemen his "Vietnam," a political quagmire that added to the economic woes of the grossly mismanaged Egyptian economy.

In 1967, Nasser declared, "All our people are ready for war," the

objective of which would be "Israel's total destruction." But the Israelis moved first. Egypt's defeat in the 1967 Arab-Israeli war, which
included the loss of the Sinai Peninsula to Israel, dealt a major blow to
Nasser's position and prestige. His great ambition to expunge "Sykes-
Picot" and redraw the map of the Middle East had come to naught. He
faced the ignominy of having to be bailed out financially by the very
Saudi and Kuwaiti monarchies that he had so reviled. Seriously dia-
betic, he died prematurely, in 1970, aged fifty-two.[14]

Chapter 27

━━ ━ ━━

IRAN'S REVOLUTION

The battle today between Saudi Arabia, the largest oil producer in OPEC, and Iran, normally one of the other major ones, is a struggle for predominance across the map of the Middle East. It is a battle shaped by the clash of religion, ideology, and national interests, and by the drive for primacy. Oil is integral to this battle, and the impacts in turn are global.

The scorn with which Iran and Saudi Arabia regard each other is hardly muted. Ayatollah Ali Khamenei, Iran's current supreme leader, denounces the Saudi royal family as "sinful idols of arrogance and colonialism," as "idiots" and "milk cows for the Americans," and as "heartless and murderous."[1] Crown Prince Mohammed bin Salman—known as MBS—in turn describes Khamenei as "the Hitler of the Middle East . . . trying to conquer the world." MBS adds, "If you see any problems in the Middle East, you will find Iran."[2]

The religious roots of this struggle go back to the seventh century and the battle that ensued after the death of the Prophet Mohammed. Who would be his successor—his father-in-law, Abu Bakr, or his cousin and son-in-law, Ali? The Sunni are followers of Abu Bakr, who became the first caliph. But his legitimacy was—and is—contested by the Shia, "the party of Ali." Each group regards the other as heretics.

For all Muslims, of course, whether Sunni or Shia, Medina and Mecca in Saudi Arabia are sacred cities. The Shia are the majority in theocratic Iran, in Iraq, and in Bahrain. They are also significant in eastern Saudi Arabia, Syria—where the ruling Alawites are deemed a Shia offshoot—Lebanon, Azerbaijan, Pakistan, and India. The Houthis in Yemen are today, if not always, considered a Shiite-related sect.[3]

In the 1960s, Shia Iran, under Shah Mohammad Pahlavi, and Sunni Saudi Arabia were allied in their opposition to their common enemies— Soviet encroachment in the region, Ba'athists, Arab socialism, and most notably, as already noted, to Nasser and his pan-Arab campaign. But at the beginning of the 1970s, when the British decided that they could no longer afford to maintain a military presence in the Gulf, the shah, with the strong support of Washington, stepped forward to be the regional power, the policeman of the Gulf. The Saudis saw him as a "megalomaniac."[4]

The shah pushed for pell-mell modernization and rapid economic growth. The flood of petrodollars that came with the four-fold increase in oil prices that followed the October 1973 war enabled him to push even faster—and to spend massively on weapons.

The oil wealth may have driven growth; it also turned Iran into a case study for the resource curse, leading to rampant inflation, mushrooming slums, wasteful and unproductive expenditures, and widespread corruption, all of which meant vast social dislocation, which fed rapidly growing discontent and opposition across the political and social spectrum.

The shah's most resolute enemy was the Ayatollah Ruhollah Khomeini, an austere, singled-minded, narrow, and intensely devout Shia cleric, implacable in his resistance and ruthless in destroying those who stood in the way of an Islamic republic ruled by the clergy. From exile, Khomeini called for an Islamic revolution. The country was thrown into turmoil by strikes and larger and larger demonstrations, marked by increasing violence. In January 1979, with his regime crumbling, the shah left Iran. Two weeks later, the seventy-seven-year-old Ayatollah Khomeini returned from exile to a tumultuous reception. Shortly after, Khomeini proclaimed himself the "Supreme Leader of the Revolution."

Critics of the shah rushed to embrace Khomeini. In the *New York Times*, a prominent Princeton professor trumpeted how Khomeini would provide "a desperately needed model of humane governance for a third-world country." Khomeini and his followers moved to consolidate power. Hundreds were shot in a few months on the roof of Khomeini's headquarters in 1979. In November, after the shah was admitted to the United States for cancer treatment, a mob of young Iranian zealots "following the Imam's line" invaded the U.S. embassy. They were to hold fifty-two American diplomats hostage for 444 days under degrading conditions. Khomeini and his allies used the seizure of the hostages as the opportunity to take complete control.

The Ayatollah introduced a new constitution that is the basis for political power in Iran today. It enshrined the *velayet-e-faqih*—"the guardianship of the Islamic jurist"—and the ultimate power of religious scholars, with Khomeini having final say as the chief jurist, or supreme leader. Since he takes his authority from Allah, his word is therefore unchallengeable. The supreme leader has control over the Guardian Council, which decides who can run for office, and has final approval over parliamentary acts. He also has control over the Revolutionary Guard, the mainstay of the region, and of the media, and judiciary. The elected president is subordinate to the supreme leader. Under this new constitution, there were few bounds on the authority of the Ayatollah Khomeini. His legitimacy, he asserted, came from the Prophet and from God and from his expertise in Islamic law. In short, one scholar has written, "Khomeini had obtained constitutional powers unimagined by shahs."[5]

While Iranian presidential campaigns, and the shifts they may portend, get global attention, these constitutional arrangements—an Islamic republic under the control of the most conservative parts of the Shiite clergy—remain the foundation for the way Iran is ruled today. This theocratic regime would take control of the economy to a much greater degree than did the shah's regime, both directly and through "foundations" and the Revolutionary Guard, which is the way Iran's economy works today.

The Iranian Revolution shook the regional geopolitical order—and cemented deep fault lines that run through the region today. Khomeini's new constitution made abundantly clear that the revolution was not just for Iran but also was meant to form "a single world community." Khomeini himself declared, "Iran has only been the starting point." The objective was to destroy "all oppressive and criminal regimes." To this end, he proclaimed that Iran must "export the revolution to other countries and reject the idea of containing it within our borders."[6]

Major targets for Khomeini would include Saudi Arabia; Egypt's president, Anwar Sadat, who had had the temerity to make peace with Israel in 1979; the "Little Satan," Israel; and "the Great Arrogance"—the United States.

The Iranian Revolution would draw the United States much more deeply into the Middle East, would shape American foreign policy to the present day, and would have far-reaching effects on the region itself.

Chapter 28

WARS IN THE GULF

B ut it was not only the Iranian Revolution that drew the United States into the Gulf. On Christmas Eve 1979, the Soviet Union invaded Afghanistan, under the mistaken belief, in a fit of collective paranoia, that the local communist leader was secretly dealing with the United States. In Washington, the invasion came as a massive shock. The Soviet Union had launched the first large-scale deployment of military forces beyond the communist bloc since World War II. And with the shah gone and the security system of the Gulf in disarray, and Iran no longer the regional policeman determined to resist Soviet moves, Afghanistan was seen as the potential first step in a Soviet campaign toward the Gulf and control of Mideast oil.[1]

In January 1980, President Jimmy Carter told aides he needed to "preach a sermon, so we let the Persian Gulf countries know that we'll be there if the Soviets invade, and let the devil take the hindmost." In his State of the Union address to Congress, Carter declared, "The Soviet invasion of Afghanistan could pose the most serious threat to the peace since the Second World War." His response was the Carter

Doctrine, building on what U.S. presidents had been saying since Harry Truman.[2]

"An attempt by any outside force to gain control of the Persian Gulf region will be regarded as an assault on the vital interests of the United States of America," said Carter, "and such an assault will be repelled by any means necessary, including military force." The United States had taken on the direct security role—protecting the Gulf and the oil—that Britain had abandoned less than a decade earlier. For Moscow, the invasion would prove much more costly than the Kremlin could have imagined, for it would help set in motion the process that would lead to the collapse of the Soviet Union. It would also unleash a new jihadism, the impact of which would reverberate around the world and would reach to the very heart of the Arab world, throwing the region into crisis.

IN IRAQ, SADDAM HUSSEIN HAD BRUTALLY CONSOLIDATED control, destroyed all potential rivals, real and imagined, and stepped forward to seek the mantle of leader of the Arab world. The Arab peoples, he said, invoking the Ba'athist ideology of his regime, were "one nation." He would aim to reshape the Middle East. In February 1980, a report from the U.S. interests section in Baghdad (the United States did not have an embassy there at that time) described the dictator as "an egoist of massive proportions. Thoroughly accustomed to adulation, obedience, unctuous publicity, slavish devotion, and servility, he acknowledges the cheers of the masses with a cool, distant smile and an upraised royal hand." To work for Saddam at the senior level was terrifying. "He looked you straight in the eye, as if to control you," said one of his officers. "One moment he would be extremely affectionate; the next moment he would be extremely hostile and cruel." Saddam's predilection for taping everything, even his meetings with his most senior confidants, created an enormous trove of recordings. Recovered after the 2003 invasion of Iraq, they provide not only Saddam's voice amid his inner circle, but also an extraordinary window into his cynical thinking and blundering decision making.[3]

Oil was the enabler that made Saddam what he became. The increase in oil prices that followed the 1973 war brought a tremendous windfall to Iraq, financing the command economy and the launch of large-scale industrialization projects. Most important, the oil revenues also paid for a modernized and vastly expanded military. Yet the 1979 Iranian Revolution and the emergence of the Shiite state posed an immediate threat for the Ba'athist regime in Baghdad. Saddam and his clique were part of a Sunni minority that ruled Iraq. Despite fierce repression, the majority Shiites were becoming more restive, and resistance groups were organizing underground.

Saddam's burning hatred for Khomeini and, as Saddam called them, his "satanic turbans," was fully reciprocated by Khomeini, who saw the Ba'athists as a secular Sunni minority repressing the Shiite majority. Saddam, said Khomeini, was not just the "little Satan" but also "a pig," and the Ba'athist regime a "blasphemy against Islam."

Everything that Saddam heard from his intelligence services confirmed that a great opportunity was at hand: Iran was fragmenting, they reported, torn up by internecine warfare, "descending into a state of chaos, crimes and law breaking," with an "increase in the disintegration and dismantlement of Iranian forces."[4]

Saddam envisioned a limited war and a quick victory that would bring great benefits at a low cost. He would conquer the Arab-speaking province of Khuzestan (part of which had been historically known as Arabistan), with 90 percent of Iranian oil reserves, free it of the "Persian yoke," and add its oil riches to the economic arsenal of what was going to be his new Arab superpower. He would undermine Khomeini and end its threat to his regime. And he would enshrine his position as the leader of the Arab world. "We have to stick their noses in the mud," Saddam said of the Iranians on September 16,

1980, in a meeting with top officials. "This cannot take place except militarily."[5]

On September 22, 1980, Iraq launched the battle with what were supposed to be knockout air raids on Iran. Instead, Saddam had blundered into what would prove a long and bloody stalemate, the longest major war of the twentieth century. The Iranian forces were bolstered by the emergence on the battlefield of a new force, the Pasdaran, or Revolutionary Guard—initially a poorly trained militia, but fanatical fighters devoted to the Khomeini revolution. Saddam's forces used what he called "special ammunition"—chemical weapons. But his drive, in his words, for the "Arabic atom" was set back and forced underground when Israeli jets destroyed the Osirik reactor that would produce fuel for a nuclear weapon. During the war, both Iran and Iraq were hard hit by the collapse in the oil price in 1986 and their respective declines in production and exports. But Iraq had "bankers." It received many tens of billions of dollars in loans from Saudi Arabia and Kuwait. Iran, however, did not have bankers; it was running out of money. In 1988, senior members of the Islamic regime persuaded Khomeini that Iran was going to be defeated. He accepted a United Nations cease-fire, though describing this decision as "more painful and deadly to me than drinking a cup of poison."[6] The war had cost half a million lives, and another million wounded.

Saudi Arabia, despite its antagonism toward Saddam and the Ba'athist regime in Baghdad, had supported Iraq with money and weapons during the war as a bulwark against what seemed the greater and more immediate threat—revolutionary Iran. The animosity between Tehran and Riyadh never ceased. Iran sponsored the creation of a new terrorist organization, Hezbollah Al Hijaz, with the aim of either toppling the Saudi monarchy or at least facilitating instability and the secession of the eastern region of Saudi Arabia. It launched attacks both within the country, including against oil installations and military targets, and against Saudi diplomats abroad.

"I don't know where it will end," said Saudi King Fahd in 1988. "Iran has harmed relations not only with us but also with its neighbors and the whole world. . . . Iran has tried many times to undermine security in the Gulf region, the Arabian Peninsula, and the world. What has Iran gained? Iran has gained nothing." Yet he would add, "We cannot change the geographic reality of Iran, and Iran cannot change our geographic reality."[7]

IN THE SUMMER OF 1990, IRAQ'S WAR WITH IRAN HAD BEEN over for less than two years. But Saddam was once again determined to redraw boundaries of the region. Early in the morning of August 2, Iraq invaded one of its own bankers, oil-rich Kuwait. It took less than two days to complete the conquest. Saddam's regime moved quickly to try to obliterate Kuwait from the map—what an Iraqi military handbook called the "Iraqization of Kuwait." The campaign was implemented with a brutal reign of terror and plunder. Saddam's half-brother, now chief of security in Kuwait, reported the process to him: "After we complete the interrogations, we treat them harshly, really harshly, then kill and bury them."[8]

In the immediate aftermath of the August 1990 invasion, Iraqi forces began to mass on the Kuwait–Saudi Arabian border. Was Iraq going to send its army into Saudi Arabia and capture the Saudi oil fields—accounting for 25 percent of world proven reserves? "They see a chance to take a major share" of world oil, British prime minister Margaret Thatcher said to President George H. W. Bush. "Losing Saudi oil is a blow we couldn't take."[9]

With Kuwait, Saddam controlled almost 20 percent of proven world oil reserves. If he went further and gained Saudi oil, he would command 45 percent. Even without Saudi Arabia, he would dominate the Persian Gulf, with its two-thirds of world oil reserves. In short, he could become the arbiter of world oil.

Saddam could achieve what had eluded Egypt's Nasser—to be the leader of the "Arab nation," expunge the borders that went back to Sykes-Picot and before to the Ottomans, and dominate the map of the Middle East. He would be launched on achieving his ambition of creating an Arab superpower based on arms and oil, and on the money and global influence that would flow from that oil. All this would change global politics in a way that was not expected of the new post–Cold War world. These were the stakes for Saddam, and, equally, they provided the compelling rationale for George H. W. Bush and Secretary of State James Baker to build what eventually became a unique post–Cold War coalition of thirty-four nations to oppose Saddam.

The war began in January 1991, with constant bombardment of Iraq that had a devastating impact on its forces. At one point, an urgent plea went out to find Iraqi soldiers with "camel grazing experience," the idea being that camel caravans were less likely to be attacked from the air than convoys of trucks. On February 24, the huge coalition launched the ground war. It lasted just over a hundred hours. Iraq was a defeated nation.[10]

It had been widely expected within the coalition that, with the defeat, Saddam would be overthrown by the army. But he maintained his absolute grip on power.

Directly toppling Saddam had never been part of the coalition's war plan. Writing seven years later, in 1998, Bush and his national security adviser, Brent Scowcroft, laid out the reasons why the coalition had not pushed on: "Trying to eliminate Saddam . . . we would have been forced to occupy Baghdad and, in effect, rule Iraq. Had we gone the invasion route, the United States could conceivably still be an occupying power in a bitterly hostile land."

The coalition had achieved its objectives. Iraq had been ejected, its military humbled; Kuwait had been liberated, and the map of the Gulf restored. Iraq was boxed in, contained within a whole mesh of sanctions and restrictions around it. And Saddam had been defeated.

But that was not how the war was portrayed within Iraq. As a

Republican Guard history described events, the Americans and the rest of the coalition were just about to "fall into the trap" of direct and bloody battle with entrenched Iraqi troops. "With the certain imminence of vast losses in lives and equipment," George H. W. Bush, according to this narrative, had concluded that a "ceasefire was the only way out."

A few days after the ceasefire, Saddam told his military staff, "The strongest scientific, technological, and military powers . . . all got together against us and they did not succeed despite what happened. They did not dare attack Baghdad." The "Mother of All Battles," he proclaimed, had been a great victory for Iraq and for his regime.[11]

IN THE AFTERMATH OF THE GULF WAR, THE WESTERN GOVERN-ments remained convinced that Saddam was determined to get the wherewithal to acquire WMD. As the chief UN weapons inspector put it, "Saddam had an addiction to weapons of mass destruction."[12]

George W. Bush became president in January 2001, eight years after his father's loss to Bill Clinton. On September 11, 2001, Al Qaeda operatives commandeered civilian airliners and attacked the World Trade Center in New York and the Pentagon near Washington, D.C., with the loss of 2,977 lives. The United States responded with the "war on terror," beginning with a counterattack on the Taliban regime in Afghanistan, which had sheltered Al Qaeda. Some in the administration of George W. Bush advocated as the next step the removal of Saddam, finishing what they saw as the unfinished business when the coalition had stopped short of Baghdad in 1991. Some argued that Iraq was allied with Al Qaeda, though that was sharply disputed by many who said there was little connection between secular Ba'athists and the fundamentalist Al Qaeda. It was also clear that the sanctions regime and containment of Iraq were faltering.

The focus increasingly became weapons of mass destruction. It was thought that Saddam had to be concocting a new WMD program.

This had been the view of the preceding Clinton administration. The conviction was buttressed by the discovery, after the 1991 war, that Iraq had been eighteen months or less away from a nuclear weapon. If Saddam had waited until 1994 or 1995 to invade Kuwait, he would have been in a far stronger position to deter or resist the coalition that swept into Iraq in 1991—with a potentially very different outcome.

There was also the rebound from the shock of 9/11, the determination to demonstrate American power anew and ensure that there would not be another, much more dreadful attack. Instead of deterrence or containment, the Bush administration adopted the policy of "preemption." For, said Bush, "If we wait for threats to fully materialize, we will have waited too long."[13]

On March 20, 2003, the Iraq War commenced with a fierce aerial barrage known as "Shock and Awe." By the end of the first week of April, Iraqi resistance was collapsing, and Saddam was delusionally issuing orders to units that no longer existed. By April 9, the "coalition of the willing," led by the United States, had accomplished what had not happened twelve years earlier—the capture of Baghdad.

What was meant to be a short war turned into a long and grueling conflict. While the war had lived up to its billing in terms of its effect, the planning for the aftermath did not. Policies thereafter put in place were improvised, contradictory, and poorly adapted to the reality on the ground.

The policy of "de-Ba'athification"—a phrase deliberately chosen to echo the denazification of Germany after World War II—was intended to remove the top of the Ba'athist system, but it was applied more deeply into the society, down to schoolteachers. Many of the civil servants who ran the ministries and the state-controlled economy were gone, as were many others who had made the country function on a daily basis.

The order dissolving the Iraqi military overturned what had been the previous U.S. policies, including prewar psychological warfare that had promised Iraqi soldiers that if they did not fight they would be taken care of, as well as plans to use many of the soldiers for reconstruction and to maintain order. One U.S. general was addressing six hundred senior Iraqi officers on what their roles would be in rebuilding the Iraqi military when word came that they all had just been dismissed. Even a plan to give officers $20 each to buy emergency supplies for their families was canceled. Altogether, more than 600,000 soldiers and other security officials were sent home devoid of compensation and pensions, no prospects, but only their guns and resentments and anger that would fuel resistance. Months later, a payment system for ex-soldiers was finally set up. "Demobilizing the Iraqi Army instead of depoliticizing it set the most capable group of men in the country on an adversarial course against us," wrote Marine general and later U.S. defense secretary James Mattis. That decision to dissolve the army, a U.S. officer in Iraq was later to say, was the moment "that we snatched defeat from the jaws of victory and created an insurgency."[14]

Nine months after the invasion, a U.S. Army soldier kicked aside a rug in an orchard, revealing under it a piece of styrofoam. He was about to throw a grenade down the hole it covered when Saddam Hussein emerged, disheveled, with a dirty, matted beard. The man who intended to make himself the dominant figure in the region, the "hero" of the Middle East and the king of oil, had been hiding in little more than a rabbit hole, protected with a few guns and in the company of a suitcase filled with $750,000 worth of hundred-dollar bills.[15]

The weapons of mass destruction that had been the rationale for the war were never found. It turned out that Saddam had indeed heeded the message of the previous war and had shut down the WMD operations— although apparently maintaining the option of a "quick restart" if sanctions were lifted. He could have undercut the entire rationale for the war had he let it be known that this time he had no WMD. But to do so would have sent a message to his own population of his fallibility. There was an even more important reason for maintaining the ambiguity about his WMD—deterrence. And whom was he going to deter? To his FBI interrogator after the war, Saddam summed it up in one word—Iran.[16]

Saddam was later convicted of crimes against humanity and executed.

Chapter 29

A REGIONAL COLD WAR

Mohammad Khatami was not supposed to win. Among Iranian clerics, he was considered a moderate and a reformer, but with little political base. But so deep was the dissatisfaction with life under Supreme Leader Ali Khamenei, the successor to Ayatollah Khomeini, and so significant the demographic changes (70 percent of the population was under thirty) that in the 1997 Iranian presidential election Khatami scored an upset victory over the candidate of the conservative religious establishment.

Khatami came in with a mandate to liberalize society, roll back strict Islamic controls, promote rule of law, and reform the economy. Shocked, the conservative clerics around the Ayatollah Khamenei and their allies in the Revolutionary Guard quickly set out to undermine Khatami's domestic reform program and Khatami himself.

Khatami had somewhat more success in foreign affairs. He proposed a "dialogue of civilizations" as a path out of Iran's isolation. There was even an outreach to the United States along those lines, but Tehran and Washington were never in the same political phase and never found a common meeting ground. From Washington's perspective, any accommodation had to deal with the 1996 truck bomb attack on the Khobar Towers in eastern Saudi Arabia that killed 19 U.S. service members and wounded another 372, which was masterminded by Iran's Revolutionary Guard.[1]

Moreover, in trying to move toward détente with Washington, Khatami was entering into what to the Islamic hardliners was a "no-go" zone. Iran's supreme leader, Ayatollah Ali Khamenei, had been one of Khomeini's chief lieutenants during the Iranian Revolution and the head of the Revolutionary Guard at the beginning of the Iran-Iraq War before becoming president of Iran. For him, anti-Americanism is an absolute necessity for the regime's survival, a foundation for both the ideology and the legitimacy of the Islamic Republic.[2]

Despite the internal opposition, Khatami moved to improve relations with Saudi Arabia, facilitated by the fact that at that time Saddam was still in power and Iran and Saudi Arabia regarded him as their common enemy. They signed a broad cooperation agreement. In 1999, Khatami himself visited Riyadh; then, in 2001, the two countries adopted a limited but painstakingly negotiated security agreement, which called for noninterference in each other's internal affairs.

This rapprochement, from 1998 through 2001, gave rise to statements about collaboration that are startling in light of today's antagonism. Prince Nayef bin Saud, the Saudi interior minister, declared that the signing of the agreement marked "a big step toward security between our two countries. We consider Saudi Arabia's security as Iran's security and Iran's security as our security." Iran's ambassador to Saudi Arabia went even further. "Iran's missile capabilities are at the disposal of the Kingdom of Saudi Arabia," he said. "Our relations with Saudi Arabia have reached a historical stage where we are complementing one another."[3]

Yet what remained immutable was the mutual suspicion based on the fundamental incompatibility of the two governments and two systems—one powered with revolutionary zeal and the other the status quo—one Shia, the other Sunni. The rapprochement foundered on Iranian intervention in Lebanon. With the fall of Saddam in 2003, the two countries no longer faced the threat of a common enemy. Thereafter, Iraq's descent into a Sunni-Shia civil war put Iran and Saudi Arabia on a collision course, all the more so with Iran's drive to assert hegemony over Iraq. Saudi King Abdullah said Iran was not "a neighbor one wants to see" but "a neighbor one wants to avoid." For, he added, "Iran's goal is to cause problems."[4]

IN 2002, THE UNITED STATES AND OTHER GOVERNMENTS were stunned to learn that Iran was secretly pursuing programs to develop nuclear weapons. Recognizing that pursuit of "dialogues" and normalization was now jeopardized, Khatami suspended the programs.

His move unleashed a torrent of denunciation from hardliners. "Those who are handling the talks are terrified, and before they even sit down to talk they retreat 500 kilometers," declared one of them, Mahmoud Ahmadinejad, who was running to succeed Khatami as president.

The end of the Saudi-Iranian détente and the beginning of the renewed rivalry can be dated to Ahmadinejad's victory in the 2005 presidential election. A Ph.D. in traffic management, Ahmadinejad had been mayor of Tehran. He was also a veteran of the Revolutionary Guard, and he had pledged to return Iran to its revolutionary course and assert itself as "the preeminent power of the region." There was to be no "dialogue of civilizations," no compromise with the United States. As for 9/11, it was a Zionist plot. The Holocaust had never happened, Israel was to be "erased from the pages of time," and a final apocalypse was to be welcomed, for it would bring back the Hidden Mahdi.[5]

Once in power, Ahmadinejad not only resumed but stepped up the nuclear and ballistic missile program, using many subterfuges to evade UN sanctions. Whirring centrifuges were enriching increasing amounts of uranium fuel—potentially to weapons grade. The Saudis, along with other Arab neighbors and the Israelis, had no doubt that Iran intended to develop nuclear weapons and seek to dominate the region. King Abdullah exhorted the United States to act against Iran and "cut off the head of the snake."[6]

Over the years 2011–12, two sets of sanctions and restrictions were constructed by the "P5+1"—the permanent members of the UN Security Council, plus Germany. Bolted together, these sanctions would act like the jaws of a powerful vise, squeezing the Iranian economy in a way that had never happened before.

One set aimed to lock Iran out of the global financial system. The United States would punish any banks that did business with Iranian banks by imposing huge fines and other penalties, including shutting them out of the dollar-based global financial system. The financial sanctions would have much greater impact than initially expected.

The other jaw of the vise was sanctions on Iranian oil. This went to the heart of the Iranian economy. Iran was producing about four million barrels per day, and exporting 2.5 million. Iran's economy was more diversified than those of the Arab oil producers on the other side of the Gulf, but oil was still the major driver of its economy and responsible for 65 percent of the government budget prior. Other Western sanctions prohibited imports of Iranian oil into Europe, insurance for Iranian oil tankers, and investments by international companies in Iran's oil sector.

But then came the arduous job of convincing the major Asian importers—China and India—to slash their imports from Iran. The message to them was that cutting the purchase of Iranian oil was much less disruptive than a war erupting in the Middle East over Iran's nuclear program. In addition, they risked financial sanctions. U.S. State Department energy envoy Carlos Pascual worked to demonstrate to major Indian oil refiners how they could diversify their supply sources, substantially reduce their imports from Iran, and thus avoid being black-balled from the international financial system. For its part, China decided on "national security" grounds to reduce its imports from Iran and concentrate what remained in a single buyer that was not connected to the international financial system and thus insulated from U.S. sanctions.[7]

Altogether, Iranian exports fell from 2.5 million barrels per day to, at their lowest, 1.1 million. The squeeze did not end there. The payments even for those 1.1 million did not go to Iran; they went into escrow bank accounts—eventually totaling well over $100 billion. Altogether, the loss of earnings, combined with reduced economic activity, hit Iran very hard across its entire economy.

Many, including the Iranians themselves, had assumed that the oil sanctions would fail because consuming countries would need Iranian oil to avoid a shortage and a price spike. But they did not see what was coming—the rapid rise in U.S. shale oil output. At the same time, Saudi Arabia and the other Gulf producers increased their exports. The global market was moving toward surplus.

In the summer of 2012, U.S. diplomats slipped into Oman, at the tip of the Arabian Peninsula. Oman was the one Arab country that maintained a dialogue with Iran. In the capital city of Muscat, they met with mid-level officials from Tehran. The discussions ended up at what one of the Americans called a "brick wall." Yet it was significant that such a secret meeting had taken place at all.[8]

THE FOLLOWING YEAR, IN JUNE 2013, A NEW PHASE OPENED, owing to a change of government in Tehran. Once again, as in 1997, a reformist, or at least a pragmatic, candidate had bested the conservative establishment's candidate in an Iranian presidential election. Once again the candidate—in this case, Hassan Rouhani—had ridden to victory on a tide of pervasive dissatisfaction, driven by the deep recession into which the sanctions and isolation had plunged the country.

Rouhani himself was no outsider to power. He had studied abroad, in Scotland, at Glasgow Caledonian University, where his Ph.D. was about the "flexibility of Shariah law." For sixteen years, he had headed Iran's national security council. But he had made his position clear during the 2013 campaign: "It is good for the centrifuges to operate," he had said, "but it is also important the economy operate as well and that the wheels of industry are turning." Now, as president, he recognized that his new administration was inheriting the "worst conditions" in the economy and that the government was about to run out of money. If Iran was to extract itself from the sanctions vise, it would have to work something out.[9]

What now ensued were secret negotiations, in a remote seaside villa in Oman, led on the American side by U.S. deputy secretary of state William Burns. Remarkable in these—and in subsequent—negotiations was how so many of the high-level Iranians had done their graduate education in either the United States or Britain, beneficiaries of the educational opportunities created by the shah whom they despised. When the secret negotiations became known, Saudi Arabia, along with the UAE and Israel, were aghast to find that their close American ally had been covertly meeting with their biggest enemy, without any signal or even hint to them. A subsequent set of negotiations was required to get all of the members of the P5+1 on board.[10]

In January 2014, an interim nuclear deal—the "Joint Plan of Action"—was announced. Given the complexity of the political bargaining and technical details, it was followed by another year and a half of wearing, frustrating negotiations. Secretary of State John Kerry led the American side. Energy Secretary Ernest Moniz, a physicist from MIT, was recruited on a few days' notice to provide the nuclear expertise that would prove crucial in the protracted negotiations. At one point, Obama had an unexpected fifteen-minute phone conversation with Rouhani as the latter drove from the UN General Assembly to the airport. The call ended with Rouhani saying to Obama in English, "Have a good day."

At last, in August 2015, the final deal emerged. In essence, it hit a giant nuclear pause button. For ten years, it restricted Iran's ability to enrich nuclear fuel and develop nuclear capabilities. It provided for "intrusive inspections" meant to ensure a year's warning if Iran did try to "break out" and restart its weapons program. The core of the Arak nuclear reactor was to be filled with concrete to ensure that it could not be restarted. In turn, the financial and oil sanctions would be lifted. Tehran would gain control of the $100 billion-plus held in the oil export escrow account. It would also get $50 billion of assets that had been frozen in the West, going back to the hostage-taking in 1979. All this was money that Iran desperately needed.[11]

Despite strong opposition in both Washington and Tehran, the agreement went into effect in January 2016. Iran began to crank up its oil exports and, at the same time, claim the billions of dollars being held outside the country.[12]

WHAT WOULD HAPPEN AFTER TEN YEARS? THE WESTERN SIDE hoped that the passage of time and generational change would mellow the government, shifting it from revolutionary zeal and hegemonic ambitions to integration with the global community. The advocates also counted on the provision that if Iran were to break the terms of the deal, the sanctions would "snap back" into place. For the critics, however, ten years was a short time; and Iran was not trustworthy, was continuing to develop ballistic missiles and pursue its aggressive designs in the region, and in due course would inevitably cheat on the agreement. By that time, the international consensus to contain Iran's program would have evaporated. "Snap back" sanctions would have lost their snap.

Donald Trump, both as a candidate and then as president, repeatedly denounced the nuclear agreement as "the worst deal ever." In May 2018, rejecting personal appeals from German chancellor Angela Merkel and French president Emmanuel Macron, who had both come to Washington to lobby him, Trump announced that the United States was withdrawing from the agreement. This meant that the United States would be reimposing sanctions.[13]

If the expectation of the Obama administration in negotiating the nuclear deal was what might be called "regime mellowing," in one way or the other the Trump administration was seeking regime change in Tehran—change either in the way it operated, in its very character, or in who occupied the seats of power. By November 2018, unilateral U.S. sanctions on Iran were put in place, including new restrictions on the purchase of Iranian oil. "Oil is in the frontline of confrontation and resistance," Rouhani said. The other signatories of the P5 vigorously argued against the U.S. withdrawal, but their companies and banks had no choice but to go along with the sanctions. Otherwise they would be locked out of the U.S. financial system and find themselves subject to U.S. penalties.[14]

ONE OF THE MAIN ARGUMENTS OF THE DEAL'S CRITICS WAS that it did not address Iran's intervention in the region. Amid all the local and national struggles, so many of the conflicts in the Middle East now fit into what has been called the new regional "cold war," the larger struggle between Iran and Saudi Arabia across the Middle East. Iran declares that it is exporting its revolution and supporting the "party of resistance" against the United States, Israel, and the Arab monarchies, and by so doing protecting the Islamic revolution at home. Its engine is the Quds (which means "Jerusalem") Force, the foreign arm of the

IRGC—the Islamic Revolutionary Guard Corps. To the Saudis and the other Gulf Arabs, Iran is striving for hegemony over the region and the establishment of a "new Persian Empire." Look at the map, they say. The reach of the Quds Force in one country after another is part of a strategy to encircle Saudi Arabia and the UAE and ultimately bring them down.

Lebanon, on the Mediterranean at the western edge of the Middle East, became the first battlefield for exporting Iran's revolution, in this case well prior to the invasion of Iraq. The country was created by the map-drawing after World War I, with a slim majority of Maronite Christians and a political order based on power sharing among Christians and Muslims. Its politics ever since independence have been turbulent, with a chronic pattern of assassinations of prime ministers and presidents. It was also a testing ground for the modern suicide bombers, including an attack that killed 241 U.S. military personnel and another that destroyed the U.S. embassy in Beirut. The political-military Hezbollah ("Party of God") was fashioned by Iran out of Lebanon's Shia communities. Its militia fighters were trained by the IRGC; its development was overseen by clerics and soldiers from Iran; much of its money and weaponry, including rockets and missiles, comes from Iran. It has become the dominant force in Lebanese politics, maintains networks in Africa and South America, and fields a formidable fighting force, both in Lebanon and elsewhere in the Middle East, in support of Iranian objectives. In 2018, it became the largest party in the Lebanese parliament. In 2020, a Hezbollah-dominated coalition took over the government. Lebanon was Iran's first great regional success.[15]

Chapter 30

THE STRUGGLE FOR IRAQ

Iraq had been in Tehran's sights from the beginning of the Iranian Revolution. Its majority Shiite population, and the religious flow between the two countries, made it the obvious target for exporting the revolution. But as long as Saddam was there, that road was blocked.

What ensued following the 2003 war and the fall of Saddam's regime was a battle for power and resources, and a maelstrom of revenge, in which politicians and Sunni and Shia militias fought a deadly struggle for power and the spoils of the war. It also, at last, provided Iran with the major opening it craved in Iraq, in conjunction with Shia politicians and militias. Iran aimed to keep Iraq fragmented and under its control and assure Shiite domination. And, critically, Iraq could be the platform in the Arab world from which to export its revolution and confront the Sunni Saudi Arabia and Gulf Arab states.

It had many tools with which to pursue its objectives in Iraq—religion and trade, bribes and subsidies to political and tribal leaders, intimidation and threats, and violence and assassination. But the most important instrument was the Quds Force of the Islamic Revolutionary Guard Corps, which fielded its own soldiers and created and directed Iraqi Shiite militias. But the Iranian campaign was counterbalanced by American power—which was reinforced in the face of Sunni insurgency by the U.S. military "surge" in 2007–2008. But, in 2011, the decline in violence and the inability to negotiate a new status of forces agreement with the Iraqi government led to the withdrawal of most U.S. forces.

After the withdrawal, Prime Minister Nouri al-Maliki, a member of the Shiite Dawa Party, shifted and become increasingly despotic. He centralized power, stoked sectarian tensions, allied with the pro-Iranian militias, and came to be seen as "Iran's point man" in Iraq. Saudi King Abdullah refused to meet him. "He's an Iranian agent," the king said.[1]

Maliki abrogated Shia-Sunni collaboration. Sunni officials and military officers were abruptly fired, without salary or pension, or jailed. The American military had collaborated with the Sunni Awakening and might now have protected its members, but the Americans were leaving. "It was my duty and honor to work with the Americans as a tribal leader fighting terrorism," said a Sunni sheikh. "But now it feels like we've been abandoned. We were left in the middle of the road."[2]

For Tehran, Iraq was the essential building block in what it would call the "axis of resistance." At the center of the axis stood Qassem Soleimani, who for two decades had been the commander of the Quds Force. He was both the architect and implementer of the axis of resistance.

Soleimani was barely in his twenties in 1980, working in a municipal water department, when Saddam Hussein invaded Iran. Dispatched on what was to be a short mission to bring water to troops at the front, he never left. War became his vocation; he would describe the battlefield as "another kind of paradise."[3]

Under him, the Quds Force became a highly organized and very efficient machine, whether for conventional battle or asymmetric war against stronger powers. Soleimani was driven in all of this by his devotion to the Islamic Republic and the Iranian Revolution.

After the U.S.-led invasion of Iraq in 2003, Soleimani oversaw the war against the coalition forces. His most important instrument was the Shiite militias that were sponsored and controlled by Iran—and specifically by him. Iran recruited fighters, trained them at secret camps in Iran and Lebanon, and indoctrinated them with Iran's ideology and the glories of martyrdom. It financed the militias and provided them with intelligence and with deadly weaponry, including heavy armor, improvised explosive devices (IEDs), and "explosively formed penetrators" (EFPs) that can pierce the metal armor of tanks and Humvees. These weapons would be responsible for the deaths of many Americans and other coalition forces.[4]

OIL WOULD BE CENTRAL TO THE FUTURE OF IRAQ, AS IT HAD been in the past; but output had collapsed—the result of lack of investment, mismanagement, and sanctions prior to the war, and then made much worse by postwar looting and lack of security and attacks by insurgents. It was not until 2009 that Iraq, on an annual basis, finally regained the production levels of 2001.

And 2009 was also the year that finally, six years after the invasion, Iraq was organized enough to hold a competitive bidding round to bring in the foreign companies with the capital and technology it desperately needed to revitalize its giant oil fields that were underdeveloped and underinvested. Some fifteen companies ended up in consortia in ten existing fields.

A *New York Times* editorial reported "understandable suspicions" that gaining control of Iraqi oil was the real reason why the United

States had invaded Iraq. But those "suspicions" were gainsaid by what actually happened. As it turned out, of the fifteen companies, just two were American. The nationalities of the other countries included Malaysian, Chinese, Korean, Russian, Norwegian, Turkish, French, British, and Dutch. Iraq needed these companies. Oil is not merely the heart of the Iraqi economy. In economic terms, Iraq *is* oil, which makes up over 90 percent of government revenues, over 99 percent of exports, and almost 60 percent of GDP. The World Bank describes Iraq as "the world leader in terms of dependence on oil."[5]

Saddam had blocked oil development in the Kurdish region. After his overthrow, the semiautonomous region gained the right to develop its petroleum resources. But were the resources there?

The Kurds recognized that they had to be highly competitive, owing to the lack of both development and knowledge of the geology, in order to bring in foreign investment. By 2016, some twenty-seven companies were active in Kurdistan, most of them independents. As they drilled in the fractured mountains, they discovered that the geology was challenging. Still, between 2008 and 2016, Kurdish production rose to over three hundred thousand barrels per day, substantial for a region that had produced virtually no oil at all prior to 2003.

Strife was constant between Baghdad and Erbil, the capital of Kurdistan, over control of the oil and revenues, partly resulting from ambiguity in the constitution adopted in 2005. That created problems in terms of hooking into the existing Baghdad-controlled northern pipeline. By 2013, a new Kurdish pipeline to the Turkish border was completed, which gave Kurdistan its own direct route to Turkey and on to the world market. That meant a route that was not subject to Baghdad's control. But it was subject to the goodwill of Turkey.

Iraq's overall output, including in Kurdistan, continued to recover. This increase at a time of rising prices provided desperately needed funds. Iraq's oil earnings rose from almost $18 billion in 2004 to over $89 billion by the beginning of 2014.[6]

———

IN THE SUMMER OF 2014, THE POWER-HUNGRY MALIKI WAS
forced out as prime minister. His replacement was Haider al-Abadi,
who had joined the Dawa Party at age fifteen. Two of his brothers had
been executed by Saddam's regime. He had spent his exile years in
England, where he had earned a Ph.D. in electrical engineering and
had run an engineering company that, among other things, managed
the elevators in the building housing the BBC's World Service. Once
in power, he modulated Maliki's harsh anti-Sunni policies, brought
Sunnis into the government, sought to assert some state control over
the militias, and began to formulate policies for reviving Iraq's economy and fighting corruption. There was no doubt that Iran remained
dominant. Still, relations with the United States improved, and Abadi
reached out to his Arab Sunni neighbors. For they could do something
that Iran could not—provide help in reviving Iraq's economy.

Whatever the hopes for Iraq's stabilization, they were soon overwhelmed by still another crisis, this one in the oil market. The oil price
collapse that began in November 2014 cost Iraq dearly. Baghdad's annual revenues fell by 50 percent between 2013 and 2016.

Abadi had wanted the American military back in the country in
coalition with other countries, both airpower and some troops on the
ground. First and foremost, it was to help defeat ISIS and win back the
country. But there was another benefit—to help offset the "unbearable
Iranian pressure on his government."[7]

Altogether, Iran has thoroughly penetrated the Iraqi political and
security structures, with Soleimani as the orchestrator of it all. In 2019,
seven hundred leaked Iranian intelligence cables provided granular
evidence of the dense network of agents and spies, facilitated by bribes
and intimidation. As one Iranian put it, "We have a good number of allies among Iraqi leaders who we can trust with our eyes closed." The
various Shiite militias, now designated as PMFs—Popular Mobilization

Forces—are integrated into the Iraqi security apparatus. About half of the militias are under the control of the Quds Force. In order to further bolster its position in the future, Iran is seeking to help turn militias into political and social organizations along the Hezbollah model. Iran has also assured itself financial benefits from its role in Iraq, including oil contracts.[8]

Despite Abadi's successes, including the fight against ISIS, his position was undermined by pervasive corruption and by the breakdown in social services. In Basra, the center for 80 percent of Iraqi production and the economic capital of the country, protests erupted in 2018 over lack of investment and jobs and shortages of water and electricity, which made life unbearably difficult in the scorching 115-degree summer weather—made worse when Iran cut its electric power deliveries to southern Iraq. The protests, spreading to other cities in the south, turned violent, including, in what was unprecedented, the torching of official Iranian offices and facilities belonging to Shiite parties, accompanied by chants of "burn the Iranian parties."[9]

The Iraqi grand ayatollah Ali al-Sistani, to whom millions of Iraqis looked, decried the "miserable" conditions in Basra and called for new leadership. Abadi was forced out as prime minister. It took almost half a year before another new government could be formed. Though the battle over the new government had been difficult, it had been the fourth transition of power in Iraq since the end of Saddam's rule. That government did not last either. In 2019, protests once again erupted over unemployment, corruption, and failed social services. Iran's domineering role became a major target of Iraqi Shiite protestors. The Iranian consulate in the Shiite holy city of Najaf was burned. Many dozens or even hundreds of protestors were shot. The government fell. It took half a year before a new one could be formed. By the spring of 2020, Iraq seemed on the verge of collapse. Finally, in May 2020, a new prime minister was confirmed, Mustafa al Kadhimi, former head of intelligence, and before that a journalist and head of a foundation devoted to chronicling the crimes of Saddam's regime. "The security, stability, and blossoming of Iraq is our path," he said. It looked to be a very difficult path. In addition to the confrontation between the United States and Iran, the new government had to deal with a raging coronavirus crisis, the resurgence of attacks by ISIS militants, and the evaporation of much of the oil revenues on which its budget depended.

Still, to a considerable degree, Iran had achieved its objectives. Iraq is both a weak state and a centerpiece in the axis of resistance.

Chapter 31

THE ARC OF CONFRONTATION

An eruption against the systematic failures of Arab regimes began in Tunisia after the self-immolation of a despairing young fruit salesman. Swelling demonstrations, fueled by social media, forced the authoritarian and corrupt dictator, Zine el Abidine Ben Ali, to flee the country in January 2011 after a twenty-three-year rule.

Over the following months, across North Africa and the Middle East, demonstrators in massive numbers, sparked by social media and satellite television, took to the streets and occupied the squares. These movements and the change in governments that resulted became known as the Arab Spring, a time of optimism and expectations for a dawn of a new era, evoking William Wordsworth's words at the start of the French Revolution—"Bliss was it in that dawn to be alive." But it was not all that long before the bliss turned bleak, and the Arab Spring would instead become something that some would describe as an Arab Winter.

———

IN EGYPT IN LATE JANUARY 2011, DEMONSTRATORS POURED into and around Tahrir Square in Cairo, eventually as many as a million people. However disparate their politics, their common objective was to force out the eighty-two-year-old Hosni Mubarak, Egypt's president for three decades. Connected by Facebook and Twitter, captured on global television, they quickly came to be seen around the world as the vanguard of a new, modern generation. The police melted away. Thugs, attacking them from camels and horses, failed to dislodge them.

Mubarak had been a resolute ally, key to peace with Israel and to the 1991 Gulf War coalition (praised by George H. W. Bush as "my wise friend"), and then in the campaign against Al Qaeda. Barack Obama's senior advisers—Secretary of State Hillary Clinton, Defense Secretary Robert Gates, and Vice President Joe Biden—urged caution in joining the rush to push Mubarak out. Gates was on the National Security Council in 1979 when, in his view, the United States had pulled the rug out from under the shah, with the expectation that a democratic revolution would follow. The result instead was the rise of the Ayatollah Khomeini, U.S. diplomats held hostage for 444 days, and the implacably hostile Islamic Republic.

The more junior advisers around the president vigorously disagreed. They were caught up by the excitement of the Arab Spring and felt an affinity for the Facebook and Twitter generation. Sure of the power and sweep of Obama's oratory, they urged the president to not hesitate in pushing Mubarak aside. They told Obama that he should be "on the right side of history."

"But how can anyone know which is the 'right' or 'wrong' side of history," Gates later wrote, "when nearly all revolutions, begun with hope and idealism, culminate in repression and bloodshed? After Mubarak, what?"[1]

Secretary of State Hillary Clinton dispatched Frank Wisner, a seasoned retired diplomat and former ambassador to Egypt, to Cairo to see Mubarak. The message Wisner conveyed from Washington was clear—an orderly transition should begin. Shortly afterward, Mubarak went on television, pledging "a peaceful transition of power." He had said, in Gates's words, "exactly what the administration had asked him to do."

But that was no longer sufficient. Obama called Mubarak and said that change had to begin "now." And what did "now" mean? The next morning, February 2, Obama's press secretary provided the answer. "'Now' started yesterday."[2]

Nine days later, Mubarak resigned. Dispatched to Egypt, Deputy Secretary of State William Burns reported back, "The not-so-good news is that expectations are unrealistically high." In rushed elections later that year, the Muslim Brotherhood was the only group that was actually organized, and its candidate, Mohamed Morsi, was victorious, though only barely. He moved quickly to try to entrench the Muslim Brotherhood permanently in power. Too quickly, as it turned out. In 2013, on the second anniversary of the downfall of Mubarak, even larger demonstrations swelled in Tahrir Square to protest Morsi's power grab. They succeeded in bringing him down. His successor was Abdel Fattah el-Sisi, former chief of staff of the military. The "right side of history" had proved elusive.

The fall of Hosni Mubarak had dramatic impact on the balance of power in the region. "Our allies in the Middle East," Robert Gates later said, "were wondering if demonstrations or unrest in their capitals would prompt the United States to throw them under the bus as well." What the Gulf Arabs saw as the unreliability of the United States in a crunch heightened their sense of vulnerability in the face of Iran's deployment across the region.[3]

LIBYA IS AN IMPORTANT BUT SECOND-ORDER OIL EXPORTER.
After Tahrir Square in Cairo, protests erupted in Libya against the

government of the mercurial dictator Muammar Gadhafi, who had been in power for forty-two years. It quickly became a civil war. In 2003, Gadhafi had come in from the cold, ending his international isolation and giving up his weapons of mass destruction. But now, in 2011, with Gadhafi preparing to launch what was expected to be a bloody attack against the opposition in the city of Benghazi, the Arab League urgently pleaded for help, and U.S. and European air forces, operating under UN and NATO authority, came to the rescue of the rebels. Here, said an adviser to Obama, the United States was "leading from behind."

This civil war had an immediate impact on the oil market. For a time, all of Libya's output was disrupted and prices headed toward $130 a barrel. In October 2011, a fleeing Gadhafi, seeking to evade air attacks, sought refuge in a drainpipe, and there he was killed. His regime was finished.

Gadhafi's highly personal rule left little in the way of institutions. Weapons were everywhere, and a civil war became instead a war of militias and gangs, while ISIS was establishing itself on Libya's Mediterranean shoreline. Vast quantities of weapons from Libya flooded down into sub-Saharan Africa. Petroleum output continued to be disrupted, as militias fought for control of oil fields and terminals and the revenues that accrue from oil. On September 11, 2012—the eleventh anniversary of the attack on the twin towers—the U.S. ambassador to Libya, Chris Stevens, and another American official were killed in an attack on the American diplomatic compound in Benghazi. Two more Americans were killed the next day in an attack on a nearby CIA annex.

Libya may still have existed on maps. But it was no longer a functioning nation-state.[4]

THE ARAB SPRING ALSO TURNED THE RIVALRY BETWEEN SAUDI
Arabia and Iran into a direct confrontation—most immediately in Bah-

rain, the island kingdom just off Saudi Arabia's eastern province. The security of oil was very much at stake. About 60 to 70 percent of Bahrain's Muslim population is Shia. In the course of the Arab Spring, Shia demonstrators, motivated by what they experienced as endemic political and economic discrimination, took to the streets to protest against the ruling Sunni royal family and the Sunni establishment.

Bahrain may be small, but it is a critical link in the standoff between Iran and Saudi Arabia. It is connected to Saudi Arabia by an eighteen-mile causeway. It is also a potential bridge for Iran to extend its writ across to the western shore of the Gulf.

Bahrain is a minor oil producer. Prior to the rise of Abu Dhabi, Dubai, and Qatar, it was the commercial and financial hub for the Gulf. In 1995, after the Gulf War, the U.S. Fifth Fleet, mothballed after World War II, was reactivated and based in Bahrain as a foundation for security in the Gulf. Bahrain's Shia population looked to religious centers in Iran and Iraq. For its part, Iran asserts that Bahrain, though across the Gulf, is a "lost province"—a claim made both by the shah and subsequently by the Islamic Republic. In 1981, two years after Khomeini took power, Iran backed a failed coup. Another coup in Bahrain was averted in 1996. Despite the fact that Bahrain had been ruled by the Al Khalifa family since 1783, Iran renewed its claims to the kingdom starting in 2007. Quds Force leader Qassem Soleimani declared that Bahrain was an "Iranian province separated from Iran as a result of colonialism."[5]

Many of the young Shia on the streets were protesting unemployment, poor housing, and discrimination, but a cadre was intent on overthrowing the government. The United States rushed to broker a new social contract that would give the Shia more say in power. From the viewpoint of Washington, Bahrain's government was harsh and obtuse in its response to the demonstrations—and in its treatment of its Shiite citizens. But the proffered deal was not acceptable to the other members of the Gulf Cooperation Council, in particular to the

United Arab Emirates, and especially to Saudi Arabia. Instead, they sent troops and police across the causeway into Bahrain to quell the protests and restore order.

For Saudi Arabia, Bahrain was of critical importance. For the island nation is forty miles from Ghawar, Saudi Arabia's largest oil field, fifty miles from Ras Tanura, Saudi Arabia's giant oil export terminal, the largest in the world, and less than thirty miles from the headquarters of Saudi Aramco in Dhahran. Moreover, Shia are predominant in the eastern part of Saudi Arabia, adjacent to Bahrain. The possibility of Shia ascendancy in Bahrain, and its subordination to Iranian policy, would be a direct threat to the Saudi kingdom.

Since the Arab Spring of 2011, parts of the Shiite opposition in Bahrain have been integrated into Iran's "axis of resistance," and Iran has stepped up its support for insurgency in the country. Scores of young Shia from Bahrain have been trained in bomb making and military tactics in Revolutionary Guard camps in Iran and by Hezbollah in Lebanon and Shiite militia in Iraq. Attacks on police and security officials have continued, as has the increasing use of IEDs and armor-piercing bombs. In 2017, the oil pipeline from Saudi Arabia to the refinery in Bahrain, a major revenue source for the country, was blown up. The pressure will continue. Qassem Soleimani prophesied that a "bloody uprising" in Bahrain will lead to "the toppling of the regime."[6]

SYRIA'S DICTATOR, BASHAR AL-ASSAD, WAS CONFIDENT, IN fact so confident in late January 2011, as people were pouring into Tahrir Square in Egypt to demand that Mubarak leave office, that he received two American reporters in Damascus. He assured them that nothing like that would happen in Syria. There was a "new era" in the Middle East, he said. "If you didn't see the need of reform before what happened in Egypt and Tunisia, it's too late to do any reform." Syria, he added, was spared that problem. "Syria is stable," he said. "Why? Because you have to be very closely linked to the belief of the people. This is the core issue."[7]

His confidence would be tested sooner than he thought.

His father, Hafez al-Assad, had seized power in a coup in 1970. Under him, Syria came to be regarded as the most rejectionist of Israel's neighbors, and the most militant. For Hafez Assad, there were many threats, from Israel and the rival Ba'athist regime in Iraq on his borders to coup plotters at home. But the biggest internal danger was the Islamic challenge arising from the majority Sunni population.

Hafez Assad was an Alawite, a Shia offshoot that comprised about 20 percent of the population. The Alawites, despite their small numbers, dominated the government, military, and security services, and increasingly the economy. Assad unleashed overwhelming force to meet any challenges to the Alawite control. The regime ruled with relentless police state surveillance maintained by at least fifteen separate intelligence agencies. Beyond his own borders, Assad used assassinations and military force to assert his sway over Lebanon, which he regarded as part of Syria that had been shorn by the "imperialists" in their mapmaking during and after World War I.

Hafez Assad also did something very important. He made Syria an ally of the Soviet Union. He forged a second alliance that, as with the Soviet Union, would prove critical, decades later, to the survival of the regime—this with Ayatollah Khomeini and revolutionary Iran. His move was bolstered by a politically convenient *fatwa* from the head of Lebanon's Supreme Shia Council, which decreed that Alawites, despite differences in doctrine and practice, were part of Shia Islam. Syria would become the staging point through which Iranian "volunteers" passed on their way to join Hezbollah in Lebanon.[8] After the fall of Saddam, a Shia belt ran from Iran through Iraq and Syria to the Hezbollah Shia in Lebanon. The alliances with Iran and the Soviet Union were essential support for the regime. In terms of economics, oil also did much to keep the Assads afloat.

———

IN 1947, AFTER A SERIES OF FAILED EXPLORATION WELLS, THE Iraq Petroleum Company announced that it was pulling out of Syria. The country, it said, had no commercial oil prospects. Syria's president at the time, desperate for revenues that oil might provide, instructed the Syrian ambassador at the UN to try to find someone somewhere in the United States of Syrian descent who knew something about petroleum. And that is how members of the Syrian delegation to the United Nations ended up, in 1948, in the tiny town of Benton in southern Illinois. They were there to seek out a certain James Menhall, who had emigrated decades earlier from Syria to the United States. Menhall had come up with several patents for portable drilling rigs, and had developed some producing oil wells in the small oil fields of Illinois and Kentucky. He was their man.

But because of political instability in Syria, he was not able to start drilling for another eight years, not until the spring of 1956. Within half a year, he had discovered commercial oil. Unfortunately, shortly after Syria merged with Egypt to form the United Arab Republic under Nasser, Menhall's concession was canceled, with no compensation.

Nevertheless, Menhall had laid the basis for what would become the Syrian oil industry. Production never made it into the big leagues. But it was highly material for the Assad regime. As late as 2010, oil revenues accounted for about 25 percent of Syria's total budget.[9]

IN THE FIRST FEW YEARS AFTER SUCCEEDING HIS FATHER IN 2000, Bashar Assad loosened the iron grip that his father had imposed. The commitment to Arab socialism gave way to the idea of a more open social market economy. Bashar permitted access to the internet and even legalized cell phones. The economy was bolstered by the influx of the middle class and money fleeing neighboring Iraq. Syrians from the international diaspora came back and bought property in Damascus.

Bashar had been an eye doctor in London. But when his elder brother was killed in a car accident, Bashar was recalled to Damascus as heir apparent. Succeeding his father in 2000, he was confirmed in his position by winning 99.7 percent of the vote in a presidential election. Bashar seemed to personify a new modernity, as did his wife, Asma, the daughter of a Sunni Syrian cardiologist in London. She had been a junior investment banker bound for Harvard Business School when she diverted to marry Bashar. She appeared to embody the new "open-minded" era that her new husband talked about. She even became the subject of an article in an American fashion magazine.[10]

Yet the economic reforms only went so far, and the repressive security and surveillance system put in place by Assad's father remained.

Some days after Assad had declared so confidently that the upheaval in the rest of the Arab world would never intrude in Syria, a gaggle of schoolboys in a town called Daraa spray-painted on a schoolyard wall: "The people want the downfall of the regime." The local secret police arrested and beat the boys. Outraged citizens took to the streets in protest. A wave of demonstrations followed across the country.

For Assad, reform was over. The regime's survival was now at stake. He vowed to destroy the "terrorists." By May 2011, tanks were pouring into city streets to quash the protests.

The demonstrations turned into a full-scale uprising. It was now a world of the internet, of cellphones and YouTube, of instant messaging and Arab satellite television. And it was a world in which well-funded Islamic jihadists were being mobilized into violent, brutal, and highly motivated fighting forces.

As the army stepped up its campaign, dissident officers defected to form the Free Syrian Army (FSA), which became a loose coalition of rebel groups. With this, the uprising was becoming a true civil war. The ranks of the defectors swelled, and came to include Assad's former prime minister and even the general responsible for prosecuting defectors. In August 2014, President Obama weighed in: "For the sake of the Syrian people, the time has come for President Assad to step aside."[11]

ASSAD WAS NOT GOING ANYWHERE. AT THE SAME TIME, RADical Islamists became a growing force in the country. One group declared itself the Al Qaeda branch in Syria. Some of the rebel fighters shifted to the Islamist militias, which were better funded and had better weapons.

The growing chaos in Syria brought in other regional players. Iran's Revolutionary Guard and militias and the Shiite Hezbollah from Lebanon were soon fighting in Syria beside Assad's own soldiers. On the other side, Sunni states, including Saudi Arabia and Turkey, were keen to see him removed and provided both funding and weapons to rebel groups. Private individuals from the Gulf, intent on the fall of Assad's Shia-linked Alawite regime, donated generously to Islamist groups.[12]

The highly sectarian nature of the Assad regime pitted Syrian Alawites, supporting the regime, against Syrian Sunnis. Meanwhile, Syrian Christians, fearful of what would happen to them under Islamist rule and sharia law if the Sunni jihadists prevailed, rallied to the Assad regime. The Syrian Kurds organized themselves into their own fighting force and carved an autonomous region out of their cantons along the Turkish border. They would have been a natural ally for the anti-Assad forces. But to Turkey they were anathema, owing to their putative relationship with the Kurdistan Workers Party, which was resisting Ankara within Turkey. As a result, Turkey would turn its weapons and planes on the Syrian Kurds as well as on radical Islamists.

The number of refugees and displaced persons from this civil war was enormous, and many were flooding into Turkey, Lebanon, and Jordan. President Obama warned that the use of chemical weapons by the

Assad regime would pose a "red line" that would trigger an American military response. In August 2013, word filtered out that Assad's forces had used poison gas against a rebel suburb of Damascus, killing as many as fourteen hundred people. This was a key moment. The United States was just a few hours away from launching airstrikes. "Our finger was on the trigger," General Martin Dempsey, chairman of the Joint Chiefs of Staff, later said.[13]

Obama decided otherwise. He concluded that airpower would be insufficient and ineffective, and he wanted congressional authorization but could not get it. He had come into office to end America's two wars—in Iraq and Afghanistan—and he was loath to slip into a third, with no clear path to success. Air power in Libya had helped remove Gadhafi, but it had left chaos behind. Obama was also demonstrating that, as he later said, he had broken with the military response "playbook" of the "foreign policy establishment." Moreover, he feared that an air strike would not eliminate all the chemical weapons, and Assad could then claim that "he had successfully defied the United States."[14]

Still, an American president had said using chemical weapons was a red line, but had not acted on that. Coming on top of Mubarak, it made leaders in other countries question the credibility of the United States and its reliability as an ally.

Then, suddenly, the chemical weapons challenge was dealt with, but in an unanticipated way. Russian president Vladimir Putin, who was already supporting Assad, stepped in to broker a deal to turn over to the international community what were thought to be all of Assad's chemical weapons. A "stunning twist" is the way one American official described the move. It helped Russia to move out of the enforced isolation that had followed Crimea and the war in eastern Ukraine.[15]

Syria was no longer a country, but rather a series of battlefields and strongholds, but one of constantly shifting lines. By 2015, the Assad regime was on the defensive. But then, in September of that year, Russia announced that it would undertake military operations in Syria to support Assad. This was another twist. It made Russia a major player in Syria, fully internationalizing the conflict and establishing Russia as more than the "regional" power that Obama had dismissed it as. There now could be no resolution to the Syrian conflict without Russia.

By this time, several thousand Quds fighters were active in Syria, and Soleimani was flying in frequently to personally direct the fight. Iran also mustered tens of thousands of fighters from Lebanon's Hezbollah and Iraqi militia, battle-hardened from battling the Americans, joined by Afghani and Pakistani Shiites.[16]

Assad, bolstered by the Iranians and the Russians, was now on the offensive. It became apparent that, notwithstanding the deal brokered by Putin, not all of the chemical weapons had been removed. For Assad's forces used them again against rebel strongholds. President Donald Trump responded with air raids. The first, in 2017, hit an airfield that was quickly repaired. In 2018, new raids, coordinated with the British and French, were aimed at chemical weapons facilities, but choreographed so as not to hit adjacent Russian military personnel.

U.S. troops in the northeast of the country worked closely with the Syrian Kurds and the Syrian Democratic Forces to defeat ISIS on the battlefield. But then, in October 2019, after a phone call with Turkish president Erdoğan, President Trump abruptly announced the pullout of U.S. forces. This facilitated Turkey's military to move into northern Syria, to push back the Syrian Kurds—America's allies—and establish a "security zone" inside Syria. Putin dispatched Russian troops to help "secure" the region. On Trump's orders, the U.S. military hastily withdrew, though with one exception—they regrouped to protect Syrian oil fields, although whether against resurgent ISIS attacks or to keep them out of the hands of the Assad regime was not clear.

OUT OF TWENTY-TWO MILLION SYRIANS, HALF A MILLION ARE estimated to have died in the civil war, over six million have been displaced internally, and five million fled as refugees—altogether more than half of the population. The effects have reverberated beyond the Middle East. The flow of refugees from Syria (and other countries) into Europe—a million into Germany alone—has transformed European politics, stimulated a new nationalistic, right-wing populism, helped to fuel anti-Europe sentiment in Britain and its exit from the European Union, and put continuing strain on the EU itself.[17]

With its success on the ground, Iran seeks to secure its land bridge from Iraq across southern Syria and into Lebanon. This route facilitates the movement of fighters, equipment, and weapons, including rockets and missiles, as well as the building of underground weapons factories along the way. To emphasize its objectives, Iran released a video showing the Quds Force commander Soleimani walking across the Iraq-Syria border. His message was clear—"uniting the Shia communities of both countries"—and, with his boot steps, marking out the land bridge.

IN YEMEN, TOO, IN THE CAPITAL CITY OF SANAA, THE ARAB Spring ignited massive street protests in 2011. But in this country located at the southwest corner of the Arabian Peninsula, it took almost a year before another "president," Ali Abdullah Saleh, was finally toppled, forced out after thirty-three years in power. The result would be a civil, tribal, and religious war. It would also become the terrain for a proxy war: Saudi Arabia and the United Arab Emirates versus Iran.

Saleh had been the ruler of North Yemen in 1990, when it merged with Marxist South Yemen to form a single country. But the united Yemen was less a nation than a fractious amalgam of tribes that Saleh managed through patronage, force, and his cynical skill at playing one group off against another—what he called "dancing on the heads of snakes."

Yemen was a latecomer to petroleum. But discoveries pushed its oil production to peak at almost 460,000 barrels per day, which provided a revenue stream desperately needed by the impoverished country, the poorest in the Middle East. An LNG project started up in 2009 with a series of twenty-year contracts, bringing in more revenues.

In the early 2000s, an Al Qaeda affiliate established a redoubt in the southern part of the country. The attack on the U.S. destroyer USS *Cole* in the Yemeni port of Aden in 2000 and then, much more so, 9/11 drew the United States into the country. It targeted jihadists, largely with drones and jets but also with special forces. In 2009, Al Qaeda affiliates in Saudi Arabia and Yemen merged to form Al Qaeda in the Arabian Peninsula (AQAP). Centered in Yemen, it came to be considered one of the most dangerous Al Qaeda affiliates, with expert bomb makers looking for ways to bring down American and European passenger and freight jets. In response to the growing jihadist threat, Yemen received large amounts of U.S. aid, as well as the support of neighboring Saudi Arabia.[18]

Saleh had his own battles to fight, most notably with a rebellious group called Ansar Allah ("Supporters of God"). Its base were the Zaydi tribes in the rugged mountains of northwest Yemen (with some spillover into the very south of Saudi Arabia). The Zaydis represent about 40 percent of the Muslim population of Yemen. They are considered close to Shia, though with doctrines different from those of Iran, and with some affinities to Sunni.

Ansar Allah emerged in the 1990s. Its founder was Hussein al-Houthi, who denounced Saleh as the living personification of the "unjust ruler" that figures in Zaydi theology. Houthi traveled to both Iran and Sudan for religious education. He had returned to Yemen to build a Zaydi militia and lead the rebellion against Saleh in the name of piety and fighting corruption—and resisting the encroachment of fundamentalist proselytizing from neighboring Saudi Arabia. His 2002 sermon, "A Scream in the Face of the Arrogant," reflected the rhetoric of revolutionary Iran. Ansar Allah found the model for its new militia in Hezbollah. Houthi studied videos of Hassan Nasrallah, the leader of Hezbollah, and celebrated how "Nasrallah's powerful words" were "shaking Israel." Ansar Allah militia went into battle with a chant from Houthi's sermon, adapted from Iran's "Death to America": "God is Great! Death to the U.S! Death to Israel! Curse the Jews! And Victory for Islam!" The Houthis began to shuttle back and forth to Lebanon and Iran.[19]

Houthi was killed by Saleh's forces in 2004. Thereafter, his fervent followers took his name—the Houthis. They also received support from Iran and Hezbollah. With the Arab Spring and Saleh's forced departure in 2012, the Houthis went on the offensive. But it was not all that long before they gained a new ally, none other than their former enemy, former ruler Ali Abdullah Saleh, who seized on collaboration with the Houthis as his best route back to power. In September 2014, bolstered with Saleh's forces and assisted by advisers from Hezbollah, the Houthis captured Sanaa, Yemen's capital. They wasted no time in establishing direct air service between Sanaa and Tehran. Having secured Sanaa, the Houthis moved on toward Aden, the most important port on the way to the Arabian Sea.[20]

The Houthi advance on Aden set off alarms in Saudi Arabia, for a Houthi conquest of all of Yemen would consolidate an Iranian ally on Saudi Arabia's eleven-hundred-mile border with Yemen. That border had always been porous, ill-defined, and difficult to regulate. For Riyadh, the specter of an entrenched militarized Iranian ally on its southern border heightened, as they saw it, the existential danger—Iranian encirclement of Saudi Arabia and the Gulf Arab emirates.

The Iranians championed the Houthis. A deputy of Ayatollah Khamenei declared that the capture of Sanaa was "a victory" for Tehran, jubilantly adding that Tehran now controlled four Arab capitals: Baghdad, Beirut, Damascus—and Sanaa. Hassan Nasrallah sent a message of his own on behalf of Hezbollah: "The Resistance Axis triumphs in Syria, and it triumphs in Iraq, and it is steadfast in Yemen where it will also win, if God wills, a great decisive victory."[21]

A Houthi victory posed a direct threat to oil. Yemen commands the eastern side of the Bab al-Mandeb, the strait at the bottom of the Red Sea, which is only eighteen miles across at its narrowest point. It is the chokepoint connecting the Indian Ocean to the Suez Canal and on to the Mediterranean. Through it pass almost five million barrels of Middle East oil per day, as well as much world trade in general. To have that strait vulnerable to a hostile power would constitute a major risk for oil-exporting Saudi Arabia and the UAE, for Egypt, which depends on the canal for substantial revenues, and for global commerce. Demonstrating the dangers, a Houthi missile struck a Saudi tanker as it made its way through the strait.[22]

In March 2015, with the Houthis about to take Aden, Saudi Arabia and the UAE launched Operation Decisive Storm with ground forces and airpower and a maritime blockade. For Riyadh, this was the first major policy decision by King Salman, who had come to the throne in January, and more so by his son Mohammed bin Salman, who had already been appointed defense minister.

What was thought would be over in weeks, if not months, became a long war. The conflict has killed many civilians, displaced millions of people, put even more at risk of starvation, disrupted basic services like water and electricity as well as the medical system, precipitated cholera and diphtheria epidemics, and then COVID-19—altogether adding up to what the United Nations has described as a major humanitarian crisis.[23]

The Saudi-led air campaign has been criticized for indiscriminate targeting that was killing civilians, a critique extended to the United States, which has been supplying ordnance to the Saudis. For their part, the Houthis have been brutal and repressive in Yemen. They expanded the war by launching Burkan missiles and drones into Saudi

Arabia, most of which have been destroyed by Saudi air defenses. The sources of the Burkan missiles and drones are Iran and Hezbollah.

By 2017, Saleh was reaching out to the Saudis to "turn a new page" and perhaps to change sides yet again. He denounced the Houthis as Iranian puppets. He sought to flee. But, at a checkpoint, he was stopped by Houthi militia and shot dead.

Despite UN efforts at mediation and on-and-off efforts at a cease fire, Yemen has become what is described as a "chaos nation," no longer a functioning country, only a "nominal entity that exists largely as lines on a map and as a concept in newspaper reports and policymaker briefings," a series of mini-states, each with its own rulers, "at varying states of war with one another."[24]

The war itself had turned into a stalemate. Opposition was mounting in the U.S. Senate against the Saudi campaign and American military support for Saudi Arabia. The Iranians were getting the benefits of the conflict on the cheap, or, as one scholar as put it, "a phenomenally high return on their investment." While the Saudis were spending many billions on the war, the cost to the Iranians was measured in tens of millions. In 2019, the UAE announced that it was leaving the coalition.[25]

Yemen may be a prototype for a chaos nation. It has also, because of its strategic location, turned into a critical battleground in the great rivalry between Saudi Arabia and Iran in the modern Middle East. "From now on, we can't speak about [the] Syrian army, Hezbollah, Yemeni army, Iraqi army, and Iranian army," Hezbollah television announced. "We must speak about one resistance axis operating in all theaters." But it fell to Qassem Soleimani to sum up Iran's strategy and why the entire region has become an arc of confrontation. Speaking at a rally celebrating an anniversary of the Iranian Revolution, he declared, "We are witnessing the export of the Iranian Revolution throughout the region—from Bahrain and Iraq to Syria, Yemen, and North Africa."[26]

Chapter 32

THE RISE OF THE "EASTERN MED"

Iran is seeking to establish a permanent military position on Israel's northern border, which, with Hezbollah in Lebanon, would provide it with a much greater capacity to confront Israel. But that is one of the most inflammatory steps Iran can take. For that increases the likelihood—or even inevitability—of an escalating Israeli-Iranian conflict. In response to rockets aimed at the northern part of the country, Israeli military forces have hit Quds infrastructure in Syria and disrupted and destroyed Iranian weapons shipments into Syria and destined for Hezbollah.

Iran has also built up the axis of resistance on Israel's southern border by supporting Hamas, the militant Muslim Brotherhood group that rules Gaza. Hamas is Sunni, but it is joined with Shia Iran in opposition to Israel on the premise that the enemy of my enemy is my ally.

A major Iranian objective is, in the words of one of the senior commanders of the IRGC, to "extend our security border to the eastern Mediterranean." That would be an immediate challenge to the major new hydrocarbon province that is opening up in the waters of the eastern Mediterranean. These resources provide unexpected opportunities for Middle Eastern countries that have heretofore lacked resources. But they also create a new sea of contention.

IN 1999, A GAS FIELD WAS DISCOVERED OFF THE SOUTHERN coast of Israel. It was small, but still, it was the first dent in Israel's total dependence on imports for its hydrocarbons, which was a source of great vulnerability and anxiety. In 2008, Israel began to import supplies of gas from Egypt through a pipeline running across the Sinai Peninsula.

In general, however, the eastern Mediterranean was considered a dead sea in terms of any significant reserves of oil and gas. But then, in 2009, an American independent, Noble Energy, and its Israeli partners found Tamar, a world-class gas field fifty miles off Israel's northern coast. Using supercomputers to analyze the data with help from faculty at Tel Aviv and Haifa universities, geologists identified a still much larger and more promising structure that no one else had imagined was there. This led to the discovery of another gas field eighty miles off Israel's coast. This was a giant, one of the largest finds of the decade worldwide. Appropriately enough, it was dubbed Leviathan.

These discoveries were just in time. With the Arab Spring and the fall of Hosni Mubarak, the pipeline carrying gas from Egypt across the Sinai to Israel was repeatedly sabotaged, and then in 2012, the Muslim Brotherhood government in Egypt canceled the deal.[1]

As timely as these discoveries were for Israel's energy security, their development was jeopardized when the Israeli government, having earlier pledged not to change the fiscal terms, proceeded to do exactly that. The resulting dispute blocked the entire development and threw its future into doubt. It became a contentious issue in Israeli politics, sparking demonstrations by groups claiming that the country would not be getting enough of future revenues. But this debate was in a

domestic bubble, isolated from the competitive realities and investment challenges of the global gas business. Further complicating matters, a government commission recommended that the financial return
to the companies be adjusted downward and treated like a virtually
risk-free bond, rather than a return appropriate for a project with extensive geological, geopolitical, and commercial risks. Then antitrust
authorities, adopting theoretical principles that did not fit in a small
nascent gas market like Israel, further immobilized development. After
five years with no progress, it looked as though the huge natural gas opportunity was not just stalemated, but dead in the water.

In 2015, Eli Groner took over as director general of the prime minister's office. His previous time as economics minister in the Israeli embassy in Washington had opened his eyes to the geopolitical significance
of energy and what Leviathan could mean. He then came across a new
book about world energy that described the growing global gas market.
It made starkly clear to Groner—indeed convinced him—that, if the
stalemate continued, Israel would be shut out from ever becoming an
exporter by the many competitive natural gas projects around the
world.

Prior to starting the job, Groner raised Leviathan with Benjamin Netanyahu. The prime minister's first question was, "Do you
know how much it costs to protect those fields?" To which Groner
replied, "The reason it costs so much is because it's so incredibly
valuable."

On his first day in the prime minister's office, Groner announced
that his number one priority was to "get the gas out of the water." It
would not only provide significant revenues to the government, but it
could possibly be one of the most important initiatives that could be
done for the future of the Israeli economy. For a country that was almost totally dependent on energy imports it would be transformative,
making Israel more secure and changing its relationship to its neighbors and perhaps more widely. If the project did not go ahead, the big

loser would be Israel itself—unable to ever monetize the gas in a meaningful way and less secure.

Netanyahu put his political capital to work. The stalemate was over. Leviathan went ahead.

Sixty percent of Israel's power is now generated with its own natural gas, rather than imported oil or coal. Israel is already exporting gas to the Palestinian Authority and to Jordan, through private companies. It could potentially now become an international exporter, either in the form of LNG or an underwater pipeline to Europe—an astonishing turnaround for a country that worried so much about the vulnerability resulting from its import dependence.

Noble discovered another major gas field, Aphrodite, in the waters of the Republic of Cyprus, eighteen miles northwest of Leviathan. Then in 2012, Eni, the Italian major, using one of the world's largest supercomputers, discovered a huge field called Zohr in Egyptian waters, a hundred miles to the west of Leviathan, which it brought on line in record time, assuring Egypt of self-sufficiency in natural gas. The entire region has now become known as the Eastern Mediterranean Basin—or the Eastern Med.

Yet while the discoveries bring new opportunities, they also carry new vulnerabilities. As Eitan Aizenberg, the octogenarian Israeli geologist who first "saw" Leviathan, put it, "We should always be aware of the risks that await us."[2]

The most immediate risks are evident in the north of that cramped region. The success in Israeli waters has been closely observed in Beirut, prompting licensing rounds for Lebanon's offshore. But these rounds have been hampered, as would be expected, by disputes between the two countries as to where the appropriate demarcation line falls between their respective offshore waters.

Iran's position in Lebanon, directly and through Hezbollah, presents it with the opportunity to participate in oil and gas development in the Mediterranean, in Lebanon's offshore. It also provides a new base to challenge Israel— in terms of the security of its offshore fields, which are now recognized within Israel as a great national asset. "I promise you, within hours," declared Hassan Nasrallah, the leader of Hezbollah, the platforms "will cease operating" in the event of conflict. To drive home its point, Hezbollah issued a video with a bull's-eye superimposed over an Israeli platform. The day after Nasrallah's threat, Egypt— now in need of natural gas—announced a $15 billion deal to buy gas from the Israeli offshore fields. The timing was coincidental, and yet it sent a message that it was not only Israel but now also Egypt that has a major stake in the security of those platforms.[3]

In response to the Iranian-Hezbollah threat, Israel is investing heavily in security for the platforms, including adaptation of its Iron Dome missile defense system to the interception of Russian and Chinese shore-to-sea missiles that have been provided to Hezbollah and Hamas.

On the last day of 2019, the giant Leviathan field started up production. Israel quickly started exporting some of its new gas to Egypt by reversing the pipeline in the Sinai that had formerly brought Egyptian gas to Israel. As Egypt is now self-sufficient in gas, the Israeli gas will help enable Egypt's mothballed liquefaction plants to be reactivated and put Egypt back in the business of exporting LNG. "Europe is our customer," said Egyptian energy minister Tarek el-Molla. "We have the ready-made solution. We have the infrastructure."

What Israel energy minister Yuval Steinitz said—"We have already discovered much more than we can consume"—applies not just to Israel. Greece, Israel, and Cyprus have signed an agreement to build a 1,180-mile undersea pipeline that would transport Eastern Med gas to Greece and then on to Italy. Turkey wasted no time in denouncing the project, which it asserted would cross its waters. To make the point, Turkey sent warships to accompany drill ships into waters that Cyprus claims as its own exclusive economic zone.

The discoveries in the Eastern Med "came for us as a complete surprise," said Steinitz. Surprising no more. Today those waters are hardly thought of as a dead sea in terms of resources. Rather, the Eastern Med is a new and dynamic element in the global energy industry and for geopolitics, and is changing the map for both. It could become a supplier of natural gas not only to Europe but also, through LNG, to world markets. But the politics of the region are such that it will also continue to be a sea of contention.[4]

Chapter 33

"THE ANSWER"

ISIS emerged out of a century of history that began with the Islamic rejection of the nation-state, fostered by the reaction to the collapse of the Ottoman Empire, European dominance, a secular world and modern culture, the imposition of borders after World War I, and detested leaders.

In the 1920s, the Egyptian city of Ismailia, on the banks of the Suez Canal, provided visible demonstration of British dominance. For it was both a colonial enclave and a company town, home to a British air force base and the operational headquarters for the Suez Canal Company, which owned the canal.

Hasan al-Banna, a school teacher and deeply pious Muslim, took to preaching in local mosques and coffee shops. In 1928, as he recounted, a half dozen workers sought him out to complain of the "humiliation and restriction" imposed by the canal company, of their being "mere hirelings belonging to the foreigners." They asked him to provide the guidance of Islam.

"We are brothers in the service of Islam," Banna replied. "Hence, we are the 'Muslim Brothers.'" This new society, he said, would rescue Muslims who had "been assailed" by "imperialist aggression" and "exploitation" and by forces that "destroy their religious beliefs." "The answer is Islam," he said, "an all-embracing concept which regulates every aspect of life." The ultimate ambition was the reestablishment of the caliphate that would lead to "dominion over the world." The concept of a "caliphate" also became the antidote to the idea of the nation-state.[1]

The Muslim Brotherhood grew rapidly, and as such became a target of the Egyptian government. The Brotherhood established a "secret apparatus" in 1943 to carry out clandestine operations, including assassinations. In 1949, shortly after the assassination of Egypt's prime minister, al-Banna himself was assassinated, probably in retaliation. By then, the Brotherhood had already found new purposes. One was fervent opposition to the new state of Israel in 1948. The other was in resisting what was seen as the overwhelming postwar secular threat—the United States.

In the year that al-Banna was killed, Sayyid Qutb, an Egyptian civil servant and essayist, was on a fellowship in the United States. On his arrival, he had been overwhelmed by fear as to whether he could resist the "sinful temptation" that surely awaited him. What alarmed him the most was what he saw as the rampant sexuality of life in the United States and the provocativeness of its women. He dismissed Americans as "a reckless, deluded herd that only knows lust and money."[2]

After returning to Egypt in 1950, Qutb became one of the leaders of the Brotherhood and, owing to his writings, one of the most visible. He infused a new militancy to the organization and would provide an agenda for violence in pursuit of its objectives.

Early on, the Muslim Brotherhood was an ally of sorts of the 1952 army coup that brought the "free officers," led by Colonel Gamal Abdel Nasser, to power. Indeed, Nasser offered Qutb a number of positions. But as Nasser embraced a more secular nationalism and Arab socialism—and consolidated his own power—the two men bitterly fell out.

In 1954, a member of the Brotherhood failed in an attempt to assassinate Nasser. The organization was banned. Qutb, charged with being a member of the Brotherhood's "secret apparatus," was jailed. There he wrote a series of commentaries that were smuggled out and eventually published under the title *Milestones*.

Qutb argued that the modern world had fallen back into the period of pre-Islamic ignorance and barbarism that had existed prior to the Prophet. Secular Muslims, or even Muslims who did not abide by strict sanctions, were subject to *takfir*, excommunication, and thus were fair game to be killed. What was required was a "vanguard which sets out with this determination" to restore Islam through violent holy war and "eventually leads Islam to its destiny of world dominion." He added, "I have written *Milestones* for this vanguard."

Qutb was freed in 1964, then rearrested for his role in another plot, and executed in 1966. But his words, as Lawrence Wright has written, "would echo in the ears of generations of young Muslims who were looking for a role to play in history."[3]

A DECADE AND A HALF LATER, ON NOVEMBER 20, 1979, SIX-teen days after the American hostages were seized in Tehran, shots rang out during predawn prayers in Mecca's Grand Mosque, the holiest place in Islam. A wild-haired, bearded zealot seized the microphone from the praying imam and started shouting orders. This was Juhayman al-Uteybi, an itinerant preacher and fundamentalist.

He had built a network of former members of the national guard and Islamic students from the University of Medina. Egyptian members of the Muslim Brotherhood were influential on its faculty, including the younger brother of Sayyid Qutb. Juhayman's movement had expanded under the benevolent eye of some conservative senior clerics. His followers were galvanized by the physical and social upheaval that came with the flood of oil money after the 1973 oil crisis and the country's integration into the world economy. Between 1974 and 1978, the nation's GDP almost doubled. Construction was booming, transforming cities. Great wealth was made, education was being rapidly expanded, Saudis were going to universities overseas, technocrats were drafting five-year plans—and more and more westerners were coming to work in the kingdom.

Juhayman's followers embraced the ideas he had laid out in a series of epistles, damning the Saudi government for heresy and corruption, for undermining Islam, for allowing women to be educated and even appearing on television, for allowing "Christian embassies" in the Muslim land, and, in a great transgression, for selling oil to the Americans.

As many as five hundred of his militants had secreted themselves with their weaponry in the sprawling Grand Mosque. It took more than two weeks and many lives before the militants were dislodged and either killed or captured. The seizure of the Grand Mosque has been described as "the first large-scale operation by an international jihadi movement in modern times."[4]

Juhayman's assault may have been a failure. Yet it was also a success, for it became a rallying point for young Islamists, and his epistles gained underground currency and influence among them.

By this time, an extensive jihadist network had been established in Egypt. Anwar Sadat, who had courted Islamists on taking power in order to break the Nasserite grip, had turned against them. His wife had gone so far as to publicly endorse the right of women to divorce their husbands. Moreover, Sadat had committed his transcendental sin—making peace with Israel—in return for which Israel returned the Sinai, captured in the 1967 war. On October 6, 1981, three soldiers who were members of al-Jihad jumped out of a jeep at a military parade, tossed hand grenades, and then started shooting, killing Sadat. The lead in this attack was a twenty-three-year-old lieutenant who had been introduced to Juhayman's writings by his brother, who had been in Mecca during the assault on the Grand Mosque. Those writings, the lieutenant would say in his interrogation, had put him on the "path to martyrdom."[5] They laid out the path for jihadism.

The Islamist movements in the Middle East cover a broad range. The most prominent and widely influential is Egypt's Muslim Brotherhood, with what have been described as "affiliates, descendants, and offshoots" in dozens of countries. A hierarchical organization, its members ascend through levels until, after several years, they become full-fledged Brothers. As one scholar has written, "Islamist groups' political concerns must compete with educational and religious activities. Millions depend on their vast social infrastructure," including health care that makes up for inadequate government services. But religion and a society based on Islam are "the foundation on which everything else is built," with a commitment to overturning existing nation-states in the region and the ultimate objective of a "a global Islamic state." The aim is not, however, to seek a return to the seventh century. Nor is it to embrace the jihadism of Qutb. But others certainly would.[6]

IN DECEMBER 1979, A WEEK AFTER THE CAPTURE OF JUHAY-man in Mecca, the Soviet military crossed pontoon bridges over the Amu Darya River and began its invasion of Afghanistan. What the Soviets thought would be a quick campaign turned into a long, wearing war; they discovered that they faced a fierce and implacable foe in the mujahedeen, the Muslim fighters whom the United States and Saudi Arabia would come to support with weapons and money.

Also among those providing money was Osama bin Laden, seventeenth son of an illiterate Yemeni migrant who became Saudi Arabia's greatest contractor, building much of the country's modern infrastructure. Osama bin Laden, however, found his vocation not in construction, but in the Islamic resistance to the Soviet invasion of Afghanistan and then jihadism and terror. And he saw the United States as the great enemy.

His chief collaborator became Ayman al-Zawahiri, a medical doctor who had founded al-Jihad in Egypt. Zawahiri saw the Afghan jihad as the way "to prepare the Moslem Mujahideen to wage their awaited battle against the superpower that now has sole dominance over the globe, namely the United States." Zawahiri also honed the arguments of his hero Sayyid Qutb about *takfir*—that Muslims who did not follow his strict strictures or who cooperated with democratic institutions were by definition apostates and thus deserved to be killed. He promoted the use of suicide bombers, though suicide is forbidden by the Koran.[7]

In 1995, bin Laden declared that Saudi Arabia's king was an infidel—an irrevocable break with his homeland. The next year, invoking the continued presence of U.S. troops in Saudi Arabia, he issued a "Declaration of War Against the Americans Occupying the Land of the Two Holy Places." In the world outside Saudi Arabia, this jeremiad issued from a cave in Afghanistan went unnoticed.

In early 1998, several of the jihadist groups merged into a single organization—Al Qaeda—which meant "the Base." The leader was Osama bin Laden; his deputy, Zawahiri. The objective was to wage war "against all U.S. interests throughout the world."

Their war began on August 7, 1998, when, within eleven minutes of each other, coordinated suicide bombers detonated massive explosions at U.S. embassies in Kenya and Tanzania. In Kenya, 210 were killed and 4,000 wounded. Two years later, the guided missile destroyer USS *Cole* was bombed in a Yemeni port. And then came the catastrophe of September 11, 2001, in which three hijacked airliners destroyed the twin towers of the World Trade Center and part of the Pentagon. The only reason that the U.S. Capitol was spared was because the fourth jet took off an hour late from Newark Airport, and the passengers aboard, learning what had happened, were able to force the plane to crash short of Washington, D.C., in a field in Pennsylvania.

The United States responded with the "war on terror." The first front was in Afghanistan, against Al Qaeda and the Taliban government that had harbored it. Al Qaeda's members were relentlessly pursued. But it took a decade, until 2011, before bin Laden was finally killed in a villa thirty-five miles from Pakistan's capital, Islamabad, where he had hidden for several years. His number two, Ayman al-Zawahiri, ascended to the top position.

Oil loomed large in Al Qaeda's playbook as a strategy to attack Arab governments, the United States, and the world economy. Bin Laden called for attacks on petroleum to drive up the price of oil and "make the U.S. bleed profusely to the point of bankruptcy." One jihadist publication presented "The Laws of Targeting Petroleum-Related Interests and a Review of the Laws Pertaining to the Economic Jihad."[8]

In 2013, jihadists invaded the Amenas natural gas plant in southern Algeria, where seven hundred employees were held hostage, and forty killed. An Al Qaeda publication promoted attacks on the oil industry in an article titled "On Targeting the Achilles Heel of Western Economies" and proclaimed that it was "for the sake of a divine message" to attack energy facilities. One study tabulated ninety-seven attacks on the energy sector in seven Middle Eastern and North African countries between 2001 and 2016.[9]

AL QAEDA SPAWNED AN AFFILIATE, AL QAEDA IN IRAQ. ITS practices were so brutal that even Ayman al-Zawahiri, proponent of the *takfir* doctrine that justified killing "unpious" Muslims, criticized it for bloody excesses in killing Muslims. Al Qaeda in Iraq took on a new name—the Islamic State in Iraq—with the mission to expunge the map of the preceding hundred years and replace it with a caliphate that would recognize no borders and heed no maps.

The chaos of the Syrian civil war provided the Islamic State with a new theater of operations. It rebranded itself as the Islamic State in Iraq and Syria—ISIS. Some call it ISIL—the Islamic State of Iraq and the Levant. In the Muslim world and in Europe, people instead call it by the derogatory term "Daesh." At the start, ISIS was only one of numerous Islamic groups that fought each other as they also battled the Syrian government and its Iranian and Hezbollah allies. But ISIS was distinguished by its organization, capabilities, fervor, and violence. It proclaimed that the other Islamic rebels, including Al Qaeda, were apostates who were to be killed "wherever you find them."[10]

ISIS captured Raqqa in north-central Syria, a transit point between Iraq and Syria, and made the city its "capital." It established a host of departments to run the city, demonstrating that in addition to fighting and killing, it could also administer services, down to affixing stamps to confirm that people had paid their taxes.

It also showed what life would be like under its caliphate. The *hisbah*, the religious police, prowled the streets, looking for the tiniest infraction on which to impose the most severe penalties. Music was immediately banned, and men were required to come to the mosque five times a day to pray. Punishment was dealt out to men whose beards were not long enough or women whose abayas revealed even a hint of their form. For transgressions large or small, or even on suspicion, men were taken to Naim Square, where, with the invocation of medieval rules, they would be gruesomely killed.

Throughout 2013, ISIS had carried out targeted assassinations in Iraq, eliminating key leaders who could have led opposition to it. In January 2014, still little known, ISIS swept across the Syrian border into Iraq and into the cities of Ramadi and Fallujah. These victories did not set off many alarms. But then they began to race across Iraq.

In April 2014, an ISIS leader exhorted its fighters to "not stop until the cross is removed, the pig is killed. March forward and redraw the map." That same month, as the jihadist blitzkrieg swept across Anbar Province, one tribal sheikh moaned, "Hell has come to these villages and towns." On June 6, ISIS launched a violent offensive on Mosul, Iraq's third-largest city. The much larger Iraqi army contingent, already terrorized by videos of ISIS militants beheading Iraqi soldiers, collapsed. Within five days, ISIS completely controlled Mosul. The abandoned weaponry, paraded on internet video, added mightily to its prestige and its arsenal. And so did the hundreds of millions of dollars it seized from the city's banks.[11]

Some remnants of Saddam's army joined ISIS's advance. So did a number of Sunni tribesmen who, as "Sons of Iraq," had joined the "Sunni Awakening" in alliance with the U.S. military to resist the insurgency. But now their commitment had been betrayed. Prime Minister Maliki's Shia government had stopped paying the Sons of Iraq. Instead, the Sunni tribes were being attacked by Shia militia and central government forces, their members jailed or kidnapped or simply killed. They saw Maliki as an Iranian puppet, while ISIS was seen, at least initially, in the words of a Sunni tribal leader, as "leading the battle against the Persians."[12]

As ISIS approached the gates of Baghdad, panic and confusion gripped the city. Coordinated car bombings added to the sense of impending doom. Inside the U.S. embassy in the Green Zone in Baghdad, American diplomats burned documents and prepared to evacuate. It seemed only a matter of time before the jihadists would capture the city and sack it. ISIS announced that after Baghdad, its next targets were the Shiite holy cities of Najaf and Karbala. That meant that Iraq's major oil fields in the south were now at risk. International oil companies hastened to make plans to evacuate their staffs. Petroleum prices spiked. It was only the swelling number of Shiite militias, combined and coordinated with Soleimani's Quds Force, that stopped ISIS from taking Baghdad. By July, oil prices were beginning to subside.[13]

But the consequences of the offensive were startling. With its fanatic but well-organized military drive, ISIS had redrawn, at least for a time, the map at the heart of the Middle East. It now controlled a contiguous territory that reached from north-central Syria all the way across Iraq to Mosul—almost the same distance as from London to Edinburgh—with as many as eight million people under its rule.

IN EARLY JULY 2014, ALMOST EXACTLY A MONTH AFTER LAUNCH-ing the Mosul assault, a bearded, black-garbed figure, wearing a black turban, slowly ascended the dais in Mosul's eight-hundred-year-old Grand Mosque. His name was Ibrahim Awad al-Badri, but he had taken the name Abu Bakr al-Baghdadi.

Baghdadi had earned his bachelor's degree in Koranic studies and then a doctorate in Islamic jurisprudence, focused on medieval Koranic recitation. After the invasion of "Muslim lands" by the United States and its allies in 2003, he became active in Islamic resistance groups. That landed him in 2004 in the overcrowded American-run prison Camp Bucca, which also became known as Jihadi University. For it provided the forum for forging networks among jihadists and former Ba'athist military and intelligence officers. Released after ten months, al-Baghdadi became the chief expert on sharia law for Al Qaeda in Iraq. Among his duties was to provide the medieval theological justifications for horrendous acts of terror. By 2010, he had become the leader of what was now the Islamic State. Among its other leaders were former officers of Saddam's army. They brought experience, organization, and strategy—and bitterness and an overwhelming thirst for revenge.

From the dais that day in Mosul in July 2014, al-Baghdadi proclaimed a new caliphate. Unlike a nation-state, the caliphate has no defined frontiers, but rather is the realm of the Muslim faithful, one to be protected and expanded by force of arms—jihad. As one of Baghdadi's top lieutenants put it, the objective was the creation of an Islamic state "that doesn't recognize borders" and that would vanquish the unbelievers and apostates and extend its rule beyond the present Islamic world. As al-Baghdadi told the assembled jihadists that day in Mosul, "You will conquer Rome and own the world."

That objective was far beyond its capabilities. But with its rapid conquests, it had carved out a de facto state. Moreover, it had established itself as the richest terrorist organization in the history of the world. At one point, it was earning as much as a billion dollars a year.[14]

Oil was a big part of it, for ISIS's territory gave it control of the bulk of Syrian oil production, as well as a small slice of Iraqi output. Though production had fallen, it ran a semiprofessional operation that generated revenue.

Some of ISIS's oil was sold within its realm or, despite their enmity, to the Assad regime. But a substantial part was moved out of the country and smuggled by truck, largely into Turkey. Beyond oil, the rest of ISIS's revenues came from taxation, tolls, theft, extortion, expropriation, and the looting of antiquities, as well as funds from sympathizers in the Gulf states. All this gave ISIS financial heft never seen before in a terrorist group. These revenues enabled ISIS to pay higher salaries than fighters could receive from competing groups or from jobs back home.

ISIS gained global reach from its propaganda and extraordinary mastery of social media, which became a powerful recruiting tool. The slick, highly produced videos were aimed at young, disaffected Muslim men not only in the Arab world and Central Asia, but also in the rest of Asia and, notably, in immigrant Islamic communities in Europe and North America. In "There Is No Life Without Jihad," an ISIS recruit from Wales declared, "We understand no borders," while another, a recruit from Scotland, exhorted, "Living in the West, in your heart you feel depressed . . . the cure for depression is jihad." This video was followed by "The End of Sykes-Picot," the video that celebrated the erasure of the border between Iraq and Syria. These videos were matched with grisly images of beheadings and executions, and of jihadists triumphing in battle. ISIS's recruitment, both from the Muslim world and Europe, soared. Its foreign fighter force was estimated at one point at more than thirty thousand, drawn from half the countries in the world.

ISIS continued to expand its terrain in Syria. In Iraq, it captured the area inhabited by the Yazidis, who had their own ancient religion. Those Yazidis who could flee did so; among those left behind, the men were killed, and the women and girls taken as sex slaves, with justification found in medieval texts.

It was when ISIS was on the move toward Erbil, the capital of semi-autonomous Kurdistan, that the United States finally pivoted back, launching the first of the air raids on the advancing ISIS forces.

The physical territory that ISIS held until 2017 gave it a prestige that Al Qaeda could not rival. Other jihadist groups pledged allegiance to ISIS, including Boko Haram in Nigeria. The ISIS branch in Libya appeared to be a direct export from Iraq. The ISIS faction in the Sinai claimed responsibility for the downing of a jetliner carrying Russian tourists. Jihadists would carry the war back to Europe, with terrorist attacks in Europe, including a night of horror in Paris in November 2015, and a morning of horror at Brussels Airport in March 2016. "Lone wolves," some radicalized online, some with controllers back in Syria, would carry out attacks in Europe, the United States, and Canada. Between 2014 and 2016, ISIS and its affiliates launched more than 150 attacks in Iraq, Syria, Egypt, and Libya.[15]

At the end of 2015, the United States began sending special operations troops to Iraq to support Iraqi and Kurdish fighters against ISIS. The Shiite militias, many controlled by Iran, were a significant part of the Iraqi forces. A few months later, Iraqi troops succeeded in regaining control of Ramadi and Fallujah. It took nine months of heavy fighting in dense neighborhoods before the Iraqi government could announce that Mosul had been liberated. Among the many ruins was the eight-hundred-year-old Grand Mosque, where al-Baghdadi had announced the caliphate three and a half years earlier.

In March 2019, ISIS lost its last piece of territory, in eastern Syria. The caliphate no longer existed. But ISIS was not finished. It morphed back into a guerrilla organization, once again using terror as its weapon in both Iraq and Afghanistan and around the world. In mid-September 2019, al-Baghdadi surfaced with an audio recording meant to fire up his movement. America, he said, "had been crushed and humiliated."[16]

By then, American forces, assisted by their Syrian Kurdish allies and aided by Iraqi intelligence, were closing in on al-Baghdadi. Late on an October night in 2019, eight helicopters filled with U.S Army Delta Force commandos took off from a base in Iraqi Kurdistan. They flew fast and low over Russian- and Turkish-held territory. After a little more than an hour, they descended on a compound in the ungoverned region of northwest Syria. There, sheltered by the leader of an Al Qaeda splinter group, was al-Baghdadi, along with other ISIS fighters. The commandos blew a hole in the side of the compound and overwhelmed the ISIS fighters. Baghdadi fled into a tunnel, taking members of his family with him. But there was no way out; the tunnel was a dead end. Pursued by a Delta Force dog, al-Baghdadi set off the suicide vest he was wearing. That was his end. A few hours later, planes overhead reduced the compound to rubble.[17]

But that was not the end of ISIS. There were still an estimated fifteen to twenty thousand ISIS fighters, plus another ten thousand held in makeshift prisons, plus the affiliates and followers around the world.

Yet in one significant way ISIS's impact had faded much earlier. In 2014, its lightning advance across Iraq had induced panic in the oil market and caused it to spike. But, the impact on oil prices had hardly lasted.

Chapter 34

OIL SHOCK

For three years, 2011 through 2013, the oil price had been surprisingly stable—a little over $100 a barrel. Though that was almost five times higher than it had been a decade earlier, the world had become accustomed to this new price. It was called the "new normal," and on this basis countries could make their budgets and companies would finance their projects. When it turned out not to be so normal, the oil world was rocked, countries would reel from the shock, and, out of the price crisis, new geopolitical alignments would emerge.

Yet what was so surprising was that the price had held so steady for those three years despite turmoil and instability in global supply. The reason was a coincidental trade-off: U.S. oil production was surging because of shale oil. These gains, however, were offset by disruptions and loss of supplies elsewhere—in Libya, where the chairman of its national oil company said the country had "all but disintegrated," and in Nigeria, where militants were attacking the country's pipelines. In Venezuela, the regime established by Hugo Chávez, with its mantra of "socialism of the twenty-first century," would become, under Chávez and his successor, Nicolás Maduro, a great economic and humanitarian disaster, and its oil output was plunging. The sanctions imposed on Iran to counter its nuclear program substantially reduced its exports. All this balanced the shale surge from the United States—for a time.

In fact, by the late spring of 2014, worry was mounting about the possibility of a shortfall—not enough oil to keep up with growing demand. At that moment, ISIS was advancing rapidly across Iraq, putting more barrels at risk. "Iraq Violence Lights Fuse to Oil Price Spike," headlined the *Financial Times*.[1]

Yet over that summer of 2014, the Arab Gulf producers began to pick up disconcerting signals from the market. For some unknown reason, they could not sell all their oil in Asia.

THE SIGNALS, THOUGH NOT EASILY READ AT THE TIME, POINTED to the momentous change in the world oil market and the global economy. For this was the time when the BRIC era was giving way to the shale era. The most dynamic developments in the global oil industry were no longer on the demand side, in emerging market countries, but rather now on the supply side, in the heartland of American oil.

At the beginning of September 2014, oil dipped just slightly below $100 a barrel—to $99.51. By the middle of October, it was down to $84. Forces were swiftly coming together to undermine an entire industry structure that had become anchored on the idea of $100 oil.

Partly it was demand. Global economic growth was weaker than had been anticipated—which meant a slowdown in demand for oil. Most significant was the slowing in China. In the summer of 2014, a group of economists met to discuss the country's economic prospects. Some suggested that China consider shifting from calling itself a "high-growth" country to a "medium-growth" country. But that was a bridge too far. Instead, the group came up with a compromise—a "medium-to-high-growth" country. At the same time, petroleum output was increasing from several countries—Canada, Russia, Brazil, and Iraq. But far and away the surge of shale from the United States was dominating. By mid-November, the price was down to $77.

As so often had happened at previous times of stress in the oil market, eyes turned to OPEC. But this was not the OPEC of the past. Venezuela, one of the patriarchs of OPEC, was sinking deeper into its self-induced economic collapse. People could not get their medicines, and expectant mothers were crossing the border into Colombia to give birth, as Venezuelan hospitals had run out of supplies.

The rivalry between Iran and Saudi Arabia had become even more acute. The nuclear deal with Iran was a huge jolt to the Arab oil producers. It meant that sanctioned Iranian oil would return to the world market. The Arab Gulf states were further alarmed when President Obama said that with a nuclear deal, Iran could be "a very successful regional power," and, later, that "Iran will be and should be a regional power." Moreover, tens of billions of dollars of Iranian oil earnings, held in escrow, would be released, providing Iran with additional resources to put into what these countries saw as its drive to dominate the Mideast.

The only OPEC nations that had the capability to reduce production to bolster the market and bring prices up were Saudi Arabia in particular, followed by Abu Dhabi and Kuwait. Yet the biggest beneficiary of high prices would be Iran, and helping Iran was the last thing that any of them wanted to do. For decades, it had been axiomatic that geopolitical tensions and turbulence would drive up the oil price. But here was the opposite—geopolitics was driving down the price of oil. The rivalry between Riyadh and Tehran made impossible any deal to halt what looked like the beginning of a price rout.

The Saudis had another reason for not cutting. In the mid-1980s, during a period of glut, they had learned the lesson that if they cut and others did not, they would end up losing market share—and money.

That was much in the minds of the Saudi delegation when it traveled to Vienna for the OPEC meeting in November 2014, especially the unprecedented growth of U.S. shale. "Non-OPEC producers," Saudi petroleum minister Ali Al-Naimi said, "need to come to the table." That meant, more than anyone else, what was at the time the world's largest producer—Russia.

Naimi's own life reflected the transformation of Saudi Arabia. He spent his early years as a Bedu—"desert dweller"—living in a tent, trekking across the desert, helping to tend the goats on which his family's nomadic life depended. In 1947, age twelve or so, he was hired as an office boy at Aramco. This was when Aramco was gearing up after World War II as a joint venture of American companies. In due course, Naimi was sent to the United States to study geology, first at Lehigh University in Pennsylvania, and then at Stanford. Back in Saudi Arabia, he rose in the company, and in 1983 became the first Saudi president of Aramco, which by then had been fully nationalized. Twelve years later, he was appointed minister of petroleum. He became close to King Abdullah, who relied on his judgment and entrusted him with managing Saudi Arabia's oil policy.

Naimi arrived in Vienna a few days in advance of the November 2014 meeting for a private discussion arranged by the Venezuelan minister Rafael Ramírez with two key non-OPEC producers, the Mexicans and the Russians. To avoid the ever-vigilant press corps, Naimi and his advisers were dropped off at a loading bay of the Park Hyatt Hotel that led into the hotel's kitchen. Naimi made his way upstairs via the service elevator used for delivering room service.

In the meeting suite, the Mexican minister Pedro Joaquín Coldwell explained that Mexico could not cut back, both for technical reasons and because it was launching a campaign to bring in foreign investment to rejuvenate the Mexican industry. Ordering a reduction in production would be a very bad message. The Russians—Rosneft CEO Igor Sechin and energy minister Alexander Novak—explained that Russia could not cut back either, owing to what they later described as "climatic, logistical, and technological factors."

It was clear to Naimi that these other countries had no intention of cutting output. They were trying, he concluded, to pressure Saudi Arabia to once again "swoop in and make a dramatic reduction in production."

"It looks like nobody can cut," Naimi said. "So, I think the meeting is over." He stood up, picked up his papers, and shook hands. "Thank you," he said tersely.

And then he walked out. The room was stunned. It took a moment for the others to realize what had just happened. Naimi's aides hurried to pack up their papers and run after him.

At the OPEC ministerial meeting that followed on November 26, 2014, Naimi laid out Riyadh's position. If Saudi Arabia or OPEC as an entity cut production without major non-OPEC participation, "we would be sacrificing revenues as well as market share." The problem was "too many new producers."

The OPEC ministers came out that day with their historic decision— which was, in fact, to have no decision. Instead of trying to stabilize the market, they would leave it to the market to manage itself; exporters were free to produce at whatever level they could. The market would "stabilize itself," said Naimi. But he added one extra word: "Eventually."[2]

IN THE WEEKS THAT FOLLOWED, OIL CONTINUED TO FLOOD into the market, and by mid-January the price had collapsed to less than half of what it had been five months earlier. *The Economist* captured what was seen as the new oil war with a cover depicting "Sheikhs v[ersus] Shale." In the United States, the shale producers, many having run up high levels of debt to finance their drilling programs, were hard hit. "Shale Producers Clobbered by Oil Rout" said one headline. Almost a hundred companies went bankrupt. The rest slashed their budgets, and workers were laid off.[3]

Surely, some assumed, the market would stabilize, because, it was thought, the shale producers were the high-cost producers and would largely fold at $70. "Every study we saw said the break-even for shale was $70 a barrel," a senior decision maker in one of the Gulf countries was to recall. But it turned out that shale was not high cost. The U.S. producers became much more efficient and more focused; they continued to learn about the rocks and innovate in drilling and completing wells. Also, with companies cutting back dramatically on spending, the cost of the technology and services on which they depended for drilling fell. The glut was swelling.[4]

The emergence of shale had forced the oil industry to learn a new vocabulary—"short cycle" versus "long cycle." Short cycle was, most obviously, shale. A decision could be made to drill and within half a year, a well could be producing oil. Actual drilling time itself was shortened to as little as five or six days. A single well that might have cost $15 million a year or two earlier might now cost all of $7 million. Of course, because of the high depletion rates of shale, the producer needed to continually drill new wells.

This was in contrast to long cycle—an offshore oil or LNG project that could take five to ten years to bring on, but then would produce for many years. Instead of $7 million, a long-cycle offshore project would cost $700 million or $7 billion—or much more. Many long-cycle projects were launched or planned during the BRIC era, with the confidence that $100 a barrel was the "new normal." But now, with the price collapse, new long-cycle projects were being postponed, delayed, or canceled outright. Low prices continued through 2015.

BY MAKING NO DECISION, OPEC HAD CEDED CONTROL TO what might be called the "swing investors"—the financial markets—the hedge funds and traders and other financial players that dealt in "paper barrels." "Sentiment" among these investors—whether they were optimistic or pessimistic, bullish or bearish about prices and the market—

would determine whether they went "long" or "short" on futures contracts. And their cumulative decisions would in turn accentuate price moves in one direction or the other. At this point, sheer bearishness predominated. Prices fell below $30 and seemed headed toward $25 a barrel. Some investment banks warned that oil could fall below $20.

"The oil market is much bigger than just OPEC," Naimi said in February 2016, but other countries had demonstrated "no appetite for sharing the burden." As a result, he continued, "We left it to the market as the most efficient way to rebalance supply and demand. It was—it is—a simple case of letting the market work."[5]

Iranian oil minister Bijan Zanganeh demanded that the other Gulf countries cut back while Iran regained, as he put it, "our lost share of the market." The Arab Gulf countries were adamant that they would not cut back to make room for additional Iranian oil. "We will provide oil to whoever asks for it," said Prince Abdulaziz bin Salman, Saudi deputy oil minister at the time.

By now most of the oil exporters were in extreme distress. Nigeria, with 175 million people, depended on oil for 70 percent of its budget. In Russia, wealth was fast draining from its sovereign wealth fund. Saudi Arabia, running a deficit, was accelerating the drawdown of its large foreign reserves. Iraq's oil revenues had collapsed. In 2015, Venezuela, desperate, suggested to other OPEC countries that they organize an "environmental campaign inside the U.S." against shale.

Naimi's message about "not going it alone" finally got through. Saudi Arabia and Russia were working on a deal to steady production. By April 2016, oil ministers representing about half of world production were converging on the Sheraton Hotel in Doha, the capital of Qatar. They had pretty much concluded a plan to freeze production— to hold output steady in order to stem the rising tide of inventories and buy some time to catch up.

But one country was notably absent—Iran. Determined to ramp up its output, it would never agree to be part of a freeze.

Still, Naimi had achieved the goal he had insisted on since 2014—Russian acquiescence to some kind of output restriction. But in the night, a call came in from Saudi Arabia. The message to Naimi was clear—no agreement without Iran. And Iran clearly would not participate. The deal was off.

In Doha, the others were astonished by this overnight reversal, after all the effort. But facts were facts. It was a new order in Saudi Arabia. Naimi was no longer the man in charge of Saudi oil policy.

Some days later, King Salman announced a cabinet shake-up. Naimi, heretofore the most powerful man in world oil, had been replaced. The new minister was Khalid al-Falih, a graduate of Texas A&M University who had been CEO of Saudi Aramco for seven years. During that time, he had pushed the company to "internationalize," expanding downstream business both in Saudi Arabia and outside the country in order to secure markets in the consuming countries. And then he had done a stint as health minister to handle MERS, a respiratory virus epidemic that had become a national medical emergency. But that assignment lasted just a year before his new appointment returned him to oil, but now as head of an expanded super-ministry that, in addition to petroleum, also included industry, electricity, water, and minerals, altogether representing about 60 percent of the economy.

Meanwhile, oil prices were certainly doing what prices do—rebalancing the market, reducing the gap between supply and demand. The price collapse was slowing the development of new supplies. Low prices were also stimulating demand. In 2015, the growth in world oil consumption was more than twice what it had been in 2014. With cheaper gasoline, the share of SUVs and light trucks sold in the United States rose from under the 50 percent it had been in 2012 to 60 percent in 2015.

WITH NO DEAL, DESPERATION AMONG THE OIL-EXPORTING countries was mounting. At the beginning of September 2016, on the sidelines of a G20 meeting in the Chinese city of Hangzhou, Vladimir Putin sat down with then deputy crown prince Mohammed bin Salman. They agreed that "no stable oil policy can be pursued without the involvement of Russia and Saudi Arabia." The deputy crown prince added that Saudi Arabia recognized Russia as "an important world player" and also as "a great power"—a potent rejoinder to Barack Obama's dismissal of Russia as nothing more than a regional power. For his part, Putin made clear that he was looking for a relationship that extended beyond oil to "cooperation" in "the broadest sense of the word."[6]

Later in September, the International Energy Forum, composed of seventy-two countries representing 90 percent of oil supply and demand, convened in Algiers for a ministerial dialogue. Alexander Novak, the Russian minister, signaled that Russia would consider signing on to a deal. But first, the OPEC countries had to come up with their own deal.

Over the next several hours, the OPEC ministers met by themselves. No one was let out of the room. The result was an outline of an agreement called the "Algiers Accord." It took the market by surprise, for it proposed a million-barrel-per-day cut in its total output. "OPEC made an exceptional decision today," declared Iranian oil minister Bijan Zanganeh. "After two and a half years OPEC reached consensus."

As Prince Mohammed bin Salman explained, "Oil revenues are the main factor, and the main reason for the oil agreement."[7]

Another factor facilitating a deal was that OPEC finally had a permanent secretary-general, a position that had gone unfilled since 2012 owing to discord among members. The new secretary general was Mohammad Sanusi Barkindo, a Nigerian and former head of the Nigerian state oil company. He was a consensus-builder, barnstorming from one OPEC capital to another, understated, listening carefully, sorting, trying to bridge gaps. The most difficult gap remained that between Saudi Arabia and Iran, for the differences over oil policy were a subset of their geopolitical rivalry. When Barkindo commuted between Riyadh and Tehran, he had to fly through a third country.

At the end of the November 2016 OPEC meeting, Barkindo declared, "We have against all odds reached landmark decisions." OPEC formally adopted, more or less, the Algiers accord. Iran was solved with a sleight of hand. It was officially allocated a higher production quota, but in practice that did not matter, as the other countries knew that Iran was in no position to achieve it any time soon.[8]

Two weeks later, OPEC and a group of eleven non-OPEC countries, led by Russia, met in Vienna and agreed on what was the breakthrough deal—an OPEC–non-OPEC deal. OPEC's 1.2-million-barrel-per-day reduction would be matched by an additional 558,000 cut by non-OPEC, of which 300,000 would come from Russia. The other ten non-OPEC countries, ranging from Kazakhstan and Azerbaijan to Oman and Mexico, would make up the rest. This new arrangement became known variously as "OPEC-Plus" or the "Vienna Alliance."

Of course, this grouping of "non-OPEC" countries did not include other big non-OPEC producers—the United States, Canada, Britain, Brazil, Norway, China, and Australia. But U.S. output had come down anyway, owing to the retrenchment in investment by the hard-hit domestic companies in the face of low prices.

Just as the oil price had cratered after the November 2014 meeting, so now it rebounded after the November and December 2016 meetings.

OPEC-PLUS HAD ALSO HELPED GENERATE A GEOPOLITICAL reordering—the new relationship between Moscow and Riyadh. Once, in the early 1990s, after the collapse of the Soviet Union, Naimi was asked what he thought of Russia. "I think of it as a competitor," he said. But now oil, which had been a source of rivalry, had brought them together. A state visit in October 2017 by King Salman to Moscow demonstrated a relationship beyond oil. He arrived with an entourage of fifteen hundred people, and the delegation took over the three main hotels that surrounded Red Square. Traffic and security were so intense that the only practical way to move around in the vicinity, despite unusually heavy rains, was on foot. During the visit, a number of agreements were signed that went beyond energy cooperation to investment, military cooperation and weapons sales, and technology sharing. The king also received an honorary degree from the Moscow State Institute for International Relations. At a conference held during the visit, a senior Saudi official was asked about how this burgeoning relationship would affect Riyadh's long-standing relationship with Washington. Saudi Arabia has to attend to its own national interest, the official diplomatically replied.

"If we continue to work the way we do," Putin said to the Saudi energy minister a few months later, "we will turn from rivals into partners." Facing a growing welter of Western sanctions, and increasingly isolated from the West, Moscow welcomed a partnership with Saudi Arabia, the United States' most important Arab ally. It cemented Russia's return to the Middle East and its growing influence. It was a player—or, as one Saudi called it, "a world player." It was the only major power that could talk across the divides, with Israel, Saudi Arabia, the UAE, Syria, Iran, Hezbollah, and Hamas.[9]

For its part, Riyadh could use the new relationship with Russia to bolster its own international position and hedge against what had been, under the Obama administration, an increasingly frayed relationship. Moreover, this new connection with Russia could perhaps help Saudi Arabia in its rivalry with Iran.

The oil market recovery was also facilitated by the continuing implosion of Venezuela and the resulting plummeting in its oil production. In the late 1990s, Venezuela had hit 3.3 million barrels per day. By the end of 2019, it was down to 600,000 barrels per day—half of what a single state in the United States, North Dakota, was producing. This was the result of the massive mismanagement of the economy by the Chávez-Maduro regime. In addition to imposing a dictatorship, the regime was in effect waging economic warfare on its own country, generating a breakdown in the economy, a huge debt burden, politicization of economic management, permanent shortages, pervasive corruption, massive diversion of government funds, and the impoverishment of the population. Inflation was heading toward an annualized rate of an altogether unimaginable one million percent.

Venezuela has the largest proven oil reserves in the world, larger than Saudi Arabia's, though "heavier" and more expensive to produce. But the industry was starved of investment, capable managers were driven out of the country, the state oil company did not have the money to pay for services and technology that it needed, and the Maduro government handed over control of the industry to the military. Almost five million refugees, poverty-stricken and hungry and desperate to escape violence, have fled the country, most on foot.

BY THE LATE SPRING OF 2018, GLOBAL PRICES WERE BACK TO $80 a barrel. But now a new factor emerged in the oil markets—Donald J. Trump.

On April 20, 2018, representatives of some of the members of the Vienna Alliance were meeting in Saudi Arabia, in Jeddah, to review production trends. In the middle of the meeting, a tweet popped up on the cell phone of one of the participants. Startled, he passed around his phone to the other participants, who were equally taken aback. President Trump, apparently having just seen a segment on rising gasoline prices in the United States on Fox News, had sent out a tweet: "Looks like OPEC is at it again. Oil prices are artificially Very High! No good and will not be accepted." The experts around the table in Jeddah were stunned. The president of the United States was negotiating with OPEC via Twitter.[10]

Trump had a long history of denouncing OPEC. More urgently,

gasoline prices were rising, which could be an important factor in the outcome of the November 2018 congressional elections. Tactically, it was also a message to the Arab Gulf countries to increase production to offset the coming decline in Iranian exports. For the U.S. administration was determined to reimpose sanctions on Iran without driving up prices.

A month later, in late June 2018, on the day the OPEC countries were meeting in Vienna, Trump came back with a new tweet: "Hope OPEC will increase output substantially. Need to keep prices down!"[11] But this was not just a message coming from the United States. India imports 85 percent of its oil, and high prices would stifle its economic growth. At a seminar in the Hofburg Palace in Vienna on the eve of the OPEC meeting, India's minister of petroleum and natural gas and steel Dharmendra Pradhan described the "pain" that the oil price was inflicting on India. China conveyed the same message. Even the Russian government was feeling pressure from consumers and other economic sectors over high prices. Moscow was also decidedly unhappy about giving up market share to U.S. shale.

By early October 2018, oil prices had spiked back to $86; moreover, some were even predicting that oil could once again hit $100 a barrel. The price had jumped in expectation that the United States would push a total ban on Iranian oil exports. In order to offset the expected shortfall, Saudi Arabia significantly ramped up output.

But the date for the sanctions to go into effect was November 4, which happened to be two days before the 2018 midterm congressional elections. When the sanctions were announced, they came with a surprise. Rather than "zero exports all at once," the administration granted eight countries, which represented 85 percent of Iran's sales, "significant reduction exemptions," which meant that they could still import some Iranian oil without facing penalties. Trump explained: "I could get the Iran oil to zero immediately, but it would cause a shock to the market." He added, "I don't want to lift oil prices."[12]

————

IN VENEZUELA IN 2018, MADURO WON A SECOND SIX-YEAR term in an election that was widely denounced as a sham. The results were rejected by several countries in the hemisphere. The National Assembly, which Maduro had tried to sideline, declared his January 2019 inauguration illegal and unconstitutional. The president of the Assembly, thirty-five-year-old engineer Juan Guaidó, described Maduro as a "usurper" and stepped forward as interim president, as specified by the constitution. A number of Latin American countries, as well as the United States and Canada, recognized Guaidó as the legitimate president. Now Venezuela had two presidents.[13]

But Maduro, bolstered by Cuban security forces, controlled two key assets—the oil industry and the military. The United States banned imports of Venezuelan oil, which accounted for half of Venezuela's total exports. Stalemate continued in Venezuela, while the economy and standard of living deteriorated. Black-outs became common, making daily life even more unbearable. Venezuela, once a global petroleum powerhouse, had become a negligible factor in the world market.

IN THE SPRING OF 2019, THE TRUMP ADMINISTRATION AN-nounced no more waivers for importers of Iranian oil. Zero exports was now the policy. Iran, which had been exporting 2.5 million barrels per day, was soon down to a few hundred thousand, either bartered or smuggled. Sanctions were also added on exports of other products.

The sanctions were choking the Iranian economy. In reply, Iran increased its enrichment of uranium, which had been limited under the nuclear deal. Gulf countries, while supporting the isolation of Iran, worried what a cornered Iran might do. In May 2019, the United States sent an aircraft carrier task force steaming into the Gulf. "Collision course," was the cover of *The Economist*. "America, Iran and the threat of war."[14]

But there was already another kind of war, as well—the trade war between the United States and China. And that was getting worse, with the United States adding tariffs on Chinese goods, and the Chinese responding by doing the same on American goods, including LNG and oil imports. The trade war was taking a toll on global economic growth and in turn lowering the demand for oil. And that sent oil prices down again.

AT ABOUT 3:45 A.M. ON SEPTEMBER 14, 2019, EXPLOSIONS shook the darkness and fires burst out at two sites in Saudi Arabia's eastern province that are critical for processing the country's oil. Firemen, thinking that these were accidents, rushed to put out the blazes. But they were no accidents; they were attacks by drones and cruise missiles. One site was Abqaiq, a sprawling industrial facility of pipes and processing units. It is the most critical "hardware" in the entire global oil industry, for it processes and purifies the bulk of Saudi crude oil for dispatch to domestic refineries or for export. The other site was Khurais, another major processing facility.

This was the moment that years of "scenarios" became, in an instant, reality—war games turned into an act of war. The attacks demonstrated the physical vulnerability of the vast oil infrastructure that lines the Persian Gulf. The attacks knocked out 5.7 million barrels a day of oil flow—in absolute terms, the biggest disruption in the history of the oil industry. The following Monday, the oil price spiked.

The Houthis in Yemen immediately claimed responsibility, describing it as revenge for the Saudi air campaign in Yemen. Iran denied any responsibility, which had the air of what could be called "implausible deniability." For the skill of the coordination and the precision of the targeting and execution—and the 750-mile distance from

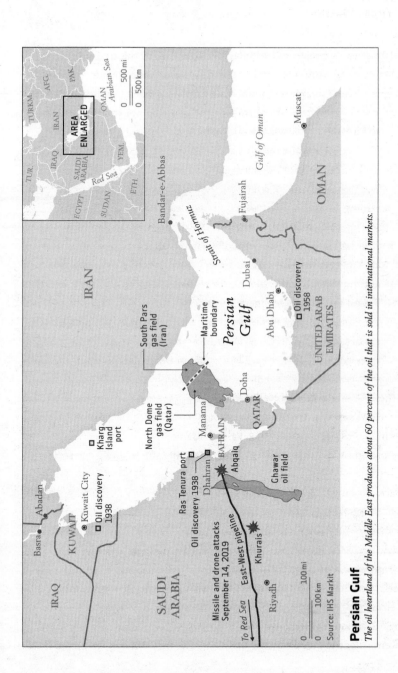

Persian Gulf

The oil heartland of the Middle East produces about 60 percent of the oil that is sold in international markets.

Source: IHS Markit

Yemen—strongly suggested Iran masterminded the attack, part of a series of escalating attacks on the oil system that responded to the U.S. drive for a total ban on Iranian oil exports. Evidence indicated that the attack came from the north—from launching points either in Iran itself or from a Shiite-controlled area in neighboring Iraq.[15]

Yet a certain degree of ambiguity on the part of the Saudis was useful. For otherwise, there would have been pressure to retaliate. But that would have risked further escalation and more destruction raining down on the region's oil and desalination infrastructure.

In previous years, something so devastating, at the heart of the world oil industry, would have radiated panic through the global market and sent prices spiraling up. What was striking now, however, was not that the price shot up, but that it did not go up that high, and that it soon came down to where it was before the attack. Part of it was that the Saudis hastened to make clear that they would maintain existing export levels by drawing down inventories and reducing supplies to their domestic refineries. They would move quickly to repair the facilities. Moreover, global inventories were high.

What also made a difference was the rebalancing of the world oil market by the continuing surge in U.S. oil. For, it turned out, shale has not only reconfigured the world oil market. It has also reconfigured the psychology of the world market, providing a new sense of security.

YET STILL TURMOIL CONTINUED IN THE REGION. FOR MONTHS, angry demonstrators in Iraq had continued to protest the breakdown of social services, lack of jobs, corruption, and the pervasive Iranian influence. Oil exports from Iraq started to be disrupted.

At the end of December 2019, rockets fired by an Iranian-dominated militia, Kataib Hezbollah, at a military base in Iraq killed an American contractor and wounded U.S. service members. The United States responded with air attacks killing two dozen militants. Members of the militia and their supporters then entered with ease into the protected Green Zone in Baghdad and counter-attacked the U.S. embassy, a spectacle that unfolded on television worldwide. Among those watching was Donald Trump.

On the night of January 2, 2020, the leader of the Quds Force, Qassem Soleimani, boarded a civilian airliner in Damascus bound for Baghdad. He had met the day before in Beirut with the head of Hezbollah, demonstrating how he seemed to be everywhere in the region. He had also been in something of a Twitter and Instagram war with Donald Trump. Dismissing Trump as a "gambler" in 2018, he had messaged the president, "You know how powerful we are in asymmetrical warfare. Come, we are waiting for you."[16] But now they were waiting for him.

At about 12:45 a.m. on January 3, as he drove away from Baghdad International Airport, missiles from an American drone struck his vehicle, killing him. Also killed was the head of Kataib Hezbollah, who had been a leader of the militias fighting ISIS and a close Soleimani ally. The outpouring in Iran was enormous; Soleimani had gone from being the shadow commander to a celebrity and national hero, even for critics of the regime. Extolling the Quds forces as "fighters without borders," Ayatollah Khamenei promised revenge. For ISIS, still on the defensive, the death of Soleimani was a great gain; its most formidable opponent on the ground was gone. Outrage among Shias in Iraq was great, and the anti-Iranian protests were silenced, for the time being. But then a few days later the Revolutionary Guard fired missiles that brought down a civilian airliner on a normal flight path out of Tehran, killing 176 people aboard, mostly Iranians and Canadians. Suddenly the streets in Iran were filled with people protesting the regime.[17]

Soleimani's death did not send oil prices spiraling. Instead, over the weeks that followed, the price plummeted. The novel coronavirus epidemic, which had broken out in the Chinese city of Wuhan, would shut down China, which had been responsible for 40 percent of all growth in world oil consumption since 2003. But now demand was going down, as people in China stayed home, traffic disappeared, and airlines canceled flights into China. Volatility in 2020 was gripping the oil market, though not in a way that had ever been anticipated nor with an impact ever imagined.

By that time, the price collapse of 2014–2015 that had come with the shale era had already helped to trigger dramatic change in the most important oil exporter of them all.

Chapter 35

———

RUN FOR THE FUTURE

The sprawling Ritz-Carlton Hotel in Riyadh, the capital of Saudi Arabia, was originally built for royal guests. It was converted into a grand commercial hostelry in 2011. In October 2017, it became the venue for a thirty-five-hundred-person futuristic investment conference, populated by leading financial and business figures from both Saudi Arabia and around the world, along with a number of friendly robots and a heady diet of virtual reality.

The centerpiece was a plenary that featured Saudi Arabia's crown prince, Mohammed bin Salman, the next in line to the king. It was not the kind of panel Saudis would have been accustomed to seeing. First, in the past, crown princes in Saudi Arabia simply did not do panels. Moreover, the moderator was a female American news anchor, who was not wearing the black hijab normally required of women in public.

Even more noteworthy than the panel itself was what the crown prince said. "The era that began in 1979 is over," he announced. From here on, Saudi Arabia would be the home of "moderate Islam." To drive home the point, he declared, "We will destroy the extremists."

To emphasize the change coming in the country for the young Saudis who were watching online, he pulled out what looked like an "old" BlackBerry—circa 2005—and a current iPhone. The difference between the two, he said, is how Saudi Arabia would change. The image went viral.

But there were also definite limits. At one point, one of the panelists, Japanese tech tycoon Masayoshi Son, became so excited by the "dream" of the plans for the futuristic new half-trillion-dollar city, Neom—with robots busily at work, even doing people's shopping— that he exclaimed that it would be nothing less than a "new Mecca."

That was too much. The crown prince grabbed the microphone. "No, no, no," he said. "There's only one Mecca."

But the overall change was undeniable. Suddenly Saudi Arabia, traditionally slow and processional in its change, was being charged to hurtle into the future. Social restrictions imposed in the late 1970s and early 1980s and regarded as immutable were now to be removed.

The discontinuity with the past was remarkable, even startling. There was more to come, as would be demonstrated a little more than a week later. Lebanon's prime minister suddenly appeared in Riyadh and abruptly announced that he was resigning as prime minister as protest against the dominance of the Iranian-backed Hezbollah in Lebanon (although he retracted his resignation a month later). The Iranian-backed Houthis in Yemen launched a missile targeted directly at Riyadh's international airport. It was intercepted over the city by a Patriot missile barrage, with explosions that shook the metropolis.

An even bigger shock, though of a different kind, reverberated across the entire kingdom when, on that same evening, over two hundred people—members of the royal family, business and government leaders, present and former cabinet ministers—were arrested, charged with corruption, and detained in that very same Ritz-Carlton, now transformed into a prison. Some of the detainees only the week before had been shaking hands with the foreign visitors at the investment conference. Over the next few months, the Saudi government would say that over $100 billion of corruptly obtained money had been recouped from the detainees.[1]

The detention would demonstrate who was in charge. Power in Saudi Arabia had formerly been shared by a collegium of the aging sons of King Abdul Aziz, otherwise known as Ibn Saud, the kingdom's founder. But now it was being concentrated in the hands of one of those sons, Salman, now king, and his own son, Mohammed.

IT WAS HARDLY FOREORDAINED THAT MOHAMMED BIN Salman—known as MBS—would be in the line of succession. His father, Salman bin Abdulaziz, born in 1935, was one of the roughly forty sons who had outlived Abdul Aziz. Salman was governor of Riyadh for almost half a century during which time the city had grown from two hundred thousand people to over five million. Salman had then become defense minister and then crown prince. In January 2015, King Abdullah had died, aged about ninety-one. Salman became king.

Mohammed, born in 1985, was one of Salman's younger sons. He had learned English as a child, he said, watching American movies. Growing up, he would be assigned one history book a week by his father, and then would be quizzed on what he had learned.[2]

One lesson he had certainly absorbed was about power, how to consolidate it and how to exercise it. He had been a private investor. He had joined the legal office advising the cabinet, where he had brashly pushed against traditional ways of doing things. He shifted over to work for his father, at that time still the governor of Riyadh. Then, when his father moved up to become crown prince, MBS headed his court.

On Salman's accession to the throne in 2015, Mohammed, heretofore largely unknown, catapulted at age twenty-nine into the position of deputy crown prince—and quickly assumed additional power as minister of defense and head of both the supreme economic council and the supreme council overseeing the oil industry. He oversaw the launch of what would prove the protracted war against the rebel Houthis in Yemen. In June 2017, MBS pushed his cousin Mohammed bin Nayef aside as crown prince. Previously interior minister, bin Nayef had led the anti-extremist campaign since the early 2000s and was well known to Washington. Now MBS was next in line to be king, with the potential for a very long timeline as his father's successor, whenever that time came. And the ministry of interior, with its responsibilities for domestic security, was also now under his sway.

For the economy, MBS promulgated Vision 2030, which promises much more than reform, nothing less than a transformation of Saudi Arabia—moving away from an economy so heavily dependent on oil and thus so vulnerable to the boom-and-bust cycles of the oil market. Instead, the economy is to be diversified, made more competitive, more entrepreneurial, and more high-tech.

By proclaiming a new era in Saudi Arabia, the crown prince is appealing to what he sees as his constituency—the social-media-savvy 70 percent of Saudis who are thirty-five years old or younger. For them, without change, the number of jobs will be insufficient, leaving them discontented and susceptible to the pulls of extremism or alienation.

Central to Saudi Arabia's challenges is its dependence on oil exports. "We will not allow our country ever to be at the mercy of commodity price volatility or external markets," says MBS. Yet he also declares that Saudi Arabia has "a case of addiction to oil." To explain the new policies, he turns to history and his grandfather. "King Abdul Aziz and the men who worked with him," he says, had "established the kingdom without depending on oil, and they ran this state without oil, and lived in this state without oil."[3]

That was certainly true. And yet it was oil, along with Mecca and Islam, that did so much to make Saudi Arabia what it is today.

———

IN 1933, STANDARD OIL OF CALIFORNIA—SOCAL, NOW
Chevron—won the right to explore for oil in Saudi Arabia. On March 4,
1938, a telegram was dispatched from Saudi Arabia to the San Francisco headquarters of Socal. It reported that in a test in the eastern
province on a well called Damman #7, at a depth of 4,694 feet, oil had
flowed at the rate of 1,585 barrels per day. The discovery of oil in Saudi
Arabia brought much relief in San Francisco, for, after several disappointing wells during those Great Depression years, the company had
been mulling over whether to give up. No longer. Still, the initial reaction in San Francisco was cautious. "While 'one swallow does not make
a Spring,'" one executive noted, "I am rather encouraged to feel that in
this instance one oil well will make an oilfield."

Several months later, the company felt confident enough to dispatch its representative to see King Abdul Aziz in Riyadh. Oil, he told
the king, had been found "in commercial quantities." The announcement appeared to come as "a complete surprise" to the monarch. "I am
glad—very glad," he said. "Some of my people have been telling me that
oil in large quantities would never have been found in my country and
others have been saying that your company had found it, but would not
admit it. But I always thought they were wrong."

The king was also "very proud" when told that Saudi Arabia had
achieved the status of the twenty-second-largest oil-producing country in the world, along with the prospect of rising even higher. The
news was particularly welcome, for his kingdom was beset with financial difficulties that greatly threatened what he had built over decades.[4]

Beginning in 1901, Abdul Aziz and his small band of followers had
set out on camels into the desert from his family's exile in the neighboring emirate of Kuwait. Their aim was to restore the Saudi dynasty
that had ruled part of the Arabian Peninsula twice before, in the eighteenth and nineteenth centuries. Over the next quarter century, Abdul

Aziz would sweep across Arabia in alliance with ferocious fighters known as the Ikhwan. They were fierce adherents of an austere Islam that traced back to an eighteenth-century cleric, Muhammad ibn Abd al-Wahhab, who had sought to purify Islam, return it to its early roots, and cleanse it of heretics and foreign intrusion. The Islam of al-Wahhab became the faith of the new state, critical to integrating the different tribes into the reestablished Saudi realm.

In 1932, the Kingdom of Hijaz and Nadj was renamed Saudi Arabia. By then the finances of the kingdom were in desperate straits, owing to the collapse in the number of pilgrims going to Mecca, the kingdom's main source of revenue. Thus the discovery of oil in 1938—coincidentally, less than two weeks after oil's discovery in neighboring Kuwait—was more than a great relief for the king. After World War II, in the 1950s, oil production began to build up and money started flowing in. Yet the true era of oil wealth only began when the 1973 oil crisis led to a quadrupling of prices. With less than eight million people at that time, Saudi Arabia became a very wealthy kingdom.

AND THIS IS WHERE THE CROWN PRINCE'S REFERENCE TO THE "era that began in 1979" becomes so significant. For that was the year in which the Saudi kingdom was suddenly challenged. The seizure by the religious zealots of the Grand Mosque in Mecca and the ensuing weeks of battle shocked the leadership of the kingdom and shook confidence. The House of Saud was the Grand Mosque's custodian, its protector; and the family's relation to Islam was fundamental to its identity.

The other major threat came from Iran, now ruled by the militant theocratic regime of Ayatollah Khomeini. He called for the destruction of the Arab monarchies, especially that of Saudi Arabia—the "arch agent of the Great Satan" (which was the United States). At the same time as the assault on the Grand Mosque, Shia protests and demonstrations were occurring in the eastern part of the country, fueled both by local grievances and inflammatory broadcasts by the Voice of the Islamic Republic of Iran.

Saudi Arabia responded to the challenge to its Islamic authority by becoming more pious, reasserting fundamentalist Islam as the fabric to bind the nation together—what outsiders would call Wahhabism, although Riyadh itself would consistently reject that term. Conservative clergy became more dominant and assertive. Religious police patrolled the streets, punishing women not properly attired or unwed couples caught holding hands. Women were segregated from men, with their decisions, education, and travel controlled by "male guardians." Pictures of women were no longer permitted in newspapers. Movie theaters, denounced as sources of "depravity," were shuttered. The educational system tilted more heavily to Islam and religious instruction.

This shift was not only defensive, shoring up the Saudi state, but also offensive. As already noted, Saudi Arabia joined with the United States in funding the mujahedeen, the religious warriors fighting the Soviets in Afghanistan. It also projected its rivalry with Shia Iran across the globe. The government spent tens of billions of dollars for religious education and proselytizing. Saudi Arabia funded imams, mosques, Islamic centers, and many Muslim schools and colleges around the world. Over time, however, the money that it spent on promoting fundamentalist Islam abroad would also foster the rise of extremism and jihadism.

The huge surge of petrodollars created a welfare state economy that has defined Saudi Arabia to the present day. Money was poured into transforming what MBS has called "a country of mud huts" into a new, modern infrastructure—houses, office buildings, hotels, schools, universities, highways, airports. The country's population grew quickly, and urbanized rapidly. The literacy rate jumped from just 10 percent to over 80 percent. The flood of money financed a new social contract between the government and the populace, with the government as the mechanism for collecting and redistributing the oil wealth. Saudi Arabia became what has been called "a nation of entitlements."[5]

This system worked well so long as the oil revenues continued to flood in. But the oil price collapse of 2014 demonstrated the vulnerability of a petro-state. Saudi revenues from oil exports fell from $321 billion in 2013 to $136 billion in 2016. But the budget—and all the commitments to spending—had been geared to a level in which oil was $100 a barrel, not $45 or $35. Moreover, from March 2015 onward, there were large new demands for money to finance the war in Yemen.

In this context work had begun to draft a different future for Saudi Arabia—to move away from an economy so heavily dependent on the export of oil and petrochemicals and thus tied to the cycles of the oil market. The goal is an economy that is more diversified, more competitive, more entrepreneurial, more innovative, more high-tech, and much more integrated into the global economy. Of high urgency is providing jobs for youth, giving them a stake in Saudi society, and thus fending off radicalism and opposition. The transformation of Saudi Arabia would also strengthen it against its regional rival, Iran. But the starting line shows how much ground will have to be covered. "Oil directly accounts for more than 40 percent of GDP," said the International Monetary Fund before the coronavirus, and "non-oil activity is highly dependent on government outlays financed by oil revenues."[6]

Diversifying the economy away from oil is a giant hedge against what has come to be seen as an existential risk for a country whose future so mightily depends upon petroleum. Previous oil price collapses had always stimulated discussion about diversification, but there had been little questioning in those times about the long-term role of oil. Demand, it was understood, was certain to continue to climb for many years ahead. Today, that assuredness is no longer there.

The collapse that began in 2014 coincided with a new global debate about the future course of petroleum demand, embodied in the phrase "peak demand." The argument is whether consumption of oil will flatten out and decline sooner than had generally been anticipated, leaving oil reserves "stranded" in the ground, without value. This debate has been hastened by climate change policies and greater efficiency in transportation, and by the rebirth, after a century of dormancy, of a vehicle that does not need oil—the electric car.

Saudi Arabia may hold the largest reserves of conventional oil and will continue to be the lowest-cost producer, and so it will have a long-term competitive advantage. Yet it also now confronts the possibility of a "world moving away from oil." Moreover, while Middle East oil has the advantage of low cost, it faces more competition, including the rise of U.S. shale oil.

All this set the stage for the Saudi government, under King Salman, to greatly accelerate the social reforms that had begun much more cautiously under King Abdullah. The crown prince announced plans for a huge theme and sports park complex so that Saudi families would no longer have to go to Dubai for holidays but rather could stay home and spend the money inside the country. As late as 2017, the grand mufti, the most senior cleric in the kingdom, on his popular weekly television show, *With the Grand Mufti*, denounced concerts and movie theaters as forbidden in Islam because they "corrupt morals and values" and "are a cause for the two genders to mix." Nevertheless, local and international entertainers began to play concerts across the country, and the government green-lighted the return of cinemas. The first one opened in Riyadh in April 2018 with a packed showing of the super-blockbuster *Black Panther*. A newly established General Entertainment Authority unveiled plans to spend $64 billion over a decade, creating a modern domestic entertainment industry that would provide young people with alternatives to the mall and the mosque—and in the process create jobs.[7]

But the most visible social reform of all has been the lifting of the ban on women driving. The prohibition had been in place for decades. In 1990 five women, arriving at a local food store, told their drivers to get out of their respective cars, and took the wheels themselves and sped off. In response, the grand mufti at that time issued a *fatwa* denouncing women driving as "a source of undeniable vices."[8]

The lifting of the ban in June 2018 removed the onus of being the last country in the world to forbid women driving. It also had major economic purpose—providing the mobility that would enable more women to join the workforce and thus spur higher productivity and economic growth. But a few weeks before the removal of the ban, several women activists, including some who had participated in the original 1990 protest, were summarily arrested. When Canada's foreign minister criticized the arrests in a tweet, spurred by the fact that one of the arrested women had family in Canada, Saudi Arabia broke diplomatic relations and recalled seven thousand Saudis studying at Canadian universities. At this writing, some of the women remain in prison.[9]

With the loosening of the post-1979 social and religious constraints comes an effort to foster an identity built more around nationalism and the Saudi state. In 2005, King Abdullah had introduced a national day to mark the establishment of the Saudi kingdom, despite criticism from religious figures who opposed a non-Islamic holiday. Under King Salman and MBS, it was expanded into a two-day holiday, illuminated with a huge fireworks show and by hundreds of drones that flew over Riyadh and Jeddah in the shape of the Saudi flag. This new nationalism undergirded Vision 2030.

SAUDI ARABIA'S VISION 2030 WAS NOT THE FIRST SUCH VI-sion in the neighborhood. It had been preceded a decade earlier by another Vision 2030, this by its neighbor Abu Dhabi, which has been a pacesetter in diversifying.

Abu Dhabi is one of seven emirates that make up the United Arab Emirates (UAE), which was formed in 1971 when the British withdrew their military forces from the Gulf. Dubai and Abu Dhabi are the two best-known of the emirates. Dubai is the region's commercial hub and a global city for the globalized world economy. Abu Dhabi, which is the largest of the emirates, is also the richest, owing to oil. The UAE's location on the sea, and historically being part of regional trading networks before oil, is often cited as explanation for its being more open and integrated with the world. The UAE may be one of the very few nations in the world, perhaps the only one, to have a minister for tolerance in the government's cabinet.

Abu Dhabi was a latecomer to petroleum. Oil was only discovered in 1958. In 1967, it was still, in the words of one of the people who worked on its first five year plan, a "developing village" that had "no road, no electricity" and "only one school." Today it is a sprawling modern city of office towers and a complex highway system, including a corniche that sweeps alongside the Persian Gulf. In terms of oil, it has the capacity to produce four million barrels a day, with the intent to raise to five, which would make it the second-largest producer in OPEC. At the same time, the state oil company ADNOC is pursuing a distinctive strategy of privatization, not of the whole company, but rather creating partnerships with international companies in different segments of its business.

Sheikh Zayed al Nahyan, who had ruled Abu Dhabi beginning in 1966 and was the founder of the United Arab Emirates in 1971, would warn that the emirate could not always depend on oil. With that in mind, he had established ADIA—the Abu Dhabi Investment Authority—considered today the second largest sovereign wealth fund in the world, with assets publicly estimated at over $800 billion. His son, Mohammed bin Zayed, became crown prince in 2004. He catalyzed the drive to broaden the economy. "In 50 years, when we might have the last barrel of oil," he said, "when it is shipped abroad, will we be sad? If we are investing today in the right sectors, I can tell you we will celebrate." One initiative was Mubadala, a second sovereign wealth fund, with about $230 billion under management, which tilts toward building and investing in companies both in Abu Dhabi and internationally. One of its companies, Strata, makes high-end components for Boeing and Airbus in Abu Dhabi. Another is a partnership with the Cleveland Clinic in a major regional medical center in Abu Dhabi. A second initiative was Masdar, which was set up to diversify in energy beyond oil and gas. It has become a major player in solar and wind both locally and globally as well as a hub for innovation and technology.

The third big initiative was Vision 2030 itself, launched in 2007, which laid out the overall strategy. The message was that the country needed to diversify its revenue base, upgrade skills, create jobs, and increase the participation of women in the economy. The results have come faster than might have been expected. Two decades ago, almost all of GDP was oil-based. Today, about 60 percent of GDP is non-oil-related. Non-oil exports have risen from just 13 percent of total exports in 2010 to 57 percent in 2018. This diversification has been facilitated by the investment climate, consistent policies, and a business culture, bolstered by inexpensive electricity that supports manufacturing and a location that has enabled it to become a trading hub. It all adds up to a playbook for economic diversification. At least its a playbook that has worked for Abu Dhabi.[10]

THE OBJECTIVES OF SAUDI ARABIA'S VISION 2030 GO BEyond economic diversification to a national transformation. Its thirteen Vision Realization Programs extend from the economy and finance to quality of life and "national character enrichment." The National Transformation Program, in its initial form, had 178 objectives, 371 indicators for monitoring progress, and 543 initiatives.

"Since the early 1970s, we've had Five Year development plans," said one official. "They all had the same objectives—diversify the economy, grow the private sector, and have less reliance on oil. We built out massive infrastructure, expansion of education, and the health sector.

But look at the three basic objectives—we did not manage to realize them. We're still dependent on oil, the private sector is dependent on the government, and we have not been able to diversify. Now we're reforming everything—society, the economy, the bureaucracy—all at once. It's truly a necessity. Look at demographics."

Vision 2030, embedded in a continuing communications program, is meant to be the roadmap for remaking Saudi Arabia. The private sector's contribution to GDP is to rise. Foreign investment is to increase substantially. Saudi Arabia and Russia are close as the third- and fourth-largest spenders on defense, exceeded only by the United States and China. Saudi Arabia has also been the top buyer of weapons from the United States since 2011. A very ambitious target is to establish defense industries that will "localize" 50 percent of spending on military equipment. Subsidies have been reduced for water and electricity, while at the same time, "citizen accounts" have been set up to provide money to lower-income Saudis to offset higher utility costs.[11]

Another goal is to greatly expand non-oil exports, including services and what MBS calls "the huge tourism strategy." Vision 2030 looks to the Hajj—the pilgrimage to Mecca—to dramatically increase the number of "religious tourists." New resorts along the Red Sea are intended to draw a growing number of "non-religious tourists." Instead of the laborious process of applying for a visa in advance, said MBS, non-religious tourists will be able to "book a room in a hotel or an apartment" and get their visa on arrival at the airport or even online. There are multiple other targets—improve access to housing and health care, create six million new jobs, increase the percentage of women in the workforce (women already outnumber men in universities), and, notably, "cut tedious bureaucracy."[12]

A top priority is to generate jobs for Saudis in a private sector that is meant to become less dependent upon government spending. Yet nothing so clearly demonstrates the difficulties than the employment structure itself. For Saudis, there are jobs, but largely in government. Jobs in the private sector are not only generally less well paid but are also considered less appealing. Those jobs tend to be filled by "guest workers" on temporary visas, usually Muslims from countries like Pakistan, India, Bangladesh, Egypt, and the Philippines. They can never become citizens, nor can their children, even if born in Saudi Arabia.

The result is a two-tiered labor force. For every two Saudis, totaling about twenty million, there is one foreigner, adding up to about ten million, in the country. But the ratios are reversed when it comes to the workforce. About four and a half million Saudis are employed—70 percent by the government. By contrast, there are twice that many foreigners, over eight million, most of them less well paid, working in the private sector.[13]

Ironically, one requirement underlies the entire economic program: In order to finance Vision 2030 and the diversification away from oil, Saudi Arabia will need a lot of oil revenues. It is attempting to do in little more than a decade what took the miracle economies of East Asia more than two decades, or longer.[14] People and resources need to be mobilized. The value system has to shift away from a system based upon entitlements and subsidies to one based upon performance, competition, and delivery, and one much less dependent upon government decision making. Pacing and balancing of choices is daunting. The very concept of "time" has to change—become less elastic and more disciplined. The buy-in of the public and the continuing support of young people are crucial, and so is cushioning against the potential turbulence, including resistance from the more traditional sectors of society, that can come from transition. And, it could be assumed, the Islamists will do everything they can to destabilize and disrupt.

Can all this be achieved in this compressed amount of time? "Even if only 50 percent is achieved," said a Saudi who was a senior official under King Abdullah, "that will be great." Even at that rate, it would still mean another transformation for Saudi Arabia, comparable to that which followed the 1970s.

———

BUT THE ROAD TOWARD REFORM—AND SAUDI ARABIA'S NEW engagement with the world economy—was suddenly disrupted by an event in Istanbul. Jamal Khashoggi, a journalist, former editor of a Saudi newspaper, and former unofficial spokesman for some members of the Saudi royal family, had moved to the United States in 2017. He became a contributing writer to the *Washington Post*, posting columns criticizing governance and clampdowns on dissent in the Arab world.

On October 2, 2018, he went into the Saudi consulate in Istanbul to do paperwork so that he could remarry. He never came out. He was gruesomely killed inside the Saudi consulate in Istanbul by a fifteen-man hit squad from Riyadh. What the hit squad had not counted on, however, was that the Turkish intelligence services would be taping the killing. Thereafter, drip after drip, they released detail after detail.

Why did the government of Turkish president Recep Tayyip Erdoğan proceed to release details in the way it did? Was it sheer fury that a fifteen-man hit squad would fly into Istanbul and, ignoring Turkish sovereignty, so brazenly kill a journalist who had cordial links with the Turkish government? Or was this meant to constrain and undercut the crown prince, who in the Turks' view was seeking to be the dominant figure in the region, and instead use the opportunity to assert Turkey's regional role, which Erdoğan describes as "a continuation of the Ottomans"? Turkey's "accretion of history, and geographic location," he says, make it "the only country that can lead the Muslim world." Or could it be that the clash between Ankara and Riyadh was one battle in the war between the Muslim Brotherhood and political Islam on one side, and the Gulf dynasties and their Egyptian ally on the other?[15]

In the aftermath of the Khashoggi affair, the reform program would continue, but no longer with the same international acclaim and excitement. Germany announced that it would no longer sell arms to Saudi Arabia. U.S. senators who had praised MBS as a reformer and modernizer now criticized him and Saudi Arabia. The same for international media. The U.S. Congress passed a joint resolution to end U.S. military support for Saudi Arabia in the Yemen war. Donald Trump pushed back against these critics and reaffirmed his administration's support for MBS and Saudi Arabia.

RELATIONS WITH THE UNITED STATES HAD ALREADY BECOME complicated on another front. In May 2017, for his first foreign trip as president, Donald Trump flew to Saudi Arabia, where he and King Salman cohosted a summit with the leaders of Arab and Muslim countries. The two also jointly inaugurated an anti-terrorism center, to which the other Arab Gulf states signed on. That included Qatar, the small gas-exporting emirate adjacent to Saudi Arabia. But there was tension between Qatar and the other Gulf countries, and it was rising.

Two weeks later, on June 5, 2017, Saudi Arabia and the United Arab Emirates abruptly broke diplomatic relations with Qatar and announced a blockade and an embargo. Phone connections were cut. Populations were ordered home. They were joined in this campaign by Egypt and Bahrain. The Saudis and Emiratis said that Qatar was continuing to fund and host Islamic extremists and dally with Iran. But the biggest critiques were interconnected—what was said to be its support for the Muslim Brotherhood and other Islamists, and its efforts to undermine its neighbors through its Al Jazeera television network, both by providing a platform for extreme imams and by its aggressive and critical news reporting.

The stand-off has persisted. Qatar is a very rich country, and it quickly replaced the food it would normally receive from Saudi Arabia, including by airlifting in thousands of cows. Turkey, enhancing its regional role and Ottoman vocation, swiftly dispatched several thousand troops to Qatar and established a base there, which infuriated Saudi Arabia and the Emirates. Further infuriating them is Qatar's relation with Iran. As one Qatari explained, "Iran is a neighbor and we have to treat it like it is." Food imports from Iran surged, and Iran opened its air space to Qatar, which was now denied access to the air space of its Arab neighbors. Qatar, for which gas revenues are much more important than oil revenues, quit OPEC because of what it said was Saudi domination.

Qatar has held three valuable cards in this stand-off and blockade. One is the scale and global importance of its LNG exports. The second is its wealth. Its $350 billion sovereign wealth fund has invested strategically around the world, including acquiring a 25 percent stake in the parent of British Air. And the third is uniquely valuable—its strategic relationship with the United States. Qatar is home to a large U.S. Air Force base with 10,000 personnel, and crucial, as one U.S. Air Force general put it, "because of its tremendously strategic location, right in the center of everything." It is also the forward headquarters of the U.S. Central Command. This is one stand-off that the United States would very much like to see ended.[16]

ON THE ECONOMIC FRONT THINGS WERE DIFFERENT. IN THE spring of 2019, Saudi Aramco launched a $10 billion bond offering to help finance the acquisition of SABIC, the largely state-owned petrochemical company. The response in the international financial community was enthusiastic. It was vastly oversubscribed, about ten times over the actual offering. The offering had revealed something that was very significant—Saudi Aramco was the most profitable company in the world.

The success of that bond restored the momentum for something that would have been considered unimaginable even a few years earlier—the IPO of Saudi Aramco. In January 2016, MBS had offered a startling preview. In an interview with the *Economist*, he was asked, "Can you imagine selling shares in Saudi Aramco?"

"This is something that is being reviewed," he replied. "Personally, I'm enthusiastic."[17]

The global oil and financial communities were stunned. Until the prince's words, this possibility was considered absolutely off the table. It was a shock for Saudis too, for Aramco was so identified with and so fundamental to Saudi Arabia. With a potential enterprise value that the crown prince posited at $2 trillion, the potential magnitude of an Aramco privatization would exceed any previous IPO in history and could change the dynamics of the world oil industry.

Saudi Aramco is the engine of the Saudi economy and one of the most important institutions in the country. Wholly owned by the state since nationalization was completed in the 1980s, it is the world's largest oil company, and everything it does is on an enormous scale. With engineers and scientists trained at universities in the United States, Britain, and elsewhere, it is generally recognized as at the forefront of technology in its industry. It ranks among the top in terms of new patents issued to the world's oil companies. It takes a twenty- to twenty-five-year perspective in its major investment programs. Sitting atop proven reserves of 261.5 billion barrels of crude oil, it will be producing petroleum long after others are running dry.

Yet much would change with an IPO. Thereafter, its management would need to report on a quarterly basis to investors. Saudi Aramco would no longer solely be a NOC—a national oil company—representing the Saudi state. It would also be a financial asset. "Oil should be treated as an investment, nothing more, nothing less," the crown prince declared. Some of those who doubted this radical break, he said, were "close to the socialist communist approach, where everything is to be owned by the state, even the bakery."[18]

But why was the idea of an IPO of Aramco so startling? After all, IPOs of parts of state companies and privatizations, either complete or partial, had been a major feature of the global economy going back more than three decades. But no other company is so identified with a country and plays so large a role as does Aramco in Saudi Arabia.

Preparing a company of this scale for privatization was a complex undertaking that took a couple years. A whole new system of internal financial controls, accounting, and economic wiring had to be implemented. Many questions had to be answered, including about dividends and taxes and governance. And there was the fundamental question for investors—what, in a world of volatile oil prices, would be the valuation? How much was the company actually worth?

The proceeds of the IPO were slated to go into the Public Investment Fund (PIF), the sovereign wealth fund. And that, in turn, was part of a larger strategy—to turn the PIF into the world's largest sovereign wealth fund.

Riyadh could look next door to the sovereign wealth funds of Abu Dhabi. Or at Norway's sovereign wealth fund, with over a trillion dollars in assets. But Saudi Arabia usually produces about three times as much oil as Abu Dhabi and more than five times as much as Norway. And thus, said the crown prince, Saudi Arabia should have a sovereign wealth fund "larger than the largest fund on earth." In turn, the money in the PIF would help launch Vision 2030, and remake Saudi Arabia. MBS wanted to build a sovereign wealth fund that would be, he said, "a global investment powerhouse." The PIF can certainly provide the kingdom with significant non-oil revenue streams. But replacing oil revenues is a high mountain to climb, as the IMF pointed out.[19]

Where was the IPO to be launched? New York or London—or Tokyo, Hong Kong, or Singapore? This became the subject of much speculation. But finally the offering was launched on December 11, 2019, on the home court—the much smaller Riyadh stock exchange. Just 1.5 percent of the company was on offer. Still, it finally topped out at $29.4 billion, eclipsing the debut value of China's Alibaba to become the largest IPO in history. The shares surged and did touch $2 trillion on the second day of trading, making it the most valuable company in the world. The value has fluctuated since then with oil prices.

SUCCESS IN THE ECONOMIC TRANSFORMATION OF SAUDI Arabia would rewrite the domestic social contract and reshape the kingdom. But the 2020 price collapse that came with the coronavirus has constrained funding across the board, including for implementing Vision 2030. Ultimately, its impact, however it plays out, will be felt across the entire region. Altogether, the stakes around change are very high for Saudi Arabia and the Middle East, and for geopolitics and the world economy. There is yet more than that.

In the den of a large house on the shores of the Persian Gulf, a ruler in a neighboring country reflected on the necessity of reform in Saudi Arabia. "What is at stake," he said, as the shadows of night fell across the desert outside, "is not what people think. It is not about the oil, and not about the territory and geography of Saudi Arabia. It is about who controls Mecca and Medina because of what that means for Muslims around the world."

Chapter 36

THE PLAGUE

I n October 2019, Vladimir Putin arrived in Riyadh for a state visit, reciprocating King Salman's visit to Moscow two years earlier. The Russian president's limousine was guided into the city by an elegantly attired horse guard carrying Saudi and Russian flags. Knowing that falconry is a favored sport of Saudi royals, he brought as a gift a rare Arctic gyrfalcon the largest of all falcons, which the king warmly received. The trip was an opportunity for signing ceremonies for what were to be deals worth billions of dollars. It also affirmed that the new strategic relationship between the two countries was taking wing. But that thought would be tested sooner and in a way that no one could have anticipated.

Two months later, in December 2019, a puzzling new respiratory illness was detected in the city of Wuhan, China. Doctors saw a resemblance to the SARs epidemic that had begun in 2002. One doctor reported that the "exact virus strain" had yet to be identified but cautioned his colleagues to take special care when seeing patients. It took another three weeks before it became evident that the new virus was spreading exponentially in Wuhan and the surrounding province of Hubei. Its official name was SARS-CoV-2, and the illness it caused, COVID-19; but it became known as the novel coronavirus, owing to the fact that there are many coronaviruses, including one responsible for the common cold.

By late January, it was a full-blown epidemic and the Chinese government imposed a lockdown on Wuhan and other cities, quarantining people in their homes. The consequences were enormous for the economy and for the energy sector. If people were not driving or flying, they were not burning oil. Chinese consumption plummeted. When the SARS epidemic had begun in 2002, China accounted for only 4 percent of the world economy, and the impact on the global oil market was negligible. But now China accounted for 16 percent, and the impact was global; for China not only had become the world's second largest oil consumer, but it also had accounted for half the total growth in world oil demand. With the shutdown in China, world oil consumption, instead of increasing as would normally have been the case, was dropping precipitously as never before—by eight million barrels per day in the first quarter of 2020.

IT WAS THIS DECLINE THAT PRECIPITATED, IN THE FIRST WEEK of March, a meeting of OPEC-Plus—OPEC and its non-OPEC partners, twenty-three countries in all—in Vienna to address what was turning into, by far, the biggest decline in consumption ever recorded.

While the countries coming to Vienna knew that the situation was bad, they did not know just how bad, nor how much worse it could get. By then, however, the common interest that the two leading countries of the group—Saudi Arabia and Russia—had forged over the last few years was unraveling. The Russian budget was pegged at $42 a barrel, the Saudi budget at $65, and, according to the IMF, Saudi Arabia needed $80 or more to balance its budget. Moreover, the Russians had seen the 2016 OPEC-Plus deal as temporary and expedient; the Saudis wanted to make it permanent and keep Russia in it.

The Saudi energy minister, Prince Abdulaziz bin Salman, sought new cuts that would be deeper, and then insisted strongly on even deeper cuts. The Russian energy minister, Alexander Novak, just as strongly resisted. He wanted to extend the existing deal and not make any further cuts for a few weeks to see the impact as the coronavirus advanced. On the morning of March 6, Novak flew into Vienna from Moscow and went to the OPEC headquarters. There, in a small fifth-floor conference room, he met privately with Abdulaziz. There was no meeting of the minds. They descended stone-faced to the first floor for the official joint meeting of OPEC and non-OPEC ministers. It was an impasse. The meeting broke up with no agreement.

"We will all regret this day," Saudi minister Abdulaziz said on the way out. Asked what Saudi Arabia would now do, he replied, "We will keep you wondering." The OPEC countries "didn't consider any other variants," said Russian minister Novak. And now, he added, since there was no agreement, all countries were free to produce whatever they wanted. An effort at calming words was made by Suhail al Mazrouei, the United Arab Emirates' petroleum minister. "They need more time to think about it," he said. But OPEC-Plus had blown up.

The failure in Vienna shocked the global oil market, with reverberations in financial markets. Saudi Arabia wasted no time in ending the "wondering" by announcing that it was going to go all-out, increasing production from 9.7 to 12.3 million barrels per day over the next month. "Increasing production when demand is falling," said Novak, trained as an economist, "is irrational from the economic theory point of view." Russia had nowhere near that extra production capacity but said it would increase as much as it could.[1]

The comity going back to 2016 was gone—in its place a price war and a battle for market share. The would-be partners had once again become fierce competitors. Some in Moscow, who had opposed a deal to restrain production, welcomed the breakdown. "If you give up market, you will never get it back," said Igor Sechin, the CEO of Rosneft and the biggest Russian critic of OPEC-Plus from the beginning. Those like Sechin opposed to any deal had been particularly loath to give up market share to the United States. In the four years that Russia had been part of the agreement and its production constrained, U.S. oil output had increased 60 percent, catapulting the United States into the number one position. Beyond markets, they regarded U.S. shale as a "strategic threat." For they saw the abundance of shale oil and gas as an adjunct to U.S. foreign policy, giving the U.S. a free hand to impose sanctions on the Russian energy sector, as it had done only a few months earlier, in forcing a halt to the almost-completed Nord Stream 2 pipeline. U.S. shale, they expected, would inevitably be a major casualty of a price war, owing to its higher costs and the constant drilling it required, compared to Saudi and Russian conventional oil.[2]

YET WHAT WAS NOT FULLY UNDERSTOOD AT THE BEGINNING of March was that this battle for market share was being launched into a market that was rapidly shrinking owing to the virus. The epidemic in China was turning into a global pandemic.

Sixteen years earlier, in 2004, the National Intelligence Council, a research organization in the U.S. intelligence community, had published a report titled *Mapping the Global Future*, which presented scenarios for the year 2020. One of the scenarios imagined was a pandemic in 2020. It was eerily prophetic, even as to the year:

> It is only a matter of time before a new pandemic appears, such as the 1918–1919 influenza virus that killed an estimated twenty million worldwide. Such a pandemic in megacities of the developing world . . . would be devastating and could spread rapidly throughout the world. Globalization would be endangered if

the deaths rose into the millions in several major countries and the spread of disease put a halt to global travel and trade during an extended period, prompting governments to expend enormous resources.

In 2015, Bill Gates, who was devoting much of his energies to health philanthropy, had warned that there was "great risk of global catastrophe" from a "highly infectious virus." He continued, "We're not ready for the next epidemic," and yet the costs of preparedness were "very modest compared to the potential harm." Perhaps a complacency developed—not among people focused on infectious diseases, but more generally after the relative success and limited numbers affected—in controlling SARS (8,098 people became sick, 774 died), MERS (Middle East Respiratory Syndrome) (2,494 cases, 858 deaths), and Ebola 2014–2016 (11,325 deaths).[3]

Only at the beginning of March did it become apparent that this coronavirus was far more transmissible. On March 6, the day OPEC-Plus broke up, 101,000 people were infected by COVID-19 worldwide, and more cases were emerging in Europe and the United States. "No handshake policy" was turning into "social distancing." Offices emptied, businesses shuttered, schools and restaurants closed, conferences were canceled, airports emptied, travel stopped, and people were told to stay home. The unprecedented drop in oil demand during the first three months of the year—eight million barrels per day—would pale in comparison to what ensued in the months that followed.

The oil war quickly became acrimonious. King Salman tried to reach President Putin by phone, but with no success. The Russians explained no disrespect was intended, but that Putin had been tied up in a six-hour meeting with Turkish president Recep Tayyip Erdoğan over the conflict in Syria.[4]

In March, China was just starting to come back to life, but other countries were closing down fast, one after another. Streets around the world were empty save for an occasional vehicle. The world almost seemed a different planet at the end of March from what it had been at the beginning of March.

This is when consumption really collapsed. Such a sudden massive decline in oil demand has never been seen before. Like so many other sectors of the world economy, the oil and gas business entered a deep crisis. In the United States, oil companies were quickly slashing budgets. But the market was moving even faster. By late March, the price, which had been $63 less than three months earlier, fell to as low as $14 a barrel. Some oil sold for less; and in Canada, where access to the market was constricted, some barrels went for single digits. "The situation is dire," said a senior U.S. official, "and it's getting worse."

Ironically, the peak of U.S. production had been reached in February 2020—thirteen million barrels per day. But in March it was clear that, with drilling rigs laid down and budgets cut, U.S. oil output was going to decline. The travails of the oil industry and the lost jobs were a major setback to the U.S. economy. The Dallas Federal Reserve warned that the price collapse had "weakened the U.S. economy" and "overall" reduced investment spending, not just in the energy sector. It was also a shock to the debt market and broader financial markets and to the industrial Midwest, which was a major supplier of equipment to the oil and gas industry.

In the United States, gasoline consumption was down about 50 percent; in Europe, by 65 percent. But nothing could be done about the collapse of demand until governments in the United States and Europe lifted their lockdowns. But, with the disease still advancing, that was not going to happen any time soon. On March 16, cases worldwide had reached 181,000; in the United States, 6,000. On that same day, March 16, thirteen U.S. Senators from oil and gas producing states, including the chairman of the Armed Services Committee, wrote to Crown Prince Mohammed bin Salman expressing dismay at what they described as Saudi policy "to lower crude prices and boost output." A few days later, when oil was heading toward $14, several of them sent a second letter warning it "will be difficult to preserve" the U.S.-Saudi defense relationship if "hardship" continues to be "intentionally inflicted" on U.S. oil and gas producers.

Nine of the senators got on a phone call with Saudi ambassador Reema bint Bandar Al Saud, a cousin of the crown prince who had gone to university in the United States when her father had served as ambassador. The senators did not spare any words—Saudi Arabia was waging "economic warfare" against the United States. The Saudi ambassador insisted that the problem was Russia and its refusal to accept the Saudi proposal in Vienna. The Senators pushed back. "Let me take you through the math in the Senate," one senator said. The math, he said, was how votes worked in the Senate regarding the U.S. military relationship with Saudi Arabia. The signatories of the letter were the key supporters of the military relationship with the kingdom. If the price war continued, their support would disappear.[5]

"I WAS ALWAYS FOR THE PERSON DRIVING THE CAR AND FILL-ing up the tank of gas," said President Trump on March 19. But a few days later he said, "I never thought I'd be saying that we may have to have an oil price increase because the price is so low." Low gasoline prices would not do much for motorists when people were not driving. Now it was a matter of national security and ensuring that a strategic industry was not, as Trump put it, "wiped out." Otherwise, the vaunted "energy dominance," which the Trump administration had championed, and which had provided foreign policy flexibility, would be gone. And there was the matter of politics in a presidential election year. Texas has thirty-eight electoral votes, second only to California, and almost as many as Pennsylvania and Illinois combined.

Trump began doing what he had done his entire career: working the phones, this time in a round-robin with King Salman, MBS,

Vladimir Putin, and other leaders. The dealmaker was now going for a mega-deal. Given what were described as the "irreconcilable differences" that had led to the breakup in Vienna between Saudi Arabia and Russia, it was also something like divorce mediation. Over two weeks or so, Trump talked with Putin more than in the entire year previous. On April 1, Saudi production rose to twelve million barrels per day. Some of the phone calls were very direct. Mention was made of those thirteen senators. After one such call, Trump tweeted, "Just spoke to my friend MBS of Saudi Arabia, who spoke with President Putin of Russia & I expect & hope that they will be cutting back approximately 10 Million Barrels and maybe substantially more." Shortly after, he raised the ante to fifteen million.[6]

Given the oil war and the animosity, his numbers were greeted with skepticism. But the wheels were grinding. Saudi Arabia called for an urgent meeting of producers—"in appreciation of the request of the President of the United States Donald Trump." On April 3, Putin told a video conference that Russia, as well as Saudi Arabia and the United States, were "all interested in joint . . . well-coordinated actions for ensuring the long-term stability of the market." He said that the price collapse was caused by the coronavirus, but pointedly added, also by "attempts by our Saudi partners to eliminate competitors that are producing so-called shale oil.[7]

But how could a deal be made? The Russians and Saudis argued that, if they cut back, they expected the United States to do the same. In other countries, the national government can order a cutback, but in the U.S. system the president does not have that authority. Individual states regulate production within their borders. Still, it took some explaining of America's federal system to get that point across. But the administration emphasized that economics and the marketplace—low prices—would enforce cutbacks in the United States. Shale oil was short cycle. That meant that, while it could go up quickly, it could also—if drilling did not continue—go down fast. And, at these prices, there was not much money for new drilling.

The fundamentals of the marketplace were already forcing a retreat from the oil war. There was no way to win the battle for market share when oil demand was disappearing. Countries were stepping up their production, but they could not sell all their barrels. Buyers were going on strike. If the oil could not go to consumers, it had to go somewhere, and that meant storage tanks. But, worldwide, storage was filling up fast, and not only on land. Every available oceangoing tanker was being chartered, not to transport oil but rather as a floating storage tank. "The supply and demand fundamentals are horrifying," said OPEC secretary general Mohammad Barkindo. And it was clear that the clock was ticking. By the end of April, or May at the latest, the world was going to run out of the last crannies of storage. And when that happened, prices would crash. In some parts of the world where storage was not available, producers could even face what had seemed unthinkable—"negative prices." They would have to pay customers to take their oil away.[8]

Such a collapse would deliver a tremendous blow to all the exporters' economies—dealing a huge hit to the funding for Saudi Arabia's Vision 2030, and to the Russian budget just when Vladimir Putin was pushing the constitutional revision that would enable him to remain president until 2036.

"I hated OPEC," Trump said, but now he needed OPEC to help get a deal done. What followed over the next several days was a dizzying whirl of day and night phone calls and video conferences among OPEC and OPEC-Plus. Of course, the United States was not a member of either, but it was a member of the G20, along with other producers like Canada and Brazil, and major importing countries like Germany and Japan, that also wanted stability. The G20 was useful now because it brought other major countries into the discussion and, specifically, put the United States into the same forum with Russia. Moreover, Saudi Arabia was the G20 chair in 2020 and thus was greatly invested in its success. And a deal was taking place in this new multitudinous configuration. But it was clear who was driving it. It was the Big Three—the United States, Saudi Arabia, and Russia—and, in particular, the United States.

On April 10, the energy ministers of the G20 assembled. "We must stabilize world energy markets," said U.S. Energy Secretary Dan Brouillette. "This is the time for all nations to seriously examine what each can do to correct the supply/demand imbalance." By then, everything was more or less in place for a grand bargain. Except one member of OPEC-Plus was holding out. Mexican president López Obrador did not want anything to do with the deal. He had his own politics; he was committed to Pemex, the national oil company, increasing production, not cutting it—even if, in fact, its actual production was in decline. More nighttime phone calls ensued, and an understanding was worked out with Mexico. That was followed by the conference call with Trump, Putin, and King Salman that sealed the deal.

The total OPEC+ deal was for 9.7 million barrels a day reduction, of which Russia and Saudi Arabia would each contribute 2.5 million barrels. Now they were on absolute parity—an agreed baseline of eleven million barrels a day each, which would go down for each to 8.5 million barrels. The other twenty-one members of OPEC-Plus agreed to their own cutbacks. So did other major non-OPEC producers that were not part of OPEC-Plus—Brazil, Canada, and Norway. But these reductions would include declines driven by economics, and those were already occurring.

The deal itself was historic, both for the number of participants and the sheer complexity. It was the largest oil supply cut in history. Nothing like this had ever happened before in the world of oil, and certainly not with the United States at the center of it.[9]

After the deal was done, Prince Abdulaziz described the oil war as "an unwelcome departure" from Saudi policy. But, he said, "We had to because of a desire to capture some revenues versus sitting on our hands and doing nothing." And the "mediation" from Washington had helped, for it had ended the rift with Russia, at least for the time being. "We don't need divorce lawyers yet," the prince said with some relief.[10]

THE AGREEMENT HAD SIGNALED A NEW INTERNATIONAL order for petroleum, one shaped not by OPEC and non-OPEC, but by the United States, Saudi Arabia, and Russia. In the future, markets would shift; it would be a different planet again after the coronavirus; politics and prices and personalities would change over the months and years ahead. But the sheer scale of their resources, and the dramatically changed position of the United States, guaranteed that these three countries, one way or the other, would have dominant roles in shaping the new oil order.

The deal was indeed historic, but it turned out to be not enough, not when measured against the ever-deepening collapse in demand—twenty-seven-million barrels down in April, more than a quarter of total world demand. After the deal, prices slid into the high teens and, in some places where oil could not be stored or transported, a lot lower. The world was now running out of storage. Owing to an anomaly in the way the futures market worked, the price dropped to one cent and then, on April 20, went "negative." That meant that a financial investor selling a futures contract, who would be obligated to take physical delivery of oil for which they had no storage place, had on that day to actually pay a buyer to take the oil. That, too, was historic—the lowest price ever recorded for a barrel of oil—minus $37.63. But that was not a price in the oil field, but a one-time fluke in financial markets, an aberration in a futures contract.[11]

Meanwhile, the global calamity continued. On May 1, coronavirus cases in the world exceeded 3.2 million, with more than one million in the United States, where more than twenty-five million people had lost their jobs over five weeks. The IMF, which at the beginning of the year had predicted solid global growth of 3.4 percent, announced that the world had already entered the worst recession since the Great Depression.

May 1 was also the day that the mega-oil deal, the OPEC-Plus agreement, went into effect; and Saudi Arabia and Russia and the other producers began to sharply reduce production. At the same time, the brute force of economics was forcing companies to curtail output or shut down wells altogether. Why sell oil for less than it cost to produce—assuming you could find a buyer or storage—when you could, in effect, store it in the ground—allow the oil to "shelter in place"—and wait for prices to recover. The biggest market-driven curtailments by far were in the United States, followed by Canada. In May the global combination of OPEC-Plus cuts and market curtailments took thirteen million barrels per day of crude oil off the world market. The planned spending by the larger U.S. oil upstream companies was slashed in half, meaning many fewer wells would be drilled in the months to follow, ensuring that U.S. production would slide significantly over the next year. The United States would certainly remain one of the Big Three, but not as big.

By the beginning of June, the number of coronavirus cases worldwide was over six million, more than double what it had been a month earlier. Yet the economic darkness was beginning to lift. China, the first country to lock down, was the first to unlock, and it was mostly back in business. European countries were at different levels of increased activity, and the United States was opening up in stages, albeit with considerable variation among states. With economies coming back, oil demand was increasing. Consumption in China was almost back to pre-crisis levels, and the streets in Beijing and Shanghai and

Chongqing were once again gridlocked as people who had the option chose to drive rather than take public transportation. Gasoline consumption in the United States, which had fallen by half at the beginning of April, was now growing again. All this pulled oil prices back up higher—to levels that not so long ago would have been considered a low-price scenario, but now a relief.

With prices rising, would OPEC-Plus stay together and the cutbacks hold? Key would be the restored relationship between Saudi Arabia and Russia. But also of importance would be how quickly. U.S. producers, who had shut down their wells, would turn around and open them again, which could renew the oversupply and deliver another blow to prices, as could low economic growth or a persisting recession—or a resurgent virus.

And there were many perspectives on what lay ahead. Looking beyond the crisis, some thought that market cycles were over and that, even with economic recovery, oil prices would be low for a long time. Others though otherwise—more likely that the slashing of investment in new production would lead, with renewed economic growth, to a tightening in the balance between supply and demand that would send prices higher. And some thought entirely differently. They sought a "green recovery"—governments' taking advantage of the crisis to reorient their energy mix away from oil and gas and hasten what they saw as the coming energy transition.

ROADMAP

Chapter 37

THE ELECTRIC CHARGE

The lunch in a Los Angeles seafood restaurant in 2003 was not going well. Two engineers, J. B. Straubel and Harold Rosen, were pitching Elon Musk. An entrepreneur of iron-man determination, Musk was already known as one of the original members of the "PayPal Mafia," who had launched the online payment system, and then as the founder of SpaceX, which was aiming to undercut the government's cost for space transportation and pave the interplanetary path for travel to Mars. The engineers were pitching Musk on something that would operate at a lower altitude—an electric airplane.

"That's not going to work," interrupted Musk. "I have zero interest."

An awkward silence ensued. They all went back to poking at their fish. Then Straubel thought, "Why not?" He might as well tell Musk about his obsession—the electric car.

"It's amazing how lithium batteries have really advanced what electric cars can achieve," said Straubel. He explained his idea of "stringing together about ten thousand laptop battery cells and jamming them into a car."[1]

Musk became excited—much more excited than anyone else with whom Straubel had ever shared his idea.

At the time, for anyone other than a small coterie of enthusiasts, an electric car would have been a fanciful idea. It was widely agreed that transportation was the one market in which oil was firmly entrenched, specifically in autos.

That was then. But today the electric vehicle (EV) has become an existential question for the global automobile industry, which is rushing to make sure it can ensure its future. And equally so for the world oil industry, which for the first time in a century faces a potentially serious competitor for autos and "light trucks," which represent 35 percent of world oil demand (cars alone are 20 percent). Will the power that moves people still emanate from oil wells, or will it be from electric power lines? The answer will affect how billions of people move around and will have a profound impact on geopolitics, on jobs, on national economies, on the global economy—and the vast flows of money within it. Climate policies, more than anything else, are today providing the great motive force for the adoption of electric vehicles. Cars generate about 6 percent of energy-related CO_2 emissions. But back in 2003, at the time of that lunch, the rationale was not climate. It was electric cars for the sake of electric cars.

The lunch with Musk certainly ended on a much more positive note than it had begun. He even picked up the check. And a few weeks later, after more meetings, he wrote a larger check—to launch the venture.

Straubel had been gestating the idea for a long time. "I love batteries, I love power electronics and motors," he said. At age thirteen, he built a sort of souped-up electric golf cart. At Stanford University, he invented his own major—energy systems engineering. He was the major's only student. After college, he fooled around with various ideas for solar-powered and electric cars.

But now, with Musk's backing, Straubel had his main chance. Out of the Silicon Valley ecosystem, he pulled together a team. Still, however great the passion, the enterprise looked irrational. No new car company had been started in the United States since 1925.

Yes, advocates of the battery-powered electric car could point to a long paternity. In 1900, electrics far outnumbered gasoline cars on the streets in New York City. No one was a more powerful advocate of the electric car than the great inventor Thomas Edison, who poured a lot of his own money, along with his reputation and effort, into trying to perfect an electric vehicle.

But two things killed that first generation of electric cars. One was Henry Ford's Model T and the mass production of the assembly line. The other, though less well known, was the electric starter, invented by Charles Kettering in 1911 for Cadillac after a person died from cranking a car. Kettering's invention eliminated the need for someone to stand in front and crank. Over the next several years, electric cars faded away.

The beginning of the modern era of electric cars was in the 1990s in California. The reason was smog. The California Air Resources Board—appropriately known as CARB—is a little-known regulatory agency but one with outsized worldwide impact because of the size of the California auto market and because CARB's emission standards are followed by several other states. In the 1990s, CARB began to require, in order to reduce smog, that a certain number of new cars sold in the state had to be "zero-emissions vehicles"—ZEVs. The only way to achieve that was either through electricity or hydrogen. But the regulations generated much controversy, owing to the fact that such vehicles did not exist.

General Motors did try, spending a billion dollars to develop the two-seater electric EV1. Introduced in 1996, the vehicle was not exactly compelling—owing to its shape, it became known as the "Egg on Wheels"—and its range was limited. Aside from a handful of aficionados, the EV1 failed to gain traction and ended up in the junkyard. The battery just wasn't good enough. Moreover, how many people really wanted a gasoline-free car when gasoline was at that time only $1.30 a gallon?

Yet now, just a few years later, a group of geeks in Northern California was going to try to reinvent the electric car. They called their company Tesla after electricity pioneer Nikola Tesla. The goal was to develop a high-performance "cool brand" that people would feel excited to drive. No more eggs on wheels.

Despite the formidable obstacles, the Tesla project had a number of positives. One was the advance in batteries using the element lithium. The lithium-ion battery was first invented in an Exxon laboratory in the mid-1970s, during a time when it was thought that the world would run out of oil and Exxon would need to find another way to stay in the mobility business. Then oil prices collapsed in a glutted market, and the incentive disappeared. Over the next decade the battery was improved, and in the early 1990s, Sony commercialized the batteries, which became the power source for laptop computers and cell phones. "We had this harebrained idea that we could use what were then basically only laptop cells and power a car," said Straubel. But the lithium battery did have four times the energy density of the conventional lead-acid battery. Tesla's idea was to package thousands of these cells into a battery pack. By this time, the door had already opened to the use of electricity in autos, owing to the uptake of hybrids, which combined an electric motor with a gasoline engine, stretching out the miles per gallon.[2]

Meanwhile, interest was growing in "clean tech" and renewables among Silicon Valley venture capital firms, in response to rising concern about climate change and the conviction that much money could potentially be made from them. That made it easier for Tesla to raise investment. The rapid rise in oil prices in those years reinforced the fears of "peak oil"—running out of petroleum—which made non-gasoline-powered transportation look increasingly interesting.

And of critical importance were government policies—incentives, subsidies, and regulations on fuel economy and emissions. In California, CARB increased the targets for zero-emission vehicles and, with climate moving up the agenda, introduced limits on CO_2 emissions. If automakers could not make the targets, they had to buy credits from companies that did—which meant that established auto companies would eventually end up paying a tithe to Tesla. As gasoline prices surged, the federal government offered tax credits to consumers for buying electric cars and for installing electric chargers. So did a number of states.

Ironically, the start-up also benefitted from the fact that it was not taken very seriously. "It was almost a decade since everyone had looked at it and said, 'Nope it doesn't work,'" said Straubel. "And they put on blinders and forgot about it. That really gave us a huge window of opportunity."

Even so, at the beginning, Tesla seemed wholly impractical. Many times the company was within months of running out of money. "We had to keep going right up to the brink to try and hit milestones and demonstrate things that would convince people to give us enough money," said Straubel. The stress was further amplified by personnel conflicts and feuds within Tesla.

Short of funds, they found that they could bring down the cost of solving myriad design issues by simulating and testing them on computers. Some things, however, had to be done in the real world. They could not afford a multimillion-dollar crash-test facility. Yet they needed to test battery safety. "We realized we could basically simulate this if we just held things up in the air and dropped them," said Straubel. And so they rented a crane and dropped batteries. "It was really a shoestring way of making it work," said Straubel. "We were innovating at an amazing pace with a very small team with few resources." Every system needed to be redesigned and then redesigned again.[3]

The world finally saw what Tesla was up to when the prototype Roadsters were revealed in 2006—two exciting sports cars, one black and one red. "Until today," Musk announced, "all electric cars have sucked." But that was only a teaser. The first cars were not actually delivered until 2008. The Tesla Roadster was an arousing, dashing sports car that could go from zero to 60 in under four seconds. It quickly became an iconic status symbol. The Roadster may have been described as a "limited edition," but it demonstrated that the lithium-ion battery pack would work in an automobile. "I basically said, 'Now, wait a minute,'" recalled the then–vice chairman of GM, Robert Lutz. "I've accepted everybody's arguments of why we can't do this. But here's this small start-up company in California" that was doing it. Tesla, he added, was "the crowbar that helped break up the logjam."[4]

Starting at $109,000, the Roadster was not exactly a mass-market car. And it was not the only new electric vehicle. In 2010, Chevrolet came out with the Volt, a plug-in hybrid, and Nissan introduced the all-electric Leaf. Sales, however, were disappointing.

Tesla leapfrogged in 2012 with the Model S sedan, which started at about $65,000 and went up from there. *Motor Trend* made it "car of the year." An even greater honor was that virtually every major automaker was buying a Model S to tear it down to understand how it worked.

The year 2012 was certainly a big one for Musk. It was not only the Model S. It was also SpaceX. His Falcon rocket had successfully rendezvoused with the International Space Station and delivered its cargo and then had been recovered with a return cargo. These were both industrial accomplishments of enormous scale, both the result of Musk's unusual drive and determination. "My drive is sort of disconnected from hope, enthusiasm, or anything else," he once said. "I just give it everything I've got—irrespective of what the circumstances may be. You just keep going and get it done."[5]

Still, the Model S was pricey—definitely not a people's car. Next on the road, in 2016, was the Model 3. This aimed at the mass market—$35,000. By 2016, Renault-Nissan had sold half a million electric cars—three times as many as Tesla, but only a third of what that company had forecast. By then, however, Tesla and Renault-Nissan were no longer lonely outsiders. Indeed, the on-ramp to electric mobility was starting to get filled with new vehicles and promises of multitudes more to come.

It was in the mid-2000s that GM, under then-CEO Rick Wagoner, had reengaged with EVs and launched the Volt. But then, with the 2008 financial crisis impending, the economic outlook was darkening quickly, casting a deep shadow over everything, even the humor. To a visiting speaker at a management meeting, Wagoner joked that they used to give speakers a car as an expression of gratitude, but now the company could only afford a pen. Still, it was a nice pen.

Then came the full fury of the 2008 global financial crisis: the factories shut, the layoffs, bankruptcy, and the massive U.S. government bailout. Daniel Akerson, who succeeded Wagoner as CEO, faced a formidable challenge in reviving the company. As he put it, "We had a monkey on our backs to deliver." That didn't leave a lot of time for reinventing the business.[6]

But in 2013, General Motors committed to the development of a wholly new electric car, the Bolt. The most challenging aspect of the Bolt would be its heart—the battery. Batteries can be particularly sensitive to changes in temperatures, and the designs were subjected to grueling simulations ranging from 85 degrees below zero to 185 above. The goal was to bring down the battery costs so that the Bolt would be competitive in the mass market. And the battery had to be able to deliver the all-important "200-mile range" on a single charge, which had become the benchmark for allaying consumers' "range anxiety"—the fear of running out of charge with no recharger nearby.

The person in charge of GM's global product development at the time was, for the first time, a woman—Mary Barra. And the Bolt was her top priority.

Barra was a rebel from within. Her father had been a die maker in a GM plant. He was, she once said, a "car buff," and so was she. As a child, she would visit the plant with him and walk around, fascinated by the factory floor. When her father drove a new model home, it was, she said, "a neighborhood event." She and the other kids would climb through the car and take turns getting rides in it. When the new models were launched in the autumn, she would go down to the dealership and watch excitedly as the paper was pulled off to reveal the gleaming arrivals. There was no question—she was going into the auto industry.

Barra first went to GM's technical institute—"I had a love of science and math," she said—and then to Stanford Graduate School of Business. She rose through the company, from plant manager to head of human resources and, eventually, head of global product development. Delivering the Bolt was her responsibility. At a key decision point, her team presented two options for the all-critical battery: On one timeline, they could get a certain distance on a single charge. Or they could get twice the range, but that would take longer.

"Let's get twice the range," Barra replied. Then she added, "But in the original timeline."

Along with the rest of team, the head of electric vehicles was stunned. "I'll never forget the look on her face," recalled Barra.[7]

Once they had recovered, however, the team executed, and the Bolt rolled into showrooms in December 2016, beating Tesla by seven months with a moderately priced model that could go two hundred miles on a single charge.

But it was only after Barra became CEO of GM that the company had to confront head-on what was coming to be seen as a revolution in mobility. Much effort over the next few years went into grappling with the external trends that would shape the industry—regulatory and public policy, climate and environment, changing consumer demands and needs, new business models, an onrushing flood of new and converging technologies—and the emerging challenge from Silicon Valley. And "this was only the beginning," Barra later recalled. By 2023, she pledged, GM would have at least twenty electric models on the market.[8]

Mark Sykes, left, the "Mad Mullah," was Britain's Middle East expert during World War I. He and French diplomat François Georges-Picot, right, drew a map for the postwar Middle East following the collapse of the Ottoman Empire. Contentious ever since, the map was the basis for the modern nation-state system of the Middle East.

An ISIS jihadist in a 2014 propaganda video declares, "We've broken Sykes-Picot," as a bulldozer erases the border between Iraq and Syria. The U.S. defense secretary said ISIS's lightning offensive across Iraq was "beyond anything we've seen."

In 1918, chemist and Zionist leader Chaim Weizmann, left, traveled into the Arabian Desert to meet Prince Faisal, right, son of the sharif of Mecca, to discuss a Jewish "homeland" in Palestine and an Arab "nation." Faisal was as "handsome as a picture," said Weizmann.

T. E. Lawrence went out to the Middle East in 1910 to work as an archeologist. As a British intelligence officer during World War I, he helped organize the Arab Revolt against their Ottoman rulers. He also became "Lawrence of Arabia."

Hasan al-Banna, an Egyptian schoolteacher, founded the Muslim Brotherhood in 1928 to rescue Muslims from "exploitation" and "humiliation." He said, "The answer is Islam."

While in an Egyptian jail, Sayyid Qutb, militant leader of the Muslim Brotherhood, wrote *Milestones*, which advocated "violent holy war" and has influenced generations of jihadists.

In July 2014, in the Grand Mosque of Mosul in Iraq, ISIS leader Abu Bakr al-Baghdadi proclaimed a new caliphate with himself as caliph. "You will conquer Rome and own the world," he declared to his followers.

U.S. secretary of state John Kerry has a word with Iranian foreign minister Mohammad Zarif (Ph.D. from University of Denver) during negotiations over Iran's nuclear program in 2015. Looking on are U.S. energy secretary Ernest Moniz, on the left, and, on the right, Ali Akbar Salehi (Ph.D. from MIT), head of Iran's atomic energy agency.

General Qassem Soleimani, commander of the Quds ("Jerusalem") Force, the international arm of Iran's Revolutionary Guard Corps, pulled the strings in Iraq and masterminded the "axis of resistance" across the Middle East. The battlefield, he said, was "another kind of paradise." He was killed by an American drone in 2020.

Houthi militants in Yemen cheer the launch of a ballistic missile aimed at Saudi Arabia. The Yemen civil war led to a huge humanitarian crisis.

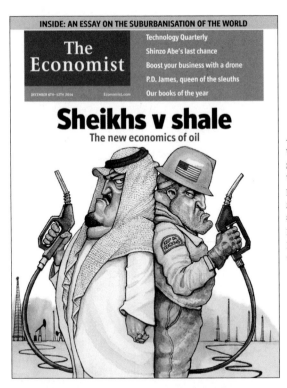

INSIDE: AN ESSAY ON THE SUBURBANISATION OF THE WORLD

The Economist

Technology Quarterly

Shinzo Abe's last chance

Boost your business with a drone

P.D. James, queen of the sleuths

Our books of the year

DECEMBER 6TH–12TH 2014   Economist.com

Sheikhs v shale
The new economics of oil

A surge in U.S. shale oil—the biggest and fastest growth ever registered in the history of oil production—created a global surplus that set the stage for the 2014–15 oil price collapse. Both U.S. oil producers and oil-exporting countries were hard hit.

The failure of OPEC and non-OPEC countries to agree on production cutbacks triggered the 2014–15 oil price collapse. "We left it to the market," said Saudi petroleum minister Ali Al-Naimi.

Iran's supreme leader, Ali Khamenei, embraces Syria's Bashar al-Assad in Tehran. Iran's Revolutionary Guards and Shiite militias secured the survival of Assad's regime in the Syrian civil war, while over half the population has either been killed, displaced, or fled the country.

Iranian missiles and drones hit Saudi oil processing facilities at Abqaiq and Khurais in September 2019, causing the biggest disruption ever in world oil supplies. But panic in the oil market proved very short-lived, as Saudi Aramco quickly repaired facilities. And U.S. shale had changed the world oil market's psychology.

Israel's giant Leviathan natural gas field, started up at the end of 2019, marked the rise of the "Eastern Med" as a new energy province and possible exporter to Europe—and is critical to ending Israel's almost total dependence on energy imports.

King Abdul Aziz was elated when told in 1938 that oil had been discovered in Saudi Arabia, making it the twenty-second-largest producer in the world. Some people had insisted that oil "would never have been found," the king said. "But I always thought they were wrong."

"The era of 1979 is over," said Saudi crown prince Mohammed bin Salman in 2017. He launched the ambitious Vision 2030 to modernize the country and reduce its dependence on oil. Here he sits below a photo of his grandfather King Abdul Aziz.

Saudi King Salman examines an Arctic falcon, the largest of all falcons, a gift from Russian president Vladimir Putin on a 2019 state visit to Saudi Arabia.

Thomas Edison championed the electric car but it never gained traction. A century later, below, Elon Musk introduced the Tesla Roadster, saying, "Until today all electric cars have sucked."

Sent to the countryside as a laborer during China's Cultural Revolution in the 1960s, Wan Gang took apart tractor engines. Years later, as minister of science and technology, he seized on the "strategic window" for the electric car. Today, half the global market for electric cars is in China.

"People are still physically going to need to move from Point A to Point B," says Mary Barra, chairman and CEO of General Motors. "But they're going to have multiple ways to do that." Her goal is "zero emissions, zero crashes, and zero congestion."

Sebastian Thrun and his team celebrate victory by "Stanley," their self-driving Volkswagen, over 132 miles of Nevada desert in the 2005 Grand Challenge. "The first time ever," said Thrun, "that the machine made all the decisions."

Garrett Camp and Travis Kalanick rode up the Eiffel Tower during a Paris snowstorm and hammered out the idea for a "better cab"—a ride-hailing company based on the smartphone. It would become Uber.

Bill Ford, executive chairman of Ford Motor Company, stands next to a Model T built by his great-grandfather at the one-hundredth anniversary of the company. "I'd like Ford to be around another one hundred years," he says. "You'll live in a world where you'll have internal combustion engines, plug-in hybrids, and pure electrics. Over time, the shift will take place."

A road trip to Tibet, crossing a fifteen-thousand-foot elevation, convinced Jean Liu that she made the right choice in leaving investment banking to join a start-up called DiDi. She is now president of DiDi—which means "beep-beep" in Chinese and is the largest ride-hailing company in the world.

A key moment for the first "energy transition" was in 1709, when Abraham Darby, a metalworker in the English village of Coalbrookdale, replaced wood with coal for making iron and "paved the way for the Industrial Revolution."

The Paris climate compact, announced to cheers in December 2015, aims to prevent temperatures from rising more than two degrees from pre-industrial levels. Now there are two eras for energy and climate—"Before Paris" and "After Paris."

Representative Alexandria Ocasio-Cortez and Senator Ed Markey unveil the "Green New Deal" in 2019 that would use "massive federal investment" to eliminate oil, gas, and coal and make the United States carbon free by 2030.

In August 2018, Greta Thunberg, then aged fifteen, skipped school to protest on climate. In little more than a year, she became a global climate phenomenon, with millions of Twitter followers, a speech to the United Nations Climate Summit, and selection as *Time* Person of the Year.

Photovoltaic technology "is a bit magical," says pioneer solar researcher Martin Green. "Sunlight just falls on this inert material and you get electricity straight out of it."

Solar panels being installed on an office building in China. Solar costs have fallen an extraordinary 85 percent over a decade, and deployment has grown enormously. China represents half the global market.

Workers assemble solar cells in a factory in China, which supplies 70 percent of the world's solar cells.

Almost three billion people do not have access to commercial energy, but instead depend upon gathering wood and crop and animal waste for cooking and heating. The resulting indoor air pollution, says the World Health Organization, is "the greatest environmental health risk in the world today."

Offshore is the new frontier for wind power. Here a supply ship services one of the 150 turbines in the Gemini Wind Park, which covers twenty-six square miles off the Netherlands' coast.

OPEC secretary general Mohammad Barkindo, left, with Saudi petroleum minister Prince Abdulaziz bin Salman and Russian energy minister Alexander Novak, the two leaders of "OPEC+." The new group fell apart, but then came together for the biggest oil output cut in history, brokered by the United States.

Members of the U.S. House of Representatives wearing masks, including Speaker Nancy Pelosi, center, came back to the Capitol during the 2020 pandemic to vote on the $484 billion financial aid package.

Hours after his inauguration, Joe Biden signed an executive order making good "on the commitment I made to rejoin the Paris climate accord"— making clear that climate will be a pillar of his administration.

———

AROUND THIS TIME, WHAT BEGAN AS AN OBSCURE EVENT
would become a giant turning point for the global auto industry—and
a big accelerator for the EV. In Europe, motorists traditionally bought
more diesel than gasoline cars, mainly because they were more fuel-
efficient and less taxed, which mattered a lot in a region where prices
at the pump could be three times higher than in the United States.
But in the United States, diesels were rare and came equipped with
a bad reputation for performance left over from the 1970s. Around
2005, however, Volkswagen, the leading proponent of diesel technology worldwide, decided to push the sale in the United States of "clean"
diesel cars with a new engine. A decade later, it was making substantial headway with diesel sales in the United States.

In 2013, researchers at a fuels and emissions center at West Virginia University, in the town of Morgantown, rented two Volkswagen
cars to test their nitrogen oxide (NOx) emissions in on-road conditions. The work was funded by a tiny research contract—just $50,000
from a European "clean transportation" group—that had to be supple-
mented by scrounging around for another $20,000. Nitrogen oxide
causes smog and reacts with other chemicals in the air to create par-
ticulates and ozone that affect the respiratory system. In the course of
twenty-four-hundred miles of driving, the West Virginia researchers
discovered that the VW cars put out far higher nitrogen oxide emissions than were recorded in laboratory tests. Puzzled, they came to
realize that the company had been using special software that demonstrated in the laboratory that emissions were lower and thus in compliance with regulations. When the cars were on the road, the software
deactivated, and the cars were not compliant.

The software was installed to solve a problem. The company was
unable to develop a small diesel engine that could combine good fuel
economy and lower CO_2 emissions with, at the same time, low nitrogen

oxide emissions. Larger diesel cars had the space to fit alternative and less expensive emission-reducing technology. But for the small cars it was a trade-off—lower NOx emissions versus greater fuel efficiency and lower CO_2 emissions.

This revelation stunned the West Virginia researchers. After all, they were, as they later said, "diesel guys" who believed that "clean diesel technology is real and available." In 2014, they presented their conclusions at a conference in San Diego. That, said Daniel Carder, the leader of the West Virginia group, "kind of opened the can of worms."[9]

CARB and the U.S. Environmental Protection Agency, having validated the West Virginia research, entered the fray, along with the U.S. Justice Department. The result was what became known as "Dieselgate," with manifold consequences. The CEO was forced out. Some VW executives went to jail. The company itself faced fines, settlements, and remediation costs of upwards of $30 billion.[10]

Dieselgate fed into a 180-degree turn in thinking that was in process about diesel fuel and urban transportation in Europe, where diesel cars have been popular. But anti-diesel sentiment was a big threat to Germany's auto industry, which looms large in the country's economy. German chancellor Angela Merkel decried the "demonizing" of diesel cars. Diesel, she said, was essential for combating climate change, owing to its lower CO_2 emissions and greater fuel efficiency. She convened "diesel summits" to try to head off urban bans on diesel cars. But it was all to little avail. European cities, concerned about the higher levels of nitrogen oxide emissions from diesel, began to introduce limits for diesels. The aim for many is an eventual ban. This has become part of a larger movement to ban all cars from streets and parts of city centers. France's ecology minister anointed the bicycle as "the little queen of deconfinement" post–COVID-19 quarantine.[11]

———

IN 2015, HERBERT DIESS WAS RECRUITED AWAY FROM BMW
to lead the Volkswagen car group with a new agenda. One of his first
questions was, What is Volkswagen's electric car strategy? His question
went beyond diesel versus gasoline to a challenge that would be closing in on all European automakers. And he asked it with some urgency.

That year, after years of discussion, the European Union adopted
new standards that require a steep decline in CO_2 emissions, averaged
over the entire fleet, for cars sold in Europe. They were to go into effect in 2020 and 2021. "The only way to make this target is with zero
emission vehicles" as a growing share of new car sales, said a senior executive at a major European company. Failure to meet these new standards could cost European automakers as much as $40 billion in fines.
Given the half-decade lead time to bring on wholly new models, it
meant starting to pivot to EVs right away.

With the new regulations now upon them, European carmakers
are racing to announce plans for electric vehicles. At the forefront is
Volkswagen. The company faces a "new world," said Diess, who is now
CEO of the entire Volkswagen group, and a "new era" for Volkswagen.[12] The company announced that it would launch at least seventy-
five all-electric vehicles by 2028. "The future belongs to electric drive,"
said Diess. "Without EVs, we can't win the battle against climate
change." He also pledged that the company would become "carbon neutral across the whole supply chain."

Volvo got headlines when it was reported to be "the first mainstream automaker to announce the death knell of the internal combustion engine." But that's actually not quite what it announced. It
was moving away from diesel. While shifting to electric vehicles, it
was also moving to hybrids and "mild hybrids" and "plug-in hybrids,"
and it would continue to produce existing models with conventional

engines. Volvo is owned by Geely, a major Chinese automaker, and this positions Volvo for the new electric vehicle game in the world's biggest auto market.[13]

The announcements kept coming from automakers around the world. Even Toyota, deeply committed to the hybrid and the fuel cell, said it would bring out its own EV. "As laws and regulations come into effect in places like China and the U.S., carmakers will have no choice but to roll out electric vehicles or risk going out of business," explained Takeshi Uchiyamada, Toyota's chairman. This was in the face, he added, of Toyota's being "skeptical there would be a rapid shift to pure electric vehicles, given questions over user convenience." In the United States, Ford announced that it would spend $11.5 billion on the production of electric vehicles by 2022. "Electric vehicles make sense," said Ford's executive chairman, Bill Ford. "We are betting very heavily on it."

And the list went on and on—and on. "There have been so many announcements that I'm waiting for my mom to announce one," said Elon Musk. He could afford to joke. For in 2017, in terms of stock market valuation, Tesla, producing around just 100,000 cars, overtook General Motors, which that year sold 9.6 million vehicles worldwide.[14]

But one year later, Tesla was again caught up in another swell of turbulence. Production of the Model 3 was going much more slowly than anticipated. Musk himself was working 120-hour weeks and often sleeping at the Tesla factory, as the Model 3 went through what he called "production hell." Then he tweeted, "Considering taking Tesla private at $420. Funding secured." Tesla's stock price went up—then down. The Securities and Exchange Commission launched an investigation, which ended up with a $20 million fine, Musk giving up the chairmanship of Tesla, and his agreeing to have "material" tweets pre-approved by company lawyers. But by the end of the year, Tesla was back showcasing its new, all-electric military-looking "Cybertruck" against the Ford F-150, the best-selling pickup in America. By July

2020, Tesla's market cap was three times the *combined* market cap of General Motors and Ford.[15]

The electric car differs in many ways from an internal combustion car. It has a screen showing the charge and how much electricity it is using. It has many fewer parts. It has strong environmental appeal, although in some places it is running on coal-generated electricity. This would make those EVs what some call "EEVs"—"emissions elsewhere vehicles." It is enticing for early adopters of new technologies. It accelerates more quickly. It replaces the gasoline pump with a plug, although a charge-up can take much longer than a fill-up. Yet at the end of the day, it is still a car, and the experience is largely the same.

In 2019, EVs were less than 3 percent of new car sales in the United States. Some 189 zip codes—0.2 percent of the 43,000 in the country—represent 25 percent of EV sales, and all are in California. The main driver is government policy, as it is around the world. Carlos Tavares, the chief executive of the newly combined Peugeot-Fiat-Chrysler company, summarized the matter succinctly: "The rise of the electric vehicle is very much dependent on the subsidies and support governments will be able to give this technology."[16]

The country with the highest penetration of pure electrics and plug-in hybrids is Norway—45 percent of vehicle sales in 2019—and it certainly has government support. The subsidies are so extensive, along with special preferences on the road, as to make one wonder why anyone would *not* buy an electric car in Norway. Some Norwegians have speculated on the psychological motives, owing to the fact that Norway is a very wealthy country because it produces oil and gas and possesses the world's largest sovereign wealth fund, funded by revenues from oil and natural gas. Another reason is that virtually all of Norway's electricity is generated by ample low-cost hydropower. A third explanation, a very simple one, was offered by a senior Norwegian. "We have the money."

In the United States, the most significant incentive is a federal tax credit that goes up to $7,500, depending on the size of the battery, for the first two hundred thousand EVs sold by a company. Why $7,500? A Senate staffer explained, "We didn't think $5,000 was enough, and we didn't think we could get $10,000. So we decided to go for $7,500." Some states and cities offer their own incentives, ranging from additional tax credits to access to high-occupancy highway lanes and free parking. As part of the fiscal stimulus launched in response to the financial crisis of 2008, the Obama administration provided a $465 million loan to Tesla and $1.2 billion to Nissan for electric car development.[17]

There's yet another powerful driver for EV adoption. As part of a legal settlement with California, VW is required to spend $800 million in the state on electric vehicle infrastructure and promoting "awareness" of EVs through "public education and marketing programs"—with CARB approving each program. The settlement with the state also requires Volkswagen to spend another $1.2 billion outside California promoting electric vehicles. Across America, for instance, the Volkswagen settlement helps pay for installing electric charging stations.[18]

IN 2009, CHINA OVERTOOK THE UNITED STATES AS THE world's largest auto market, and the gap continues to grow. Beijing is determined that one out of every five new vehicles sold in China by 2025 should be a NEV—a "new energy vehicle." China has its own great champion of electric vehicles—Wan Gang. He ranks with Elon Musk in terms of impact on the advancement of the EV and as one of the most consequential figures for the global auto industry.

During the Cultural Revolution in the late 1960s, Wan was exiled to the countryside. He relieved the harshness and the tedium by spending hours in a tractor shed, fascinated by the tractor's engine, disassembling it and reassembling it. After the Cultural Revolution, he went to university and then did an engineering Ph.D. in Germany, and then stayed to work for Audi before returning to China. As early as

2000, he was invited to make a proposal to the State Council of China, "Regarding Development of Automobile New Clean Energy as the Starting-Line for Leap-Forward of China's Automobile Industry."

First as head of China's "Key Electric Vehicle Project" and then as minister of science and technology, Wan led China's EV drive. "There will be a strategic window for developing electric vehicles," he said. "We have to take action."

And China did. Today, more than a hundred Chinese-made EV models are on sale in China. The development of electric vehicles meets three major objectives for Beijing. The first is to reduce the often-choking air pollution (although the gains will be somewhat mitigated by the amount of electricity produced by coal). Secondly, electric cars promote energy security. "China's oil demands are increasing day by day," Wan warned. With overall auto ownership growing rapidly, Beijing counts on electric cars to dampen the continuing increase in oil imports.[19]

The third reason is global competitiveness. As a latecomer to auto production, China would have a hard time catching up and becoming a major competitor in the global market for conventional cars. But the electric car is a new game, and there are no big established EV players. Rapid growth of the national EV industry would deliver not only jobs domestically but also the platform to become a formidable exporter and a major force in the global auto industry. China already manufactures almost three-quarters of the world's lithium batteries. The electric car also fits into a larger vision of an electric transportation network inside China—high-speed electric rail connecting cities where people move around in electric cars, or on electric buses and electric bicycles. By the end of 2017, over 50 percent of China's city bus fleet was already electric.[20]

The New Energy Vehicles program is one of Beijing's national industrial strategic priorities. It is switching from a "consumer-centric" model—subsidies for buyers—to a "producer-centric" one, establishing quotas for manufacturers. That means a rising share of automakers' total production must be EVs. China also offers an enormous carrot. In the crowded megacities, one can only get a license plate through a lottery. In Beijing, the odds of winning a license plate for an oil-fueled car are tiny—1 in 907. Even then, the winner must pay a fee, which can be quite substantial—as high as $13,000. But there is a big exception: An electric car buyer can bypass the lottery and get the license automatically (although there may still be a waiting period), without having to spend money for the license.[21]

In 2019, almost a million EVs were sold in China—4 percent of new car sales there and more than half of all the EVs sold in the world.[22]

IN INDIA, THERE IS ALSO ENTHUSIASM FOR GREATER ELEC-trification, which in the country encompasses the ubiquitous two-wheelers and three-wheelers as much as four-wheelers. Today, there are only a few domestically produced EVs, but the intent is to build up India's manufacturing capabilities in this sector and use electric vehicles as an opportunity for value-added domestic production.

EVs have also become a hot topic. Transport minister Nitin Gadkari declared that he would "bulldoze" automakers into making electric vehicles. He was, as he put it, "crystal clear" as to why—urban pollution and the economic strain of the country's importing 85 percent of its oil.[23]

Others, however, asked how a country that is constantly short of electric power and suffers hours of rolling blackouts—and depends on coal for most of its power—can switch to EVs. Responding to the clamor in India for electric vehicles, India's petroleum minister, Dharmendra Pradhan, described the enthusiasm about electric cars as a "fashion." He compared it to the excitement generated by Alia Bhatt, an alluring young Bollywood film actress.

From the viewpoint of the Indian auto industry, government policy is key to determining what happens. "Unless there is a mandate, people are likely to continue using diesel vehicles," said Ajit Jindal, vice president of engineering and head of electric vehicles at Tata Motors. "There has to be some kind of mandate." Since then, India has adopted a more measured view. The new approach is one of promoting EVs for specific segments, such as public buses, taxis, and to replace oil- and human-powered rickshaws. But the focus has recently shifted from EVs to CNG—compressed natural gas—to reduce urban pollution, and to biofuels that can be made from agricultural waste.[24]

In Britain, the Conservative government in the midst of the painfully protracted negotiations for Brexit—its exit from the European Union—proposed a ban on the sale of new gasoline- or diesel-fueled cars by 2040, this despite the fact that Britain could be short of the generating capacity to meet the electricity demand and that the Conservative Party's own manifesto opposes new onshore wind farms. In 2020, the government announced that it was considering accelerating the ban to 2035. The French government, promising a new French "veritable revolution," also says it aims to impose a similar ban on new gasoline and diesel car sales after 2040.[25]

BUT HOW FAST WILL ELECTRIC VEHICLES PENETRATE THE global auto fleet? In 2019 they constituted less than 3 percent of total new car sales worldwide. But ambitions are high. Will EVs dominate new sales in the decades ahead, or remain a fraction?

"Consumers compare electric vehicles with other motor fuel options they have, such as petrol and diesel," said the top auto official in the Japanese government. "They have two options. Either government regulations put a ban on vehicles other than electric vehicles beyond some time-frame. Or the electric vehicle proposition needs to be extremely compelling to consumers. It is a matter of both safety and price."[26]

At this point, sales are still being driven largely by government policies—regulations and subsidies—and the increasingly determined response of automakers to those policies. But the costs were making it challenging for governments to provide subsidies for a mass market for electric vehicles. It would be far too expensive. The working assumption, however, was that the current incentives would be enough to promote scale and technical advances, bringing down costs. Still, the data from Denmark, the Netherlands, and Hong Kong—and in 2019 from China—indicate that when subsidies are removed, EV sales abruptly go down. But for the post-coronavirus recovery, some governments will extend incentives to promote EVs as a "green recovery" measure to support the auto industry.[27]

There are two critical gating items for mass adoption. The first is the battery itself. "Nothing else has more leverage to the electric vehicle business than the cost of batteries," said someone who knows—Tesla's J. B. Straubel. If one only looks at operating costs, running a car on electricity per mile driven is cheaper than gasoline. But it is much more expensive to store the energy. In an internal combustion car, the energy is stored in the gas tank. In the EV, it is stored in the battery. Costs for batteries dropped significantly—by over 50 percent—between 2015 and 2019 to about $180 per kilowatt-hour. This is largely the result of redesigns and manufacturing scale and reductions in weight. Still, a battery pack that will go two hundred miles or more on a charge costs around $11,000, which is expensive and not yet competitive without subsidies. It is thought that at around $100 per kilowatt-hour, the battery would be competitive with the internal combustion engine. The recent MIT study on mobility posits that the gap may not be closed until 2030.[28]

There is much debate as to how much further these costs can come down without technical breakthroughs. One path is through greater scale in manufacturing, and new factories are being added at impressive rates. But there are also concerns around the supply chain of some battery materials, which has added to the uncertainty over future costs. Under an aggressive scenario, EV demand for lithium would rise 1,800 percent by 2030 and would represent about 85 percent of total world demand. Demand for cobalt, another essential element in batteries, would rise 1,400 percent. More than 50 percent of global cobalt supply comes from one province, Katanga, in the Democratic Republic of the Congo. Overall, the performance required of batteries in cars means that these materials will have to be of the correct, high-grade specifications, adding potential bottlenecks to the supply chain. China has already staked out a key position in these industries.[29] At the same time, the potential scale of the market is stimulating research aimed at improving and developing new battery technologies.[30]

An electric car eliminates the need to go to the gas station, but it does require charging, and the second big obstacle is charging time and the availability of chargers. It still takes longer to charge a car, although using a super-fast charger—which is significantly more expensive—could bring down the time to around ten to fifteen minutes. The chargers will have to be much more widely available, numbering in the many millions. Someone will have to pay for them, and ultimately there will have to be a standardized business model. And something will have to be done for people who live in apartments and high-rises and park on the street.

MANY FORECASTS HAVE BEEN MADE FOR THE ELECTRIC CAR over the last several years. Some have been very optimistic compared to what has actually happened, and what will happen.[31] The electric car may appeal to buyers in terms of its "coolness" or "newness" or status or because it delivers a message about climate change or values or brand or because of the quality of the vehicle and the fact that it can go from zero to 60 in four seconds. Or because buyers receive financial or regulatory incentives from governments. But it is not clear how the overall utility of the electric car, on its own, is superior to that of the gasoline car. The electric car is still a car.

What then will determine the rate of adoption? Electrification "still has to be driven by customer demand, and we don't know that," said Bill Ford of Ford Motor Company. "It will be a transition. . . . You'll live in a world where you'll have internal combustion engines, hybrids, plug-in hybrids, and pure electrics. Over time, the shift will take place."[32]

The range of predictions can be very wide. A leading investment bank forecast that electrics could constitute anywhere between 10 and 90 percent of new car sales in 2050, depending on regulations and "technology development." Governments will certainly have overriding and even decisive impact. They can require that increasing shares of new car sales must be electric or meet increasingly tough carbon reduction targets. Automakers will have to comply or face increasing penalties. Either governments or consumers would absorb the extra cost until the battery comes down in price. Governments can keep tightening the screws on emissions from gasoline-powered cars and on their efficiency, both of which would drive up their cost and thus by comparison make EVs more economically attractive; they can also adjust the tax code to favor electrics over gasoline-powered cars.

So far this is a story of "car versus car"—gasoline versus electricity. But a further revolution is potentially at hand.

Chapter 38

—————

ENTER THE ROBOT

Victorville, a small town in Southern California on the western edge of the Mojave Desert, had its beginnings in the nineteenth century as a way station to provide water and supplies to those trekking across the desert. In more recent decades, the stark high-desert scenery and the good weather made it a favored location for shooting movies, including the 3-D science-fiction thriller *It Came from Outer Space*. In 2007, an unusual scene was being played out there. It might have seemed like some new science-fiction thriller. For zombie vehicles—with no one in the driver's seat—were making their way down streets of an abandoned Air Force base, evading obstructions, changing lanes, parking, halting at stop signs, turning corners, and avoiding crashes.

But these zombies were not from outer space, and this time the producer was not a Hollywood studio but rather DARPA—the Advanced Research Projects Agency of the U.S. Department of Defense. In Iraq and Afghanistan, American soldiers riding in tanks and trucks were being maimed and killed by IEDs. In response, the Defense Department was determined to develop vehicles that would not need drivers—what would become known as autonomous vehicles. And that's why these zombie vehicles were crisscrossing the streets in an abandoned Air Force base in the California high desert.

Yet, as with electric cars, this was not the first go at driverless vehicles. In 1925, an engineer named Francis Houdina set out to demonstrate his driverless "phantom" car on the streets of New York City, controlling it by radio from a car immediately behind. Unfortunately, things went a little awry. As the *New York Times* reported in a story headlined "Radio-Driven Auto Runs Down Escort," the driverless car "careened from left to right down Broadway, around Columbus Circle, and south on Fifth Avenue." It barely missed two trucks and a milk wagon, collided with a car filled with photographers, and then almost hit a fire engine at 43rd Street. Decades later, in the 1950s, Nebraska's Department of Roads experimented with driverless cars that would be guided by electric circuits embedded in the road. In the 1980s and 1990s, research groups in the United States and Europe investigated driverless concepts to improve road safety. But all to no effect.[1]

DARPA had been established in 1958, in response to the Soviets beating the United States into space with Sputnik, the first satellite. Its mission was to ensure that the United States would from then on be "the initiator and not the victim of strategic technological surprises." Working with universities and industry, it funded frontier research "in breakthrough technologies for national security." Its extraordinary record extended from advanced computing and stealth bombers to GPS systems. At the beginning of this century, DARPA decided to offer cash prizes to fire up competitive juices for developing "autonomous ground vehicles." But the first Grand Challenge, held in 2004, was a flop. The best that any vehicle could manage was just 7.5 miles on the 142-mile course in the rugged desert on the California-Nevada border. Yet this failure was also a success. "The first competition created a community of innovators, engineers, students, programmers, off-road racers, backyard mechanics, inventors, and dreamers," said a DARPA official. "The fresh thinking they brought was the spark."[2]

The advances in machine learning and technology were already evident the next year, in the 2005 Grand Challenge. This took place along a 132-mile course in southern Nevada. Almost two hundred teams entered. Five teams managed to complete the whole course. But from the beginning, it was clear that there were two top contenders: Carnegie Mellon University (CMU) from Pittsburgh and Stanford from California. It was something of a grudge match.

CMU's team was led by William "Red" Whittaker, who headed the robotics center at CMU and was a legend in the world of robotics. Decades earlier, in the aftermath of the accident at the Three Mile Island nuclear plant in Pennsylvania, there had been no way to know what was happening inside the containment building, which was flooded with radioactive water and nuclear debris. No human dared enter. It was too dangerous. Whittaker, who had just completed his Ph.D., designed special robots that could enter the building, analyze what was happening, report back, and help with the cleanup. Altogether, robots were his life's work. He not only invented and built robots, over sixty of them, but also invented a research discipline called "field robotics"—robots that would move and accomplish tasks over difficult terrains.

Stanford's team was led by Sebastian Thrun, a German-educated computer scientist who had been a colleague of Whittaker's at Carnegie Mellon University until lured to Stanford to direct its artificial intelligence laboratory. While still a graduate student in Germany, Thrun and colleagues had designed robots to give museum tours. He had gone on to pioneer what became known as "probabilistic robotics," which combines statistics to deal with the uncertainty that robots face in real-world settings. Thrun's goal now was to create a vehicle with "the brainpower to make all the decisions along the way." His interest in autonomous vehicles arose, he said, because he was "curious about human intelligence." He also once explained that his "passion for cars that drive themselves" arose from traffic jams. "I feel I've lost a year or two just in traffic," he explained. Thrun's quest for an autonomous vehicle that would make driving safer had a personal dimension; his best friend had been killed in a car crash at age eighteen.[3]

Carnegie Mellon fielded two burly military-style vehicles. The CMU team had spent twenty-eight days laser-scanning the Mojave route to create a computer model of the topography. Stanford's riposte was "Stanley," a midsized Volkswagen SUV. To Whittaker's distress, Stanley came in first, completing the course in six hours and fifty-three minutes, eleven minutes ahead of Carnegie Mellon.

It was a key moment. "This is the first time, ever, that the machine made all the decisions," said Thrun after the race. He himself was so drained from the intensity of the competition that he fell asleep during an interview with CNN.[4]

The Stanford team walked off with the $2 million prize. That was not the only win that day. As Thrun later said, the Grand Challenge proved "life-changing. There was Sebastian Thrun before, and there was Sebastian Thrun after." Among those unobtrusively watching the race was Larry Page, cofounder of Google. After the race, disguised with sunglasses and a hat, he approached Thrun. Page really wanted "to understand what's going on," Thrun later said. "Larry had been a believer in this technology for much longer than I even knew." Indeed, Page had even toyed with doing a Ph.D. on autonomous vehicles.

Sometime later, Thrun received an email from Page, who said he was having problems with a robot that he had built to enable him to attend meetings at Google without being physically present. The robot wasn't working. Thrun met up with him in a parking lot. Page opened up the trunk of his car and pulled out the robot for Thrun to examine. Thrun quickly pulled together a team, and the robot was fixed.[5]

The decisive race was two years later—that 2007 Grand Challenge on that deserted Air Force base in Victorville, California. An empty desert was one thing. But could a self-driving car navigate the streets of an American city, even if it was a ghost city?

Eleven teams made it into the 2007 competition, but once again it was Carnegie Mellon versus Stanford. Stanford's team was back with a Volkswagen named "Junior," after Leland Stanford Jr., for whom the university was named. Thrun was part of the Stanford team, although by this time he was already working at Google. CMU's entry was a Chevy Tahoe that was named "Boss" in honor of General Motors' Charles "Boss" Kettering, the inventor of the electric ignition in 1911, who had gone on to run GM research for more than a quarter century.

The invocation of General Motors' legendary research chief was not accidental, for this time CMU's partner was GM. For several years the auto giant and CMU had been collaborating on research on advanced technologies. The connection had been fostered by Larry Burns, GM's head of research and development and of strategy. Burns worried about GM's future—in particular, what could make the automobile obsolete. Once, GM's then-CEO Rick Wagoner and Burns had got to talking about the fact that while many other industries had changed over a hundred years, the basic model of the auto industry was the same as it had been since Henry Ford's Model T—"gas-fueled, run by an internal combustion engine, rolling on four wheels."

"What's the car of the next hundred years going to look like?" Wagoner asked Burns. "If the automobile were being invented today, then what form would it take?"

Burns reflected that "there haven't really been any disruptive innovations in that time." And he kept thinking, "That's true of very few industries." Autonomous cars could be a big part of the answer to Wagoner's question—if they could work.

Burns was also gripped by what he saw as the "most important sustainability issue faced by automobiles"—not energy or emissions, but a deadly epidemic in which 1.2 million people a year globally died in auto accidents. Autonomous vehicles might be able to virtually eliminate crashes. That was one of the main reasons why Burns hooked GM up with Whittaker and the robotics group at Carnegie Mellon.[6]

And in this third Grand Challenge, held on the deserted air base in Victorville, Carnegie Mellon won—by twenty minutes.

"That day in 2007," Whittaker later said, "was the moment when concepts that had been around for years suddenly came out of the laboratory and into the world." His life's work had been vindicated.[7]

THE ROAD RACE IN VICTORVILLE CAUGHT THE INTEREST OF the major automakers. But at that point they had something more immediate to worry about—survival. This was the beginning of the financial crisis. Automakers were running out of cash and hurtling toward bankruptcy. Many things were thrown overboard, including, at GM, the joint research program with Carnegie Mellon.

But there were others, across the country in Silicon Valley, who were not short of money. Google was already at work on autonomous vehicles, with Sebastian Thrun in the lead. Google's effort—"Project Chauffeur"—was housed in a separate building. "No one at Google had a clue we existed for a year and a half," said Thrun. As part of their work on autonomous vehicles, they posted 360-degree cameras on the roofs of the cars. This generated the idea for Google Street View, with the ambition of photographing every street in the world.

In 2010, with a blog post from Sebastian Thrun—"we have developed technology for cars that can drive themselves"—Google went public with the stunning news that it was working on the autonomous car. "Our automated cars" had just driven from Silicon Valley in northern California "to our Santa Monica office and on to Hollywood Boulevard," Thrun reported. The cars had also "crossed the Golden Gate Bridge," and "even made it all the way around Lake Tahoe"—altogether 140,000 miles of automated driving. While operators were in the cars just in case, the key interaction was through the cars' informational tether back to Google's enormously powerful data centers.[8]

The word was out. Google—with its scale and capabilities—had put a stamp on the viability of the driverless car. Google was hardly alone. By then, a host of other tech companies and entrepreneurs were jumping in to tackle the driverless car. Technological advances and falling costs were making it all possible. "There was no way, before 2000, to make something interesting," Thrun said. "The sensors weren't there, the computers weren't there, and the mapping wasn't there. Radar was a device on a hilltop that cost two hundred million dollars."[9]

Even with all the competitors and competing visions, there is at least a consensus on the benchmarks for defining a self-driving car. The Society of Automotive Engineers has classified cars by level of automation. The first three levels go from "no automation" at Level 0 up to Level 3, which is cruise control and autopilot that controls acceleration under the supervision of the driver. Level 4 is "high automation"— capable of driving and monitoring the environment without human supervision but only in what is called a "geofenced area," which might be a college campus, a central business district, or using the "pods" to go from Terminal 5 at Heathrow Airport to the business car park. Level 5 is "full automation"—capable of doing all the tasks of driving under all conditions. Getting to Level 5 requires the deployment and continued improvement of a host of complex capabilities.[10]

To give a sense of what's involved, start with "sensing," which in itself depends on several technologies. "Lidar" uses lasers that detect objects and send back that information at the speed of light. It is backed up by radar, which uses radio waves, less capable, but more reliable in bad weather. Surround-view cameras constantly capture digital images, which are identified at high speed by machine vision algorithms. There is also ultrasound, along with infrared to pick up thermal imaging, plus GPS for location and inertial navigation systems.

All of this is amplified by three-dimensional maps, down to centimeter-level accuracy, which have been preinstalled in the car. At present, there is no consensus on the optimal mix of these technologies or the requisite standards—or on who will provide all the ingredients.

And this is only the beginning. The sensor data needs to be constantly processed with instant decision following, and that requires a very big brain, exponentially more powerful computing power than found in today's cars. Software is key; and here the rapid advances in AI—artificial intelligence—and machine learning for image recognition have propelled the driverless car forward more rapidly than might have been expected a few years ago. At lightning speed, the computer has to barrel through vast amounts of data coming from the sensing devices to identify a stop sign or a dog and instantaneously come to a stop. This big brain computer also has to manage the electronic systems units that control braking and steering and ensure that the car swerves around pedestrians who obliviously step into the street without looking up from their smartphones. The telematics have to be developed between the car's computer and other computer systems to provide for information and software updates. The speed with which such information and updates can travel will depend on fast internet access and the availability of 5G wireless technology.[11]

And then there is the question of the interior of an autonomous vehicle—will it be a traveling office, a communications hub, a living room, or a wraparound entertainment center? Hardly to be overlooked, of course, is the human-machine interface—that is, the ability of the computer to communicate with (and reassure) the vehicle's passengers. Indeed, one big barrier may be psychological—the willingness and even ability of people to "let go," to completely surrender control to a robot.

It's become a fevered horse race with a swelling number of participants. Increasing elements of automation will be added to cars over the next few years, but, in terms of scale, will likely be well short of that holy grail of 4 or 5. At this point, a reasonable estimate is that all of the technology to be fully autonomous would add $50,000 to the cost of a car. But with scale and convergence on technologies and the beginning of production in the early or mid-2020s, the additional cost of a fully driverless car over a regular car could fall toward $8,000 to $10,000. At least that's the assumption.

A host of other obstacles will have to be addressed. Reliability has to be assured. What happens if a vehicle goes to the wrong destination? Or if there is construction or an accident en route? Or weather causes a malfunction? Or the car has to make a decision whether to hit a person or crash on the side of the road? Since people are handing over control to a machine, in order to instill confidence these self-driving cars will have to function at much higher levels of performance than cars driven by humans. While millions of miles have now been driven by test driverless cars, humans drive more than eight billion miles every day in the United States. Will some groups create havoc by hacking the software in tens of thousands of cars? In response, in addition to the obvious emphasis on cybersecurity, the concept of "graceful degradation" in the face of hacking is also being quietly tested—a sort of glide path to a minimal level of functionality, which would allow the driver to take over. Of course, that assumes that people in an autonomous age will still need to learn how to drive, let alone get a license. And how does someone take control if there is no steering wheel, no accelerator, no brake pedal, and only a computer to talk to?

What about insurance? Currently, drivers are insured because they have personal liability. But if an accident happens with a driverless car, will it be a matter of product liability, and is the automaker or software supplier responsible, not the person in the car who is its owner? And by the way, who owns all that data the car is generating about the occupants, their interests, proclivities—and destinations? And regardless of who owns the data, who can access it—and for what purposes?

In short, the prospect of autonomous vehicles is already creating major new regulatory puzzles. A host of issues will have to be addressed by both regulators and legislators—including who will be the regulators and who will "own" which part of the puzzle. Disputes continue over safety, security, and privacy, as well as the role of the federal government versus the states. Meanwhile, within the U.S. federal government, a number of agencies—as many as thirty-eight—are trying to figure out their varying roles in regulating different aspects of autonomous vehicles—safety, privacy, and connectivity—while about thirty states have already passed legislation related to autonomous vehicles. The likely result will be some mix of federal and state regulations.[12]

And then there are the social impacts. Some envision that delivery trucks will become autonomous and that large trucks will drive in convoys down the highway—"platooning," as it is called—perhaps the first truck with a driver, all the rest driverless. That would mean that a significant percentage of the 3.5 million people who now drive trucks in the United States would need to look for other work. Finally, there's the puzzle of the "intermediate period"—how does an insurgent tribe of self-driving cars get integrated into the massive current flow of traffic composed of cars with drivers?

One likely path is that cars move toward self-driving incrementally, with more and more parts of the driverless package, such as crash-avoidance and acceleration-management technologies, introduced over time. That introduces yet another concern. Will drivers be even more distracted and preoccupied with other things, such as texting on their smartphones, when their attention is suddenly and urgently required in vehicles that are only partly autonomous?

The race continues to heat up in the ultimate grand challenge—the commercially viable self-driving vehicle. The players range from the major automakers to the big Silicon Valley tech companies to venture-financed tech start-ups. At some point, there will be a giant shake-out and agreed-upon standards. But before then, many competing technologies will seek to carve out their place in the marketplace. "With different cars operating with different algorithms, it'll be interesting to see how the social interaction between them happens," observed the CEO of one start-up. "Will they be polite and say after you, or cut each other off?" She added, "It'll be AI versus AI, but ultimately this is a so-cietal problem that needs to be solved."[13]

Some are already moving on to yet newer frontiers. Sebastian Thrun cofounded Udacity, an online tech education company, and then established a new company with Larry Page of Google—Kitty Hawk—that aims to introduce autonomous flying taxis. He is con-vinced that they will preempt autonomous ground taxi service.

The technology advances required for self-driving cars still have many hurdles to overcome. Yet there is one more key element that could partner with the driverless car to usher in a decidedly new and different world of mobility—the now-ubiquitous ability to order a car on your smartphone. In short, ride hailing.

Chapter 39

——— ———

HAILING THE FUTURE

n the summer of 2008, a Canadian software engineer named Garrett Camp, already fed up with haphazard taxi service, was standing
on a San Francisco street. He was late for a date but unable to wave
down a taxi. In his hand he held his new iPhone. And that, he realized,
could be the alternative to having to stand on street corners depending
on overwhelmed taxi dispatchers and unpredictable taxi service. And
perhaps it could provide him with a better social life. Apple had released the iPhone just a year before, but it was only in 2008 that it
allowed non-Apple apps and had opened the App Store. Why not,
Camp thought, replace waving your arm on a corner with an app button on your iPhone and let an algorithm connect together passenger
and driver? "At the time," he would later say, "I was thinking 'better
cab.'" On August 8, 2008, he registered a website, www.ubercab.com.

Camp hardly imagined transforming transportation and challeng-
ing an entire business and way of life based upon people owning a car.
The initial aim was far more modest. The first pitch book described
"UberCab" as a "Next-Generation Car Service" aimed at improving on

cabs. "Digital Hail can now make street hail unnecessary." Instead, "Mobile app will match client & driver." It would be "members only—respectable clientele" matched up with drivers operating Mercedes and other high-end cars owned by UberCab. The "use cases" ranged from "trips to/from restaurants, bars & shows" to "elderly transport."

The optimistic case was "market leader" and a "billion dollars in revenue." The realistic—5 percent of taxi business in the top five U.S. cities and $20 to $30 million profit. Worst case? "Remains a 10-car, 100 client service" in San Francisco offering a "time-saver for San Francisco based executives." And next step? "Buy 3 cars" and "raise a few million."[1]

One night the following December, Camp was in Paris with Travis Kalanick. The City of Lights was shut down by a big snowstorm. The bars and bistros closed, and even the taxicab drivers went home. With nothing else to do, Camp and Kalanick trudged to the Eiffel Tower and rode up to the top. There, for the next two hours, they talked through their idea for ride hailing—the algorithm would become the outstretched arm.

AFTER DROPPING OUT OF COLLEGE SEVERAL YEARS EARLIER, Kalanick had founded one failed start-up and another that had modestly succeeded. With several million dollars from the latter, he had become a small-time angel investor and brash personality around San Francisco's tech scene. Now, atop the Eiffel Tower looking out on a silent Paris draped in white snow, Camp and Kalanick discussed this Uber idea, made all the more compelling by their own experience that night. Contrary to the pitch book, Kalanick argued that the company should not own the vehicles. It should be "capital-lite"—a twenty-first-century middleman with no investment tied up in the physical autos, but rather a facilitator, linking driver and passenger and taking a cut of the fee. "You don't need to buy cars," said Kalanick.

Uber went on the road in May 2010, offering limousines and town cars. Its slogan was "everyone's private driver." Kalanick came on as CEO in October 2010. The month before, the company had racked up a grand total of 427 rides. On the day that Kalanick started, a representative of the local transportation authority showed up at the office with a letter ordering it to cease operating, as it was an unauthorized taxi company. Kalanick's reaction was to drop the word "Cab" from the company's name and ignore the order. Over the next year, Uber gained traction in San Francisco, launched in New York City and then in London. But what it was offering was access to limousines and town cars and professional drivers. It was still "everyone's private driver."[2]

That abruptly changed in 2012. A competitor with a different approach suddenly appeared. Logan Green, who had grown up in Los Angeles, had long been appalled by the almost constant congestion of single-occupant cars on the city's freeways. On a trip to Zimbabwe, he saw local minibuses that picked up people who hailed them along the road. That was the nugget of the idea. He came back to the University of California at Santa Barbara and started a carpooling service for students going home that he whimsically named Zimride, in honor of his recent visit to Zimbabwe. In New York, a Facebook posting by Green caught the eye of a young Wall Street analyst named, by odd coincidence, John Zimmer, who could not help but note the new service's name, although obviously it had no connection to him. Perhaps it was fate. He had gone to hotel school and looked at transportation as a "hotel occupancy" problem. Hotels seek to be 80 or 90 percent occupied; personal cars are "occupied"—that is, used—only 5 percent or maybe 10 percent of the time. A more efficient transportation system would raise auto "occupancy" to a much higher level.

Green and Zimmer connected, and in 2012 began to offer short rides in San Francisco. They called the new venture Lyft. Anyone could be a driver. In contrast to the upmarket "private driver" black car of early Uber, they provided Lyft drivers with pink mustaches to affix to the front of their cars. "Friendliness" and fist bumps were Lyft's mode, a studied contrast to Uber's ersatz limousine. Uber wasted no time in striking back, launching its own service with ordinary drivers. "We chose to compete," Kalanick wrote in a blog post.[3]

And compete Uber did, and fiercely so. Its new business model was UberX, which adopted Lyft's model and enrolled nonprofessional drivers who could work as little or as much as they wanted. They would be contractors, not employees. In other words, it's a BYOC model—Bring Your Own Car. Uber drivers, 60 percent of whom have other jobs, have become prime examples for what became known as the "gig economy." Both Uber and Lyft also rolled out modern versions of carpooling services that match up a rider with another rider in close proximity headed to nearby destinations.

Uber and Lyft rolled forward, opening in city after city. Customers, initially many of them millennials, were quickly won over. In its quest to expand, Uber went to war with local taxicab drivers and owners and transportation regulators, all of whom opposed it as an unregulated taxi company. It called its approach "principled confrontation." Others called it outright aggression. Uber did not wait for permission to enter a city. It would just appear and start demonstrating the value it delivered. In the face of the inevitable counterattack, it would mobilize riders and drivers to bombard the regulators and politicians with phone calls and emails.

In London, thousands of the drivers of London's black cabs, who had spent years mastering "the Knowledge" of London's intricate maze of streets, protested by blocking roads, paralyzing parts of central London. In France, thousands of angry taxicab drivers (who paid up to $270,000 for their licenses for the right to pick up people) did the same, also blocking routes to railway stations and the airports, in some cases with burning tires. In 2017, regulators in London banned Uber on grounds that it was not "fit and proper" to operate, owing to its failure to meet "public safety and security" requirements—and because it was using a special software to evade regulators. In June 2018, Uber won an appeal and was granted a probationary license for London, which was revoked in 2019.

After app-based ride hailing took off, the number of vehicles used for ride hailing in New York grew to seventy thousand—well above the city's twenty thousand yellow taxis. In the face of traffic-clogging congestion and declining wages for drivers—"too many cars, too many drivers"—New York City capped the number of for-hire vehicles it will permit. Taxi drivers continue to be on the defensive. Unlike Uber drivers, New York City taxi drivers have to take driving courses, go to "taxi school," and get a medical examination. If they want to own the cab, they also have to go into debt to pay for a taxi medallion.

THE SEARCH FOR "BETTER CAB" WAS NOT LIMITED TO THE United States. Cheng Wei, an engineer at the Chinese tech giant Alibaba, missed several flights in China because of his failure to get a taxi in time. Fed up, in 2012 he founded DiDi, which means "beep beep" in Chinese. Now called DiDi Chuxing after a merger, it has become the largest ride-hailing company in the world.[4]

That need for "better cab" was all too painfully obvious to a banker named Jean Liu when she found herself stranded with three unhappy children on a street corner in Beijing, midst a heavy rain, unable to flag down a taxi. Liu had been raised on tech—her father founded Lenovo, which purchased IBM's personal computer business and is now the world's largest PC maker—and she had done postgraduate work in computer science at Harvard. After twelve years at Goldman Sachs, she was well along on the partner track. Her own frustrations with Beijing taxis helped fire her interest in Goldman's participating in a financing for DiDi. But Goldman lost out to Chinese venture capitalists, who had moved quickly to snap up the deal. A deflated Liu met Cheng for lunch. But she did have another idea.

"If you won't take our money," she said, "why don't I join you?" Cheng was taken with the idea. Afterward, Liu thought, "What have I done? This is not my professional journey."

To settle the matter—and test the chemistry between them—she and Cheng decided to take what they called "a social security trip," a road trip from Beijing to Lhasa, the capital of Tibet. The sixteen-hundred-mile journey, with four people—Liu and three men—crammed into the car, turned out to be more arduous than expected.

At one point, during a blinding rainstorm, the designated driver became fevered, which was a major challenge since neither Liu nor Cheng actually have a driver's license, nor even know how to drive. But they managed to truck on. The most difficult moment occurred as they crossed the huge, sparsely populated Tibetan Plateau in western China. At the fifteen-thousand-foot elevation, the three male passengers in the car became ill, short of breath, gasping for air, and urgently in need of a local hospital for an infusion of oxygen. In contrast to the men, Liu was okay. That was the moment she said to herself, "I can do this."[5]

Liu joined Cheng in 2014. Two years later, Uber went on the attack in China, seeking to dethrone DiDi in its home country. After spending $2 billion, Uber surrendered. On its exit, however, it received shares in DiDi as a consolation prize. DiDi was moving twenty-seven million people a day in China (and another three million in its Brazilian subsidiary). Its services extend from ride hailing to much-improved taxi service and bicycle sharing, to a limo service, as well as food delivery.

The market in China is very favorable to ride-hailing services. For China has only about 160 cars per 1,000 inhabitants, compared to 867 in the United States. There are many obstacles to owning a car in China. As already noted, in Shanghai, in response to urban congestion, the cost of a license for a conventional car is often more than that of the car itself. Even if you have a car, parking is a challenge in cities that were not built for cars. All of this gives an added boost to the ride-hailing business in China.

With its sprawling business in over four hundred Chinese cities, DiDi generates a vast amount of transportation data, which, with its algorithms, enables it to work with governments to improve traffic flow and reduce congestion. Operating at its scale also means that it faces passenger safety issues. DiDi now requires drivers to take courses and go through more extensive examination, and it has stepped up the monitoring of trips. It is applying artificial intelligence and machine learning to shape its business for the future. "The transportation industry will be transformed," said Liu. Or already is. As for DiDi, a company that was valued at $700 million in 2014, when Liu and Goldman lost the chance to invest, was valued in 2019 at $62 billion.

IN JUST HALF A DECADE, KALANICK HAD, AS MUCH AS ANYONE, redefined transportation globally. He did it his way and on his own aggressive terms. As the *New York Times* put it, Kalanick "built Uber in his own image." But as the company grew and became more visible, Kalanick's own image was becoming more problematic. It was coming to be seen as not only aggressive, but also incendiary and disruptive. By 2017, the woes were piling up. The company was accused of sexism and harassment. Its relations with its drivers were abrasive. Google sued it for theft of intellectual property related to self-driving cars. It was also charged with deploying software covertly and illegally to deceive regulators and undermine Lyft. In June 2017, Uber retained former U.S. attorney general Eric Holder, who ended up recommending forty-seven actions to improve Uber's "workplace" culture. A week later, five of the biggest investors in Uber dispatched a letter to Kalanick, who was on a trip to Chicago. The message was simple. He was fired. He was replaced by Dara Khosrowshahi, who had been CEO of the online travel company Expedia.[6]

By then, the ride-hailing industry was already well established; Uber alone had two million drivers worldwide, and "Uber" had the status of a verb. The growth in ride hailing had proved exponential. In San Francisco alone, Uber's revenues were in the billions, compared to less than $200 million for taxis. By 2017, Uber was operating in 540 cities around the world; Lyft, 290 in the United States. In the United States, Uber had about 70 percent of ride hailing and Lyft, 30 percent. Internationally, in addition to DiDi, other major players have emerged, including Gett in Europe and Ola in India. Overall, ride hailing could be a very big industry.

But the industry still had a major challenge—to prove that it could be profitable. In May 2019, Uber went public with an $82 billion valuation. But the costs of running it were greater than its revenues. What became apparent in the IPO was that it was losing money—billions of dollars. Half a year after the IPO, its valuation had fallen to $47 billion, still a lot for a company that had not existed a decade earlier, but not exactly great for those who had invested in the IPO. Lyft's market cap had similarly fallen, from an IPO value of $24 billion to about $14 billion.

Ride hailing on a massive scale could disrupt the century-old model of selling and servicing cars that run on oil fuels for personal use. The traditional model would give way to a whole new business model and indeed way of life—"mobility as a service" (MaaS). Instead of buying a car, keeping it in the garage, driving it to work, leaving it in a parking lot, and so on—altogether only using the car that 5 or 10 percent of the day—people would not own a car at all. Instead, they would buy mobility as they need it.

But the coronavirus and social distancing delivered an unexpected blow to ride sharing. Would people want to share vehicles, or would they rather own their own mobility? In one indication, by June of 2020, less than three months after the reopening of China, DiDi was back to 70 percent of its pre-crisis level. But the answer elsewhere would only become clear after the crisis.

Chapter 40

—— —— ——

AUTO-TECH

Much will change for the auto industry and the oil industry if the future is no longer about owning a car—"mobility as a product"—but rather hailing a car—"mobility as a service."

To begin with, what does the rapid rise of ride hailing do to automobile ownership? This is a critical question for the automobile industry and for the seven and a half million people in the United States who work in and support the industry and tens of millions more around the world. There is already a preexisting trend: In the United States, the percentage of those between sixteen and forty-four holding a driver's license has been continuously declining since the early 1980s, especially among younger people. In 1983, 92 percent of those between the ages of twenty and twenty-four had a driver's license; by 2018, that number was down to 80 percent.[1]

One reason for this decline is that owning a car is less urgent than in the past in terms of staking out identity, status, and coming of age. It is no longer the emotional signifier of freedom and autonomy. The digital world and social media now provide that platform, and the automobile becomes more a utilitarian tool, rather than an expression of aspiration, achievement, and "self," no longer the distinctive "coming of age" vehicle in which young people liberate themselves from their homes and their parents. Once upon a time, automobiles were central to romantic life. It was once estimated that almost 40 percent of marriages in America were proposed in automobiles. Today, a third of marriages result from meeting up online and through dating apps.

The second reason is cost. An automobile ties up capital with the purchase and entails significant additional annual costs in terms of fuel, parking, insurance, and repairs. Young people with college debts or "gig" jobs may not want the added burden of ownership.

Compare the economics. Let's say the average number of miles driven in a year in the United States is twelve thousand. Owning a car for that year would cost around $7,000, including the proportionate cost of car ownership, fuel, and other operating expenses. Given the average ride-hailing trip, $7,000 would equate to around six hundred separate trips per year, or twelve per week—almost two per day. Of course, on the other side of the ledger, there's no residual value from Uber or Lyft rides, as there is when selling a used car. And no pride of ownership.

It's possible, of course, that driver's licenses and car purchases are not being avoided but rather delayed. As today's under-thirties get older and their incomes rise, they will be in the market for a car. People marry later and have children later, but eventually they do, and they move to the suburbs, and then they are buying SUVs, ferrying kids, and driving more. Moreover, many people may prefer to "own" their own mobility. One other factor to note: Ride hailing does not necessarily mean few total miles driven. On the contrary, it can well mean increased mileage driven, as the accessibility and convenience stimulate more usage of vehicles—fewer people taking the bus or the subway and more people in individual cars, albeit driven by someone else.

Autos, like oil, is a global business. So what counts is not just what happens in the United States, or Europe, or Japan. Despite all the efforts in China to manage congestion, owning a car continues to be a powerful aspiration. And then there is the other giant—India. There are only 48 million cars in India compared to China's 240 million, despite similarly sized populations. But India is also a very big emerging market, and the share of young people in India's population is much greater than in China's, and its road system is far less developed. But economic growth will raise incomes and finance new infrastructure, and India's massive cohort of young people will end up having a huge impact on the global auto and oil industries.

THE "TRIAD"—THE CONVERGENCE OF ELECTRIC VEHICLES, RIDE hailing, and self-driving cars—is far from sure. It will take electrics a long time to catch up with gasoline-powered cars as a share of the fleet. People may continue to want to own cars and drive themselves. Autonomous vehicles at scale are far from proved.

Still, the new convergence is prefigured in the dizzying pace of moves—partnerships, acquisitions, and investments. The players include the likes of Apple and Google, the new ride-hailing companies, established technology companies, venture-funded start-ups, universities—and, of course, automobile companies seeking to ensure their central place in the future.[2]

In Larry Burns's view, traditional auto companies were half a decade late in waking up to the mobility revolution. "The car companies make hardware," he says. "They design steering wheels and headlights and door handles, and they're really good at getting it all in the same building at the same time in order to assemble an automobile that's going to work in hot and cold and night and day for hundreds of thousands of miles. But the self-driving question is essentially a software and mapping problem. It requires writing lots of computer code, which isn't a car company's strength. The reason the auto industry was slow was that they didn't have a bone-deep understanding of digital technology, or the full capability of computers and big data."[3]

There was another big factor. Automakers are highly regulated and can be subject to huge penalties and class action suits; and, owing to the nature of their product, they are careful and cautious and risk averse.

"A few years ago there were a lot of articles being written saying, 'Hey, tech companies are going to win in this, the old-line players are going to lose—end of story," said Bill Ford. "But it's not that simple. A vehicle's architecture has to marry up with the brains, the self-driving system, in a way that works. If you don't iterate in one aspect, you can't iterate in the other. Then there's a mismatch. You need both working together."[4]

The whirlwind of hookups demonstrates the urgency to be ready for this new world, but also the complexity. "It's a matter of surviving or dying," explains Toyota's chief executive, Akio Toyoda.[5] Toyota creates a billion-dollar partnership with MIT and Stanford for autonomous research and then invests in Uber. Google spins off its autonomous vehicle group into Waymo, a separate subsidiary of Alphabet, Google's parent company. Waymo, in turn, establishes a partnership with Lyft and launches a driverless taxi service in the Phoenix area. Ford spends a billion dollars to acquire Argo AI, an artificial intelligence company, on top of a series of other investments. Apple puts a billion dollars into China's DiDi. Audi, Daimler, and BMW spend $3.1 billion acquiring mapping capability from Nokia. General Motors puts $500 million into Lyft. GM also buys a start-up barely three years old called Cruise Automation for a reported billion dollars.

It's not only technology but also the scale of investment that is pushing different entrants into partnerships. Ford established a partnership with Volkswagen that includes developing autonomous and electric vehicles. "We both came to the same realization," said Bill

Ford. "The world we're heading into is huge. The market is potentially huge, but the capital requirements are potentially huge. As big as our balance sheets are, no company can do this alone."[6]

Economics is a big reason for the convergence of autonomous vehicles and ride hailing. The biggest cost for ride-hailing companies is the driver. Eliminate drivers and the cost of providing mobility drops. Of course, the nonpayment to human drivers, who were BYOC (Bring Your Own Car), will be offset to some degree by the need to buy cars. But the ride-hailing companies will gain greater leverage from the fact that self-driving cars will be on the road most of the day—and night. They won't need coffee breaks, let alone sleep.

Here is where the electric car can gain traction. While an electric car may cost more, its operating costs will be lower because the costs of electricity per mile will be lower than that for gasoline (unless internal combustion engines become much more efficient). So if you're running a massive fleet of cars that is working most of the time, the electric car becomes compelling. Moreover, the recharging conundrum can be solved with a central charging location.

"People are still physically going to need to move from Point A to Point B," Mary Barra of GM said. "But they're going to have multiple ways that they can do that." Her ultimate goal, she said, is "a world with zero crashes, zero emissions, and zero congestion." And that means largely a world, in her view, of autonomous electric vehicles. But, as it is for other companies, getting there is an immense challenge for GM, which sells many millions of vehicles a year around the world, virtually all of them powered by oil fuels and with a driver at the wheel.[7]

The world's major automakers will be digging in to protect their positions in the new mobility—and ensure their survival. For automakers in countries like China and India, this could be the avenue to break out of their national boundaries and become global competitors.

But the tech giants, looking for the next trillion-dollar market, could wield their mastery of software and platforms and their capital hoards to become dominant players—not necessarily in manufacturing but in the overall industry. After all, Apple does not manufacture its own phones.

And here is where we could see the emergence of new types of companies—"Auto-Tech." These would either be vertically integrated or strategically allied companies, from vehicle manufacture, to fleet management, to ride hailing through their own platforms. They would be the master coordinators of multiple capabilities—manufacturing, data and supply chain management, machine learning, software and systems integration, and the delivery of high-quality "mobility as a service" to customers around the world.

At this point, there is still no tipping point where the benefits of new technology and business models prove so overwhelming that they obliterate the oil-fueled personal car model that has reigned for so long. The bulk of cars are still bought for personal use. The traditional companies are still in business. And the empire can strike back—internal combustion engines that become much more efficient.

Altogether, the world of autos—and their fuel suppliers—has become the arena for a new kind of competition. It is no longer just about selling cars to consumers for personal use. No longer just automakers versus automakers, no longer gasoline brands versus gasoline brands. It has become multidimensional. Gasoline-powered cars versus electric cars. Personal ownership of cars versus mobility services. And people-operated cars versus robotic driverless cars. The result is a battle among technologies and business models, and a struggle for market share. Change does happen, just not overnight. With that said, electricity is advancing. Oil is no longer the unchallenged king in automotive transportation. But for some time to come, its writ will still extend quite widely across the realm of transportation.

———

THE NEW MOBILITY WOULD CREATE MAJOR DISLOCATIONS.
A shift from "mobility as a product" to "mobility as a service" would
result in a significant drop in new car purchases by individuals. What
would grow instead would be fleet purchases. Since the cars would be
used not 5 percent of the time, but 70 or 80 percent of the time, the
numbers of fleet sales would not compensate for the lost personal sales.
The traditional automotive supply chains, involving thousands of companies around the world, could be disrupted not by trade wars but by
innovation and technology, including robotics and 3D manufacturing.

While mobility as a service delivers value to users, it would, if it
includes autonomous vehicles, have a wrenching impact on the labor
force. Taxi drivers, Uber, Lyft, and DiDi drivers, people who work in
gasoline stations and auto dealerships, autoworkers, workers in public
transportation systems that see ridership decline—this adds up to very
big job losses. What new jobs do they find—or can they find? Whom
do they blame for their lost vocations? What happens to their pensions?
A sign of what could be ahead is evident in what has happened to taxi
drivers who have purchased medallions for the right to operate. Traditionally, they would sell their medallions at a higher price when they
stopped driving, helping to support them in retirement. But now the
price of medallions has plunged far below what drivers paid for them,
and no money for retirement.

For the global automobile industry itself, the future is perplexing.
The paradigm on which it is based is the premise of growth in emerging markets and replacement in mature markets. The planning horizon
for a new model is typically five to seven years. Yet the future is unfolding faster than that.

Henry Ford, more than anyone else, created the current business
model for the auto industry. As he once said, if he had asked his customers at the time what they wanted, they would have said a faster horse.

This business model has lasted for more than a century. "That was a long run, wasn't it?" reflects his great-grandson Bill Ford. But now "everything" about the business model "is changing. The ownership model is changing, the propulsion system of a vehicle is changing. Every bit of our business is being disrupted.

"A couple of things are very clear," he adds. "Number one, there's a real push to electrification. Number two, autonomous driving is going to happen, although the timing we can debate. What's unclear, though, are all the other ancillary businesses that will be developed around this new world. Those are still in the 'what if' and experimental stage.

"I'd like Ford to be around for another hundred years. But we're not the kind of business that can pivot on a dime. The more certainty we have, the better off life is. Unfortunately, right now, we seem to be in a world that doesn't have a lot of certainty."[8]

Automakers will struggle over what to build and at what pace and what to fund. They have to cope with policies that restrict and raise the cost of cars with internal combustion engines, while subsidizing EVs or setting quotas for EV production. But when electric vehicle sales reach higher levels, will the sheer scale of the subsidies force governments to whittle them down, especially after the massive government debt burden arising from the coronavirus? And the companies have to worry what that will do to the anticipated consumer appetite for EVs. But at least for now, the "demand" for EVs is largely coming not from consumers, but from governments whose evolving policies are shaped by climate concerns as well as by urban pollution and congestion.

There will be some kind of meshing of automobile manufacturing with fleet management, ride-hailing, and software platforms. It could take many different forms. But the result could well be the rise of a new breed of firm—Big Mobility companies that embody what would be the transformative world of Auto-Tech.

CLIMATE MAP

Chapter 41

———— ———— ———— ————

ENERGY TRANSITION

Mapping the path to a lower-carbon world will be a defining challenge in the decades ahead. Climate change caused by humans has been a topic of serious study for four decades. But the mobilization of public opinion on climate is more recent, driven not just by studies but by an increasingly intense focus on events around the world—forest fires, droughts, torrential rainfalls, coastal flooding, heat waves, melting ice, and hurricanes.

This alarm about climate is the great motivator for the "Energy Transition." The term is widely embraced—possibly the two most used words in talking about the future of energy. It aims to limit temperature rises to less than two—or 1.5—degrees centigrade above pre-industrial levels, but beyond that there is no clear consensus. Is it to be a transition to a "lower-carbon energy" system—that is, one in which CO_2 emissions from human activities go down over time? Or is it to "deep decarbonization," in which emissions go down much faster? Or is it a "zero carbon energy" system—no human-related emissions? Or a "net zero carbon" system, in which emissions are canceled out by mechanisms that absorb the carbon? There is certainly no consensus as to the speed of the transition, nor as to what the transition will look like decades from now, nor as to the cost—nor as to how it is all to be achieved.

Energy transitions are not new. They have been going on for a long time and unfold over time. Previous energy transitions have primarily been driven by technology, economics, environmental considerations, and convenience and ease. The current one has politics, policy, and activism more mixed in.

The first energy transition began in Britain in the thirteenth century with the shift from wood to coal. Rising populations and destruction of forests made wood scarce and expensive, and coal came to be used for heating in London, despite fumes and smell. The need for coal for warmth became more urgent during Europe's several-century-long Little Ice Age, from which the world has since warmed. It was so cold that the Thames froze over; and it was said that Queen Elizabeth I strolled on the ice. Coal's advantage was price and availability, not superior or differentiated performance.

For a specific date in the first energy transition—coal's becoming a distinctive industrial fuel, superior to wood—January 1709 could well do. That month, Abraham Darby, an English metalworker and Quaker entrepreneur, working his blast furnace in a village called Coalbrookdale, figured out a way to remove impurities from coal, thus turning it into coke, a higher-carbon version of coal. The coke replaced charcoal, which is partly-burned wood, and had been the standard fuel for smelting. Darby was convinced, he said, "that a more effective means of iron production may be achieved." He was also ridiculed. "There are many who doubt me foolhardy," he said. But his method worked.[1]

Though it took a few decades to spread, Darby's innovation lowered the cost of smelting iron, making iron much more available for industrial uses, helping to spur the Industrial Revolution. Coal was the fuel source for Thomas Newcomen's steam engine, developed around the same time as Darby's innovation to pump water out of coal mines, and for James Watt's much-improved engine, the commercial introduction of which in 1776—the same year as the outbreak of the American Revolution and the publication of Adam Smith's *Wealth of Nations*—was a decisive moment in the Industrial Revolution. But as energy scholar Vaclav Smil observes, "Even with the rise of industrial machines, the nineteenth century was not run on coal. It ran on wood, charcoal, and crop residues." It was not until 1900 that coal reached the point of supplying half of the world's energy demand. Oil was discovered in northwest Pennsylvania in 1859. But it took more than a century—not until the 1960s—for it to supplant coal as the world's number one energy source. Even so, that hardly meant the end of coal, for consumption has continued to grow. As for natural gas, global consumption has increased 60 percent since 2000.[2]

THE FRAMEWORK THAT HAS SHAPED THE GLOBAL DISCUS-sion of climate change has been the periodic reports of the Intergovernmental Panel on Climate Change, known as the IPCC, under the auspices of the United Nations. This is a self-governing network of scientists and researchers that issues periodic reports, with each one raising further the crescendo of alarm. The first, in 1990, said that the earth was warming and that the warming was "broadly consistent with the predictions of climate models" as to largely "man-made greenhouse warming." But the changes, it added, were also broadly consistent with "natural climate variability." By 2007, in its fourth report, the IPCC was much more categorical—it was "very likely" that humanity was responsible for climate change. The actual report was not as categorical in all dimensions as the summary for policymakers. "Large uncertainties remain about how clouds might respond to global climate change," it said.

That same year, the Nobel Peace Prize was awarded to Al Gore, the former U.S. vice president who had become a leading climate activist and who declared that the world faced a "planetary emergency." Sharing the prize was the IPCC, represented by Rajendra Pachauri, its chairman for thirteen years. Shortly thereafter, he told the CERA-Week conference in Houston that the IPCC's warning is "not based on theories and supposition. It's based on analysis of actual data which is now so extensive and overwhelming that it leaves no room for doubt." He would later describe "the protection of Planet Earth" as "my religion."[3]

The fifth IPCC report, issued in 2014, was the starkest yet. "Human influence on the climate system is clear," and "emissions of greenhouse gases are the highest in history. Recent climate changes have had widespread impacts on human and natural systems. Warming of the climate system is unequivocal, and since the 1950s, many of the observed changes are unprecedented over decades to millennia." Some raised questions about aspects of the IPCC report—disagreements among the several dozen different models, observations about the frequencies of hurricanes and the rate of ocean rising, understanding of feedbacks, and underestimation of natural variability. But they were a distinct minority.[4]

The 2014 IPCC report set the stage for what was to unfold in Paris a year later, which would give a whole new import to "energy transition" and make it a central global topic.

THE PARIS CLIMATE CONFERENCE—OTHERWISE KNOWN AS United Nations COP 21—convened in the northern Paris suburb of Le Bourget at the end of November 2015. Just two weeks earlier, an ISIS jihadist assault had savaged the city, leaving 130 people dead and hundreds more injured. And so security was now extremely tight as fifty thousand people descended on the French capital to debate climate policy.

The organizers were determined that this meeting be decisive after the chaotic COP 20, held in Copenhagen six years earlier, which then–secretary of state Hillary Clinton had at the time described as "the worst meeting" she had attended "since eighth-grade student council."

The essential formula for avoiding another "Copenhagen" had really been set out in the Great Hall of the People in Tiananmen Square in Beijing a year earlier, in 2014. The United States and China—together responsible for over one-third of global greenhouse gas emissions—had until then been adversaries on climate. China and other developing nations had asked why they should "pay" for all the emissions that the developed nations had been putting into the atmosphere for a century by restraining their own energy use and thus holding back their own development. But in November 2014, standing together in the Great Hall, Barack Obama and Xi Jinping announced a joint commitment that their two countries would adopt significant new measures to reduce emissions. But their respective commitments had different timelines. The United States, Obama promised, would reduce its CO_2 by more than 25 percent in 2025 compared to 2005, much facilitated by increased natural gas use in power generation. China's carbon emissions could continue to rise, peaking only by 2030.[5]

Altogether, representatives of 195 countries and the European Union, joined at various times by 150 leaders of countries, attended the Paris conference, which began on November 30, 2015, to be followed by almost two weeks of arguing and grappling.

Just after seven in the evening on December 12, after an unexplained delay that left nerves fraying in the audience, French foreign minister Laurent Fabius strode out to announce a final agreement. The room erupted in cheers, thunderous applause, ovations, whistling, embracing, and even weeping. The UN secretary-general called it "truly a historic moment." There was, he said, "no Plan B."

What they had adopted was not a treaty but rather a compact to take actions that were intended to prevent temperatures from rising to two degrees centigrade above preindustrial levels in this century—and, it was hoped, no more than one and a half degrees. It was up to each country to come up with its own "nationally determined contribution"—what became known as NDCs—based upon its particular situation, laws, regulations, volition, and mood. These NDCs would not be binding, but rather voluntary. "Nonbinding" was crucial for Barack Obama, for a treaty would have to be submitted to the U.S. Senate, where it would never get the votes required for ratification. While not mandatory, these NDCs would have the power of declarative policy and the compelling force of the global consensus. Developed countries promised $100 billion a year in aid to developing countries to help them meet climate targets. "Make no mistake," Obama said. "This gives us the best possible shot to save the one planet we've got."

The agreement went into "force" a year after the conference, on November 4, 2016. As events would turn out, however, Donald Trump was elected president just four days later. He viewed the agreement decidedly differently. The compact, Trump said, "gives foreign bureaucrats control over how much energy we use on our land, in our country." He called climate change a Chinese "hoax." In the spring of 2017, Trump took to Twitter to announce that he was beginning the process to withdraw the United States from the agreement.[6]

NOTWITHSTANDING TRUMP'S ANNOUNCEMENT, WHICH IN ANY event would take three years to implement according to the agreement, "Paris" changed the global debate. For the most part, the time was past for discussing uncertainties about aspects of climate change—rising sea levels, intensity of hurricanes, or climate models. The subject now was the warming planet, and now there were two distinct political eras when it came to the politics of climate: "Before Paris" and "After Paris."

While the degree of "confidence" rose with each new iteration of the IPCC, the basic argument was consistent. Here is how the logic works,

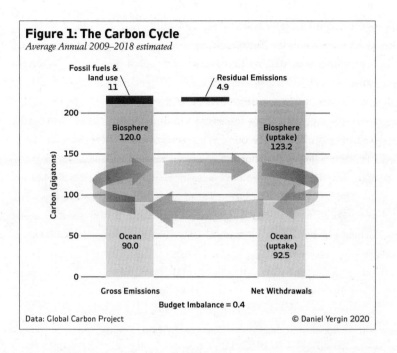

Figure 1: The Carbon Cycle
Average Annual 2009–2018 estimated

Fossil fuels & land use 11

Residual Emissions 4.9

Carbon (gigatons)

Biosphere 120.0

Biosphere (uptake) 123.2

Ocean 90.0

Ocean (uptake) 92.5

Gross Emissions
Net Withdrawals

Budget Imbalance = 0.4

Data: Global Carbon Project

© Daniel Yergin 2020

using average annual data from the years 2009–2018 (see Figure 1): Some 210 gigatons of carbon were annually, on average, naturally released by such processes as the decay of plants and breathing by people and animals. But 9.5 gigatons came from fossil fuels and 1.5 from land use. This added up to a total of 221 gigatons released. But only 215.7 were captured in the natural annual cycle—that is, absorbed by vegetation and the ocean—leaving a residual 4.9 gigatons in the atmosphere uncaptured. (There is also a budget imbalance factor.) That uncaptured 4.9 gigatons is only 2.2 percent of the naturally captured CO_2. That may seem a very small amount in any given year. But, over the years, it accumulates and builds up in that band of gases known as earth's atmosphere. Water vapor is the most prevalent greenhouse gas. Others include nitrous oxide and methane. Some of these gases dissipate after a year or ten years; others last much longer. Some are more potent than CO_2. These greenhouse gases become a shield of sorts, a global "greenhouse" around the planet, retaining more of the sun's heat, which otherwise would flow back into space. The result is greater warming for the earth—thus known as the "greenhouse effect."[7]

As the climate consensus has crystallized, concern and fervor have risen, fueled by the fear that an approaching "tipping point" will lead to "runaway climate change." The growing dread is reflected in the vocabulary; "global warming" and "climate change" have given way to "climate crisis" and now "climate emergency" and "climate catastrophe."

The Swedish activist Greta Thunberg became the voice of this urgency, beginning when, in August 2018, as she put it, she "school-striked for the climate" outside the Swedish Parliament. Her message became zero carbon. "Expansion of airports," she told the British Parliament in the spring of 2019, "is beyond absurd." At the U.N. Climate Summit the following September, she said, "You have taken away my dreams and my childhood with your empty words," adding, "How dare you?" She warned that a new "mass extinction" loomed unless climate was quickly addressed. Not long after, she elaborated in a coauthored article her thinking as to the sources of the global warming: "Colonial, racist, and patriarchal systems of oppression have created and fueled" the "climate crisis," adding, "We need to dismantle them all."[8]

Finance and energy investment have become a new arena for climate. The claxon was sounded by Mark Carney, the then-governor of the Bank of England, in 2015, several weeks before the Paris conference in a speech to the venerable insurance organization Lloyd's of London. Climate, he said, has become "a defining issue for financial stability" and created "systemic risk" for the world's financial system, which, in central bank language, harked back to the global financial crisis of 2008. He warned that investors and insurers faced the growing risk that oil and gas companies' reserves in the ground would remain in the ground—"stranded"—unable to find their way to market because demand had faded away, or, as Carney put it, be "literally unburnable" over thirty years because of policies imposed to achieve the "two-degree world." That would mean that the value of the companies would plummet, he argued, perhaps even becoming worthless, leaving investors holding equity that had also become worthless. He called for a "sweeping reallocation" of investment away from traditional energy companies to finance "the de-carbonization" of economies.

Some, in response, pointed out that companies' oil and gas reserves are not valued over thirty years by investors, but only about ten years. In any event, most of the world's oil and gas reserves are owned by national governments, and not shareholders in Britain or the United States.[9]

Thereafter, the Financial Stability Board, whose members are central banks, focused on "climate-related financial disclosure" that requires companies to disclose how their investments and strategies comport with achieving the objectives of the two-degree world.

Pension funds and other investors are now pressing energy companies to explain how their strategies and profitability would fare under the terms of the 2015 Paris Agreement. In his 2020 annual "Letter to CEOs," Larry Fink, head of BlackRock, the world's largest investment company, declared, "Climate change has become a defining factor in companies' long-term prospects" and that "in the near future—and sooner than most anticipate—there will be a significant reallocation of capital." BlackRock, he said, will "place sustainability at the center of our investment approach" and will require companies to "disclose climate-related risks." When BlackRock—$7.5 trillion under management—speaks, companies listen. One example of the "reallocation of capital" is the growth in "green bonds." These provide financing for infrastructure related to renewables and infrastructure. From $50 billion issued in 2015, the total reached $257 billion in 2019.

Divestment—the movement to get investors to sell their shares in energy companies, and banks not to lend to them—is gaining momentum. There is also pushback. Microsoft founder Bill Gates, who is investing billions in seeking technology breakthroughs for lower-carbon energy, has said, "Divestment, to date, probably has reduced about zero tons of emissions." Consumer demand still has to be met. There is no obvious way that people around the world can any time soon dispose of their 1.4 billion cars that run on oil, and people will still need to heat and air-condition their homes. There are other aspects as well. Dividends from BP and Shell were funding 20 percent of all pensions in Britain.[10]

On many college campuses, divestment has become a contentious issue. One of the great traditions in American football is "The Game"—Yale versus Harvard—which has been played since 1875. During half-time at the November 2019 game, hundreds of students carried the fight against climate change to the football field, suddenly pouring onto it, delaying the second half. Their targets were the investment offices of Yale and Harvard, which they wanted to divest of their energy holdings. One student warned, "Life at Yale cannot go on as usual until Yale divests."

Their particular ire was directed at David Swensen, the legendary head of the Yale endowment, whose investment returns had, among other things, financed the scholarships of many students. "If we stopped producing fossil fuels today, we would all die," Swensen had recently said. "We wouldn't have food. We wouldn't have transportation. We wouldn't have air conditioning. We wouldn't have clothes." He added, "The real problem is the consumption" and "every one of us is a consumer." The president of another major university was surprised when told that the financial loss from divesting energy would be greater than the university's entire budget for undergraduate scholarships.[11]

Pressure comes in other forms. Annual stockholder meetings of banks and energy companies have been disrupted by activists rappelling down from the ceiling, and opponents of hydrocarbons have stepped up their efforts—both physically and in courts—to block pipelines and other projects. A plan was developed at a meeting in La

Jolla, California, in 2012 to plot out a "tobacco" strategy—that is, to brand oil and gas companies as peddlers of a dangerous and addictive product, like the tobacco companies. The difference, of course, is that tobacco is a habit, while oil and gas are enablers of modern life.

This strategy has played out in the years since. In line with the spirit of La Jolla, the British newspaper *The Guardian* announced that, as a self-described climate campaigner, it would no longer accept advertising from oil and gas companies. It added, however, that it would have liked to accede to demands by Greenpeace and other "readers" that it also reject advertising from automobile and travel companies. But if it did so, it explained, it would be a "severe financial blow" that would force it to fire many of its journalists. But it did promise, from thereon it would no longer use the term "climate crisis" in its news columns, but now all references would be to "climate emergency."[12]

"Fighting climate change" has now become a broad social movement, engaging people not only in terms of policy and business decisions but also increasingly in their personal lives and sense of personal responsibility. In Britain, the Royal Shakespeare Company terminated an eight-year gift from an oil company because, it said, of the "strength of feeling" among young people. Some people have become vegans so as to give up meat and dairy products from methane-producing cows. Invoking "flight shaming" that has emerged in Scandinavia, a headline in the *New York Times* asked, "How Guilty Should You Feel About Flying?" The answer seemed to be if you did more than six flights a year. So significant has this personal dimension become that one of the major U.S. television networks invites "those who care deeply about the planet's future" to go to its "confessions" page on its website to share how personally "you have fallen short in preventing climate change."[13]

Chapter 42

GREEN DEALS

C limate has risen to the top rung of policy in a number of nations. Of the G20 countries, fourteen deploy or have announced plans to deploy carbon pricing mechanisms or some kind of carbon tax. The United Kingdom announced that it will legally commit to zero carbon emissions by 2050. Two dozen other countries are promising the same, though the path for most is far from clear.

Europe, more than anywhere else on the planet, is seeking to build an "After Paris" world. And, more than anywhere else, it is seeking to use government policy to drive this energy transition. Declaring that climate is Europe's "most pressing challenge," Ursula von der Leyen, president of the European Commission, has pledged to turn all of Europe into "the first carbon neutral continent in the world." The head of the European Investment Bank, in announcing an end to finance for natural gas projects by 2022, went even further, saying, "Climate is the top issue on the political agenda of our time."

The EU's "green deal" aims to make net zero carbon legally binding for the Continent by 2050. "Net zero" requires a further word of explanation, for it will be fundamental to future discourse. The World Resources Institute explains, "Net zero carbon" is not the same as "zero carbon." "Net" means minimizing "human-caused emissions" to "as close to zero as possible," with "any remaining" emissions balanced out by the "equivalent amount of carbon removal"—for instance, by "restoring forests" or with carbon capture. In other words, carbon can be released, but in some way an equal amount of carbon must be captured. Today, Europe is responsible for about 12 percent of the CO_2 emissions released by burning carbon (see Figure 2).

A basic tool for Europe's achieving net zero is the *Taxonomy*, a 66-page report, backed up by a 593-page technical analysis, by dozens of "leading thinkers" that evaluates sixty-seven economic activities as to "environmental friendliness" and "sustainability." It is meant to direct investment flows. The EU will require investment managers to label how "Taxonomy-compliant" their funds are. The *Taxonomy* will be used to guide new regulations and government programs for "green investment." While "very clean" natural gas may be acceptable, most natural gas and all of nuclear are problematic under the *Taxonomy*, coal is to be eliminated, and all coal mines to be shut down. In addition, six thousand firms with over five hundred employees in Europe will be required to identify which of their activities are environmentally sustainable. The EU is also considering "border taxes"—otherwise known as tariffs—on goods from other countries that do not have equivalent carbon-pricing programs to Europe's. This is sure to create contention with Europe's trading partners.[1]

Altogether, the EU has staked out a position on the parapets of "green commanding heights." For Europe's 2050 goal is breathtaking: nothing less than reshaping economic activity, directing investment, and rebuilding Europe's economy over the next three decades. The program will aggregate power to the European Commission in terms of regulating businesses and allocating capital. As to why this should be the EU's "first priority," as von der Leyen calls it, given all its other issues, including its own future, one European businessman close to the European Commission hypothesized that also "they are looking for a new narrative for the European ambition."

"The costs of the transition will be big," von der Leyen said, "but the costs of inaction will be much bigger." The EU has established a 100-billion-euro "Just Transition" mechanism to help buffer the impacts for countries still reliant on coal. Yet, at this point, the costs of "net zero carbon" are murky. As a paper from the Peterson Institute for International Economics explained, "Whether the transition to a climate-neutral economy will improve or hurt growth is a quantitative issue. Unfortunately, we know too little about it." While arguing that prosperity depends long-term on decarbonization, it said that over the

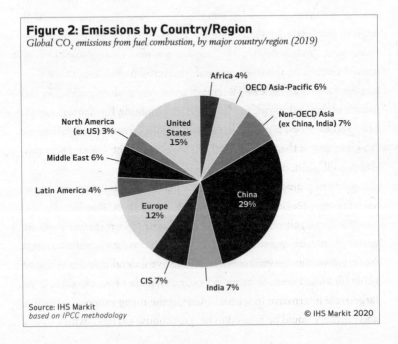

Figure 2: Emissions by Country/Region
Global CO$_2$ emissions from fuel combustion, by major country/region (2019)

Africa 4%
OECD Asia-Pacific 6%
Non-OECD Asia (ex China, India) 7%
North America (ex US) 3%
United States 15%
Middle East 6%
Latin America 4%
China 29%
Europe 12%
CIS 7%
India 7%

Source: IHS Markit
based on IPCC methodology
© IHS Markit 2020

next five to ten years, "decarbonization will inevitably reduce economic potential."

The green deal got a jump start in an $825-billion anti-crisis package that von der Leyen introduced in May 2020. The presentation described "the European Green deal as the EU's recovery package" and as "Europe's growth strategy," with a substantial share of that $825 billion going for wind, solar, "clean hydrogen," renovating buildings, "clean mobility in our cities," and the installation of a million charging points for EVs.[2]

The overall objective—net zero carbon by 2050—is a daunting ambition. How daunting is underscored by the estimate that, for Europe to achieve its target, per capita emissions will have to decline to the level of India, where the per capita income is about $2,000 a year, compared to Europe's $38,000.

FOR SOME, THE YEAR 2050 IS TOO FAR AWAY AND A TRANSItion over thirty years is too long. That is the essence of the Green New Deal launched on the steps of the U.S. Capitol by the left of the Democratic Party in 2019, led by Representative Alexandria Ocasio-Cortez, who had decided to run for Congress after joining the protest against the Dakota Access pipeline. In its content, this Green New Deal synced up closely with the Green New Deal platform of the Green Party candidate, Jill Stein, in the 2016 presidential election.

The talking points released just prior to the official release of the Green New Deal called for the United States to be powered by 100 percent clean and renewable energy by 2030. The role of the private sector would be secondary in this endeavor, as "government is best placed to be the driver," mobilizing "massive federal investment." The program would seem to ground existing airplanes because there is no large-scale alternative to jet fuel. Among the many proposals, farmers and ranchers would be pressed to be "greenhouse gas free," which would mean doing away with all their cows, owing to their bovine methane emissions. The non-energy proposals included government-guaranteed jobs. "The world is going to end in twelve years," Ocasio-Cortez said, "if we don't address climate change."

The talking points reflected the viewpoint of some of the advocates, but not all. They were pulled back just prior to the official launch of the Green New Deal. Ocasio-Cortez's partner in launching the program, Senator Edward Markey, a veteran of decades of legislative battles, explained that the 100 percent was not a forecast, but rather was "aspirational." The actual congressional resolution was more general, calling for a ten-year "new national, social, industrial, and economic mobilization on a scale not seen since World War II and the New Deal" to generate a Green New Deal that would meet a host of objectives— for instance, "counteract systemic injustices"—but principally "to achieve net-zero greenhouse gas emissions" and "100 percent of the power demand in the United States through clean, renewable, and zero-emission energy sources."[3]

In the 2012 U.S. presidential debates, not a single question was asked about climate. In the 2016 president debates, climate got a total of five minutes. In 2020, CNN hosted a seven-hour climate town hall on the subject. Climate became a major issue in the Democratic primaries. Some candidates called for a ban on fracking. Climate now polled as a major issue, especially for millennial voters.

During the primary campaigns, candidates vied over climate action plans. Joe Biden's $1.7 trillion program and Elizabeth Warren's $4 trillion paled next to Bernie Sanders's $16.3 trillion. Sanders's long list of would-be initiatives included $35 billion for people to reforest their front lawns or turn them into "food-producing spaces" as well as "ensure fossil fuels stay in the ground" and ban both exports and imports of oil. But how you ban imports and exports of oil and at the same time ban domestic production of oil—and still have a functioning economy and society—was left unexplained. Nor what happens to the 12.3 million jobs in the United States associated with the oil and gas industry.

Still, for many the determination and commitment to speed an energy transition are there and deeply felt. But will the money be there after the costs of the coronavirus crisis and the trillions of dollars and pounds and euros of government debt wracked up to deal with it?

Chapter 43

———

THE RENEWABLE LANDSCAPE

What will the energy landscape look like twenty or thirty years from now? Yes, it will be "lower carbon." But what comprises that system? At this point, it appears that the energy system in the next decades will continue to be, as it has been in the past, a mix, but a shifting mix, differing considerably among countries, but also certainly lower-carbon than today's.

One will see a multitude of solar panels and wind turbines across the energy landscape. These are the "modern renewables," as opposed to the traditional renewables of hydropower, wood, and biomass. They will be among the main engines for achieving the climate goal of a transition from CO_2-producing electricity generation to carbon-free generation. Nuclear power remains today the largest source of carbon-free generation, but the gap is narrowing with wind and solar. Although wind and solar are called "modern" renewables, neither is exactly new. Both are about half a century old.

The theoretical foundation for today's solar panels—photovoltaics (PVs)—was provided by Albert Einstein in his 1905 paper "Concerning the Production and Transformation of Light." The light arriving from the sun, he said, was composed of photons, packets of energy, which could dislodge electrons surrounding the nucleus of an atom, creating an electric current. Einstein was awarded the 1922 Nobel Prize in Physics for this paper, his "discovery of the law of the photoelectric effect." But it was not until 1953 that the photovoltaic effect was demonstrated at Bell Labs in New Jersey.

The modern solar industry really began in 1973, with the launch of two ventures. One was a unit of Exxon. The other was by two scientists who had worked on the U.S. space program—Joseph Lindmayer and Peter Varadi—both of them refugees from Europe. Over the following three decades, other ventures were launched, mainly by oil companies, hedging against an uncertain energy future, and by Japanese tech companies, sparked by Japan's alarming lack of natural resources. The appeal of solar has been enormous ever since. As Professor Martin Green, a leader in solar research for decades, puts it, "The whole photovoltaic technology itself is a bit magical. Sunlight just falls on this inert material and you get electricity straight out of it." But for many years the markets for PVs remained niche for off-grid uses—to bring electricity to isolated homes or remote locations or, for that matter, to marijuana growers, who did not want oversized utility bills to call the attention of law enforcement to their illicit businesses. The first introduction to solar power for many people was the solar-powered pocket calculator.[1]

What catapulted solar into the mainstream was the marriage of Germany's environmental politics with Chinese manufacturing prowess. Beginning in the 1990s, Germany's "feed-in" tariff laws required utilities to buy renewable electricity at high prices from generators and then spread the cost over all electricity bills. This law laid the foundation for a broad shift—Germany's *Energiewende*, the "energy turn"— which aimed to replace conventional energy with wind and solar. The generous subsidies from the feed-in tariffs speeded renewable deployment, while also leading to the highest residential energy prices in the European Union.

Companies rushed in to meet the large and growing demand for solar and wind. However, while the solar market created by the *Energiewende* may have been in Germany, the panels could come from anywhere. In time, most of them would come from the new solar juggernaut that would rise in China and eventually extinguish German manufacturers.

As late as 2006, China had barely a walk-on role in photovoltaics production. But then came the big push by Chinese entrepreneurs, backed by China's central government and regional and local authorities, in the form of land, large low-cost loans, and other subsidies. This coincided with the increased push for solar not only in Germany but also in Spain and Italy, fueled by substantial subsidies. By 2010, there were 123 solar panel manufacturers in China.

Between 2010 and 2018, China's manufacturing capacity for solar cells increased fivefold, which was well beyond the global demand. Many more solar panels were pouring out of China than the market could absorb. Prices came tumbling down. As they gained market share, Chinese companies also struggled under great financial pressure. Some went bankrupt. Over two years, the Chinese Development Bank extended $47 billion in credit to keep money-losing Chinese companies afloat.

To ease the pain of this overcapacity, as well as to support employment, the Chinese government set out to create a new market within China for the beleaguered solar manufacturers. This was also aimed at meeting the critical national need—reducing the suffocating pollution from older coal-burning plants while continuing to meet the country's surging demand for electricity. By 2013, China had overtaken Germany as the largest market for installed solar panels, and by 2017 it alone represented half of the entire global market.[2]

China now produces almost 70 percent of the world's solar panels.

Adding in Chinese companies that manufacture in other countries brings the total share up to almost 80 percent. China makes 70 percent of the photovoltaic solar cells that are the heart of the panels. When it comes to the solar wafers out of which the cells are produced, China's share is even greater—almost 95 percent. This means that, in green energy, China has already reached the "Made in China 2025" goal of a dominant role in this century's new industries.

China's overwhelming competitive advantage arises from many factors—government support and cheap financing; scale (much bigger factories); reductions in polysilicon prices; focus on costs; proximity to supply chains; standardization of products; and continuing technology improvements. Martin Green points to one other factor. "Present low photovoltaic prices," he says, are also "the outcome of serendipitous combinations of events and personalities," including that a number of leading figures in different Chinese companies worked at various times with his teams in Australia. The cost of solar panels came down an extraordinary 85 percent between 2010 and 2019, driven mainly by Chinese manufacturing and massive capacity, and by technological improvements. Like the advent of shale, a price drop of this magnitude is proving revolutionary for energy. Total installation costs have also gone down substantially, but not to the same extent.[3]

China has also established a decisive position further upstream in the supply chain for solar. It now produces almost 60 percent of the key raw material, polysilicon. It has also made a major effort to build up the domestic PV equipment industry and reduce dependence on Western suppliers.

Solar's ascent has been extraordinary. Global installed capacity in 2019 was 642 gigawatts, fourteen times what it had been little more than a decade earlier. While panels on roofs may be more visible, over half of total capacity installed between 2010 and 2019 is utility-scale—that is, solar parks that feed into the grid.

Overall, the global growth in capacity has been fueled by two things. One is that huge decline in prices and what the renewable advocacy organization REN calls "cutthroat pricing" resulting from the overcapacity in Chinese PV manufacturers. The other is a growing global system of incentives, subsidies, and mandates at national, state, and local levels, requiring increasing amounts of renewable energy in electric systems. The PV electric generating capacity added globally in 2019 was bigger than the additions from fossil fuels and nuclear. But that requires an important caveat—"operating time" is much less than "capacity." Much of the fossil fuels and the nuclear are base load or can be managed to correlate with demand for electricity at any given hour. Solar is intermittent, depending for the most part on the availability of sunlight, and actual generation may only equate to about 20 percent of capacity.[4]

THOUGH THE MODERN WIND INDUSTRY, LIKE SOLAR, GOES back to the 1970s, its real growth has only been in this century. In 2000, just 17 gigawatts of wind capacity had been deployed worldwide. By 2019, it had grown to 618 gigawatts. Over 40 percent of total installed wind capacity is in Asia, with three quarters of that in China.[5]

The growth is propelled by forces similar to those that have driven solar, beginning with technical innovation. Taller towers, longer blades, new materials, more sophisticated controls and software, better wind models and weather prediction—all these transform more of the wind into electricity. While 95 percent of total wind capacity is onshore, the industry is venturing offshore, where the winds may be steadier and stronger and the towers larger and the wind resource potentially much greater, but the technical challenges of waves and wear are greater. To date, offshore wind development is concentrated in Europe, mainly in and around the North Sea, although growing in China's offshore, and projects are being pursued off the east coast of the United States.

The second force is the incentives and subsidies, and the strong mandates requiring more renewables in electric generation. And the third is falling costs, the result of what REN, echoing its solar comment, calls "fierce competition in the industry." The last has put great pressure on companies, leading to bankruptcies, restructurings, and mergers.

As with solar, the often-cited "capacity" can be misleading, because wind, like solar, is intermittent. It depends upon the wind blowing. But capacity factors are increasing with the technological advances. Today, the global weighted average is about 25 percent, though higher with new turbines.

Europe has the highest share of wind in electricity generation, accounting for almost 12 percent of total electricity supply. China is about 5 percent, the United States about 7 percent. In the United States, the state with the most electricity generated by wind is not California, as some might expect, but Texas, at 15 percent. If Texas were a nation of its own, it would rank sixth among the countries of the world in terms of installed wind capacity. A good part of the state's wind turbines are in West Texas. It turns out that the Permian Basin in West Texas is bountiful not only in oil and gas but also, above ground, in its wind resources.

THE RAPID GROWTH OF SOLAR AND WIND IS UPENDING THE way the electric power industry has operated for over a century, changing its strategy and structure. "People understand that we need more wind, solar, and hydro," said the CEO of one European utility. "This is fundamentally challenging the model of all the energy companies." They are shifting from traditional "central" generation, based on coal and gas and nuclear power plants, to "distributed and intermittent" generation based on wind farms and solar panels that are spread across the landscape. But "distributed" systems create new challenges, especially in terms of grid stability and reliability, which is a fundamental mission of utilities. "With the advancement of distributed generation, with the monitoring of two-way flows on the system, with managing circuit overload potentials, more technology is going to have to be put into storage and control mechanisms," says Christopher Crane, CEO of the U.S. utility Exelon and chairman of the Edison Electric Institute.[6]

HOW FAST WILL THE TRANSITION BE, AND WHAT WILL IT look like on the other side? Predictions vary widely. In the IHS Markit scenarios, global electric consumption grows by up to 60 percent by 2040. Wind and solar constitute 24 to 36 percent of total generation by that date. Either is a big increase for wind and solar from today's 7 percent globally. The reasons for the variance result from what one would expect—uncertainties and varying assumptions about technology and innovation, and policies and economics.

Wind and solar together have grown dramatically, from 2 percent of U.S. power generation in 2010 to 9 percent in 2019, and they will continue to grow rapidly. Yet U.S. electricity is very unlikely to be 100 percent renewable by 2040. There's neither the technology nor the investment dollars to do that, nor the grid to support that, nor the magic wand to obliterate America's current energy infrastructure and transform the regulatory and political landscape and at the same time assure that the needs of electricity-dependent consumers for reliability are met. Further electrification of the economy will add to demand, which will make reaching 100 percent even more unlikely.[7]

The global picture underlines the same point. Even Denmark, which at times produces more wind electricity than it can consume, also depends on imports of electricity generated by nuclear in Sweden, hydropower in Norway, and coal in Germany to maintain the stability of its power supplies.

One factor to be taken into account is the huge capital investment that is in the ground today, in the long-lived investment of the electric power industry around the world—and the new investment currently being made. In 2011, following the Fukushima nuclear accident in Japan, Germany set out to close its seventeen nuclear reactors by 2022. Yet between 2011 and 2019, China added thirty-four new nuclear reactors, double the number of reactors that have closed in Germany. A few nuclear reactors have closed in the United States because of the difficulty of competing against inexpensive natural gas, but close to a hundred reactors are operating, providing 20 percent of U.S. electricity. As for natural gas, the growth of its contribution to total world energy in 2018 was more than double that of renewables. Adding it all up, energy transition is complex and requires some perspective.

THE WORLD IS INCREASINGLY ELECTRIFIED, BUT THAT ALSO increases the need for reliability and predictability of the electricity supply. The positives of wind and solar are clear. Once the capacity is in place and paid for, there is no cost for the fuel. There are costs, however, both for maintenance and for the overall electric power system in managing renewable power. The variability of wind and solar—that is, their intermittency—poses major challenges. The first is how to integrate large and fluctuating amounts of wind and solar into an electric grid that generally operates on the orderly dispatch of electricity from conventional power plants, correlated to the demand at any particular time of day, and assures reliable power to consumers. As the amount of wind and solar grows, this becomes a larger problem. In his generally positive book on solar power, Varun Sivaram warns, "Rising solar penetration could make the grid less reliable." He adds, "Much more solar is on the way, bringing with it wild swings in power output that could increase the risk of blackouts." He also cites economic risk for the solar power industry—what he calls "value deflation." When solar (or wind) floods into the grid, the swelling tides drive costs down toward zero, lowering investors' returns and potentially undermining the investment made in the solar infrastructure (unless bailed out by government).[8]

In other words, at this time at least, solar and wind cannot go it alone. They need partners. Natural gas generation is a flexible partner for solar and wind. Gas is lower-carbon and lower emissions (with methane control), and gas generation can be ramped up and down to provide balance against the fluctuations of wind and solar.

Integration of renewables will require increasingly complex management of the grid. It also depends on solving the second challenge— storage. Oil can be stored in tanks, natural gas in underground caverns. At this time, however, there is no redoubt for storing large amounts of electricity not just for a few hours, but, as former U.S. energy secretary Ernest Moniz says, for several days. The only notable capability today comes from what is called "pump storage," which is a form of hydropower. But it is very small and limited in growth.[9]

A great deal of effort is being poured into trying to develop utility-scale batteries, economically capable of storing large amounts of electricity that can be dispatched in an orderly way.

NOT SO LONG AGO, WIND AND SOLAR WERE CALLED "ALTERnatives." That is hardly the case today. They are now mainstream and will become mainstays of future electric generation. Over half of that total investment in renewables was, again, concentrated in Asia—with the majority in China. It happens to be the country that, by itself, consumes a quarter of all the electricity generated in the world. And its growing economy needs more electric generation capacity. Even as China continues to build out wind and solar at a rapid rate, it is also adding three new highly-efficient coal-fired plants a month.

Chapter 44

BREAKTHROUGH
TECHNOLOGIES

"We don't have the technologies for advancing the energy transition to net zero carbon," Ernest Moniz says. What are those technologies that will accelerate and reshape the energy transition? A new study, *Advancing the Landscape of Clean Energy Innovation*, led by Moniz and myself, conducted for the Gates Foundation and the Breakthrough Energy Coalition, identified twenty-three technologies with "highest breakthrough potential." They fall into several areas: Storage and battery technology for the intermittency that bedevils large-scale use of wind and solar. Advanced reactors and a new generation of small reactors that would revitalize carbon-free nuclear power. Today, there are more than sixty advanced private-sector nuclear research projects in the United States.[1]

Hydrogen had its false starts almost two decades ago with the hydrogen "freedom car" and a "hydrogen highway" in California. But a renewed focus has emerged on hydrogen to substitute for natural gas in heating and for fuel cells as an alternative to electric vehicles. There's no great mystery here. Hydrogen is already used extensively in oil refining and for making fertilizer. While it is the most common element, hydrogen does not naturally exist by itself, except in rare instances. It is derived by breaking up molecules. Today most hydrogen is produced from natural gas and coal. (A typical natural gas molecule contains one atom of carbon and four of hydrogen.) It can also be made with electrolysis—that is, an electric current running through water. And the source of that electricity could be renewable power, using the excess electricity generated at certain times by wind and solar. Scale will require advances in technology and cost reduction—and spending on infrastructure.[2]

Hydrogen could end up a 10 percent or more player in the energy mix in the future. Indeed, some see hydrogen today as where renewables were two or three decades ago in terms of development. It is striking, too, that hydrogen does not seem to involve geopolitical issues. It is either a tool for countries to meet ambitious decarbonization goals or an opportunity for export, becoming a globally-traded commodity.

Advanced manufacturing, including 3D printing, could have a major impact on energy use by reducing transportation costs. New technologies for buildings could make them much more energy efficient. Electric grid modernization and smart cities could apply digital technologies, increase resilience, and create two-way flows between energy suppliers and customers.

Of critical importance will be large-scale management of carbon itself. Some dismiss carbon capture because they want a world in which there are no carbon emissions from human activity. But that seems quite unrealistic given what is necessary to get to a "net zero carbon" world. The UN Intergovernmental Panel on Climate Change (IPCC) accords an important role to carbon capture, as does the International Energy Agency.[3]

Carbon capture is integral to how the natural system—the lungs of the world—works. What plants do is absorb CO_2 from the atmosphere, store the carbon in the trunk of a tree or the roots of plants, and release the oxygen back into the air for living creatures to breathe. Farmers cultivating their crops have been in the business of capturing carbon back to the beginnings of agriculture twelve thousand years ago.

A decade or so ago, there was a surge of interest in capturing CO_2 (especially from coal-fired power plants), compressing it into a liquid, and then transporting it by pipeline and storing it underground. A few projects were launched, but proved expensive and involved heavy engineering, and traction was slow in coming.

The 2015 Paris climate compact provided new impetus to develop "carbon capture and storage," or CCS. Around the same time, a "U" for "use" was added to the acronym. It became "carbon capture, use, and storage"—CCUS. That meant finding commercial applications beyond putting the fizz into carbonated soft drinks. After Paris, the Oil and Gas Climate Initiative—the group of thirteen oil and gas companies mentioned earlier—established a $1.3 billion research fund to work on energy transition technologies with a focus on CCUS. Another major impetus came from the U.S. government, which enacted what is known as "45Q." It provides a tax incentive for CCUS technologies, analogous to the tax credits that have been so crucial in the commercialization of wind and solar in the United States.

CCUS takes many forms today. For instance, captured carbon is being used to manufacture products like cement and steel. "Direct air capture"—pulling CO_2 out of the air—had seemed fanciful, but progress is being made and units are being scaled up.

And then there is going full circle, back to what are called "nature-based solutions," otherwise known as forests, crops, and other plants. It is quite possible that Mother Nature has been underestimated. Reforestation and improved cultivation practices are part of the package. Research projects are also aimed at creating super-plants that have a stronger appetite for absorbing CO_2.

The aim, says the Harnessing Plants Initiative at the Salk Institute, is to "coach plants" to "increase their carbon-storing potential." In other words, plants can play a larger role than now anticipated in closing that carbon gap and become part of the CCUS repertoire. "Back to nature" takes on a new meaning.[4]

Advancing these varied technologies will take money and time. By 2030, if not before, the signals and cadences will indicate the rate of progress on these fronts, as well as on others that may not have much visibility today.

Chapter 45

―――

WHAT DOES "ENERGY TRANSITION" MEAN IN THE DEVELOPING WORLD?

Energy transition" means different things to different nations, especially in the developing world. A billion people lack access to electricity; three billion do not have access to clean cooking fuels. Instead they burn wood or charcoal or crop waste or cow dung indoors, impairing their health. This leads to a different perspective. "We're told we have to move on beyond natural gas, to the next thing," said Timipre Sylva, Nigeria's minister of petroleum. "The reality is that Africa is not there yet on renewables. We have to overcome the issue of energy poverty in Africa. Many, many things are not being taken into account with all the talk about renewables and electric vehicles."

What those like Sylva see as not taken into account is that three billion people, almost 40 percent of the world's population—what the World Health Organization (WHO) calls "the forgotten 3 billion"— are subject to indoor air pollution caused by these poor fuels, which the WHO calls "the greatest environmental health risk in the world today." Close to four million people a year die from this indoor pollution, and many more suffer from a wide variety of illnesses. For children, it can mean stunted development.[1]

India, with almost 20 percent of the world's population—soon to be the most populous country in the world—is a case study for the challenges of the developing world. It demonstrates how different the meaning of "energy transition" can be for a developing country, compared to that for developed countries. For in a country in which almost three hundred million people live on the equivalent of $1.25 a day, poverty and economic growth cannot be separated from energy. The energy issues India faces reflect, in a giant-sized way, those of many developing countries.

The term "energy transition" has multiple dimensions for India. It is a transition out of poverty and using wood and waste and into commercial energy—and better health and reducing pollution, both in cities (India has seven of the ten most polluted cities in the world) and in village homes, where the traditional "chulha" stoves fill rooms with noxious fumes. And it means ensuring that the country achieves the growth rate required to lift hundreds of millions of people out of poverty. As the government's *Economic Survey* put it, "Energy is the mainstay of the development process of any economy."[2]

How India develops will have global impact. As its economy grows and becomes more integrated with the global economy, its economic and political weight in the world will also rise.

India has struggled with the inadequacy of modern energy for a long time. Noncommercial energy commonly known as "biomass"—wood and agricultural and animal waste—has been the fuel for more than half of India's population. In terms of commercial energy, India depends on coal for over half of its total energy, and almost 75 percent of electricity. Oil provides about 30 percent of the country's energy. But about 85 percent of the oil is imported, raising anxiety about energy security and creating vulnerabilities for the balance of payments, which morph into crises when the oil price spikes. Natural gas is 6 percent of total energy, compared to a global average of about 25 percent. Modern renewables are just 3 percent of total energy; nuclear, only 1 percent.[3]

When Narendra Modi became prime minister in 2014, his government faced a whole set of energy issues that were holding India back. It focused on energy as an essential engine for economic growth. In 2015, to jump-start energy reform, Modi convened in New Delhi the Urja Sangam, a national energy summit, at which he laid out a series of principles to guide energy development—access, efficiency, sustainability, energy security, and, since added, energy justice. He talked about adjusting the "institutional mechanisms" to be more responsive and flexible and more open to market solutions.

Implementing those principles has not been easy. It meant taking on complex, burdensome, overlapping, and often immobilizing systems of regulation, for which "timeliness" often did not seem to matter much. The "permit raj" of government control was still pervasive. The government had been managing prices disconnected from supply and demand. All of this led to inadequate supply and shortages.

Modi subsequently brought together people from government and the private sector to debate how to break the impasse in India's energy position. Some argued that the "market" was too volatile, too open to manipulation, and could not be trusted; government had to keep control and manage the market. Others said that times had changed; India could not meet its goals on growth and poverty reduction without major reform and an opening to markets and to the world. At the end, Modi looked up from his notes and simply said, "We need new thinking."

That "new thinking" underpins an energy transition across the entire spectrum. "Our energy requirements are vast and robust," says Dharmendra Pradhan, minister of petroleum and natural gas and steel. "India will have an energy transition in its own way. Mixing all exploitable energy sources is the only feasible way forward in our context."[4]

In houses and in villages throughout the country, the smoke from indoor cooking contains carbon monoxide, black carbon, and other pollutants, creating pervasive and severe health problems. In response, the government launched a "blue flame revolution" to deliver cylinders of propane—derived from oil or natural gas—to eighty million rural households for cooking. It has reformed the fiscal, regulatory, and price systems to encourage production and investment in upstream oil and gas by both Indian and international companies and has opened new areas for exploration. Overall, the government, in the words of petroleum minister Pradhan, is seeking to "usher in a gas-based economy." Some $60 billion is being spent on building a natural gas system of major pipelines and urban distribution. One focus is to replace diesel with compressed natural gas as fuel for cars and light trucks, to help reduce urban pollution.[5]

India is becoming a major player in the global LNG market. It is diversifying its sources and has become a significant buyer of both LNG and oil from the United States. This has brought a significant new dimension to relations between the two countries, one made tangible by the interdependence that comes from the scale of this trade—something that would not have been imagined a decade ago either in New Delhi or Washington. Another initiative is to convert agricultural waste in local plants into biofuels and biogas that can be fed into larger distribution systems.

And with climate change in mind, the Modi government has set out ambitious goals for renewables. It has also put tariffs on solar panels, to try to ensure that Indian companies can compete with cheap imported panels from China. As Pradhan summed it up, "India will pursue the energy transition in its own way."

What Pradhan is also pointing to is what some regard as a gap in discussion in the developed world about energy transition—underplaying the challenges and human hardship in developing countries and dismissing as "dirty energy" what many in the developing world say is the clean energy they need for healthier and better lives.[6]

Chapter 46

— — —

THE CHANGING MIX

Reading the map was more straightforward before the coronavirus. One could ascertain directions and trends, although also noting often-strong disagreement among the readers about the speed and extent. But, as the result of the pandemic, an uncharted chasm has suddenly appeared on the map, which the world is now struggling to work its way around. Yet one can see some of the features of the new topography. Some trends will remain the same, some will be accelerated, some will change direction, and some will simply play out over time.

On the premise that the coronavirus is a finite crisis, whether there are further waves of infection or not, and that science and medicine provide timely answers, what can we now see for the future of energy, trying to look beyond the global economy's recovery period?

In the years ahead, CO_2 and GHG policies will bring continuing changes in how energy is produced, transported, and consumed; in strategies and investment; in technologies and infrastructure; and in relations among countries. Established companies will be tested for their adaptability. New entrants will have to prove their business models. Partnerships and competition will characterize the relationships among different kinds of companies. Energy security concerns will expand to the supply chains supporting low-carbon industries and to the minerals on which renewable energy technologies depend. Climate change is global, but nations will respond in different ways, depending on their own particular situations. Developed countries will have more flexibility. Developing countries will struggle to balance between low-carbon and the need for low-cost solutions to promote economic growth, especially in the aftermath of the coronavirus crisis.

And aspirations will come up against an ineluctable reality—today's energy system, which is more than 80 percent based on oil, natural gas, and coal, with a huge embedded investment in infrastructure and supply chains—all of which will be required to meet the energy needed during the recovery period and get back on the economic growth track (see Figure 3). The scale of this system is enormous and cannot change overnight. So far, the energy transition has actually been, in the words of energy strategist Atul Arya, "the phase of energy addition."[1] Wind and solar have been increasing, but they were doing so atop conventional energy, which was also growing.

In the United States, no new coal plants are being built and the number of operating plants is declining. Worldwide, the picture is different. Asia is on track to substantially increase its coal consumption, with the construction of more-efficient coal-powered plants. Coal may be a declining share, but it is still a mainstay for the world's two largest countries, China and India, important not only for energy but also for employment and energy security.

As observed before, coal still represents almost 60 percent of China's total energy supply. "China is not going to abandon coal," said a senior official. "China is different from Europe. China is a developing country. We need to maintain our consumption, but it also means good use of coal, cleaner coal." China's new Five Year Plan (2021–2025) puts a renewed emphasis on coal for energy security and calls for "safe and green coal mining" and "clean and efficient" coal-fired plants.[2]

A LITTLE MORE THAN A DECADE AGO, SOME PREDICTED THAT "peak oil"—the "end of oil"—was near and the world would "run out" of petroleum. The argument has now flipped over to "peak demand": When will oil consumption hit the high point and begin to decline?

Ever since that first oil flowed up out of Colonel Drake's well in 1859, the world's demand for oil has steadily risen, though with occasional dips due to recessions, depressions, and price spikes. The great exception, of course, was when the government-mandated lockdowns shut down much of the world economy in 2020, and demand collapsed in a way that had never happened before. But, for trend, we can use 2019, when global oil consumption was more than 30 percent higher than it had been in 2000.

Yet while consumption has continued to grow, the map of consumption has changed. For decades, demand was concentrated in the industrial nations of North America, Western Europe, Japan, and Australia. The developing world's share was relatively small.

No longer. Since 2013, oil consumption in the "emerging markets" and the other developing countries has been greater than in the traditional industrial countries. Between 2000 and 2019, consumption rose a little in the United States, declined a little in Europe, and in aging Japan dropped a lot. Over that same time period, almost all the growth in oil demand has been in the developing world. China is now the world's second-largest consumer, after the United States; India, the third. And it is in emerging markets where the future growth will continue.

Of course, there was always the understanding that at some point global demand would stop increasing. But "peak demand" was something considered to be far off into the future. The reason was simple—rising population and rising incomes would continue to push up demand.

The number of autos would increase around the world, as ownership in the developing world caught up with the developed world.

For now, the gap remains very large. In 2018, there were 867 cars for every thousand people in the United States, 520 in the European Union. Compare that to the 339 in Russia, the 208 in Brazil, the 160 in China—and just 37 in India. In other words, the world's auto population will grow substantially as incomes rise and the number of people increases from today's 7.8 billion to 9.5 or 10 billion.

In "Rivalry," IHS Markit's planning scenario, the world's auto fleet grows from its current level of just over 1.4 billion to over 2 billion by 2050. Of that 2 billion, about 610 million are electric vehicles—almost a third of the total. The fleet simply does not turn over quickly. Annual new-car sales represent only about 6–7 percent of the total fleet. Most of the fleet is composed of vehicles that have been purchased

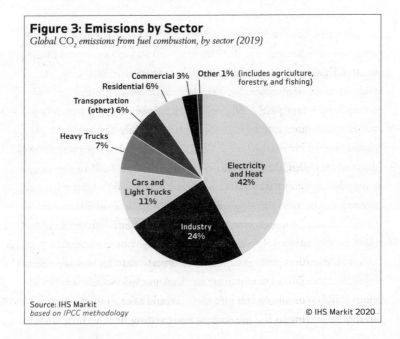

Figure 3: Emissions by Sector
Global CO_2 emissions from fuel combustion, by sector (2019)

Commercial 3%   Other 1% (includes agriculture, forestry, and fishing)
Residential 6%
Transportation (other) 6%
Heavy Trucks 7%
Electricity and Heat 42%
Cars and Light Trucks 11%
Industry 24%

Source: IHS Markit
based on IPCC methodology   © IHS Markit 2020

over the preceding dozen years—in the United States, cars on average remain on the road for 11.8 years. But EVs catch up. By 2050, in this scenario, some 51 percent of total new car sales are EVs.[3]

This is substantial change, but not as fast as some may expect. In light of the economic difficulties and job losses from the 2020 shutdown, regulators may ease up on the steep requirements for reductions in carbon emissions that are pushing a shift to electric cars. Indeed, the world would actually end up with almost the same number of oil-powered cars on the road in 2050 as today. But they will be more fuel-efficient. People may drive more miles in vehicles that require gasoline, but the amount used for each mile goes down. That, in turn, may reduce the incentive to switch away from a gasoline-fueled vehicle. In a more radical scenario, around Auto-Tech, the numbers and kinds of cars will change more rapidly, as they would with stringent climate regulations and larger incentives.

"Electric cars are not the end of the oil era," is the way that Fatih Birol, executive director of the International Energy Agency, puts it. Even if every other car sold in the world from now on were electric, he adds, oil demand would still grow. Cars and light trucks (SUVs and pickups), as pointed out earlier, constitute 35 percent of world oil demand—cars alone, about 20 percent. The rest of transportation consumption goes into heavy trucks, ships, trains, and airplanes. The global fleet of civilian airliners, while more efficient, was expected to double by 2040. That may now be pushed out a few years by the slower growth in passenger travel. Nevertheless, demand will return; over 80 percent of the world's people have never been in an airplane. "Flight shaming" may be a social mode in Sweden, population 10 million, but China, population 1.4 billion, is building eight new airports a year. One of the hardest problems is to find alternatives to jet fuel aside from biofuels, the volumes of which are small. And even if there were an obvious solution or one on the horizon, it would take a long time to have an impact, owing to the life span of the existing fleet and the time to design new planes, get them certified, and then out into the fleets of airlines. Heavy trucks, because of their weight, require the energy density of oil in order to propel their loads along the highway, although in China LNG is also being used.[4]

Oil and natural gas are also the feedstock for petrochemicals, from which chemicals and plastics are made. A growing movement focuses on limiting the use of plastic straws and single-use plastic bags, especially owing to ocean pollution and the debris washing up on beaches. In Washington, D.C., "straw cops" hand out fines to restaurants that covertly use plastic straws, which are now banned. Recyclability to replace single-use plastics has become a priority. This is seen as part of the "circular economy," where products are reused, recycled, or remade at the end of their lives—instead of going into landfills.[5]

But the plastic waste problem is largely not in the developed world. The United States generates less than 1 percent of the plastic waste in oceans. About 90 percent of river-sourced plastic pollution in the oceans comes from uncontrolled dumping into ten rivers in Asia and Africa, which, if properly managed, could dramatically reduce the wastage. Plastic bags and straws may be the most visible use of plastics, but they constitute less than 2 percent of plastics. Moreover, the coronavirus crisis did demonstrate a health advantage of plastic bags over reused cloth shopping bags.

The omnipresence and versatility of plastics make them a building block of the modern world. They are used in everything from making airplanes lighter (and thus more fuel-efficient) and manufacturing electric cars; to auto dashboards and safety glass in windshields and lenses; to bulletproof vests; to carpets, housewares, pantyhose, clothes, and shoes; to packaging (yogurt containers) and keeping food fresh (and thus preventing disease). They are used for water pipes, eliminating metal piping that rusts, and in solar panels and wind towers and blades, and in the casing of cell phones.

They also are embedded in the health system. "Petroleum products are intrinsic to modern health care," is the way an article in the *American Journal of Public Health* put it. "Plastics are central to the antiseptic model of modern health care." Look at a hospital operating room— gloves, tubing, the bags for intravenous liquids, instruments, and the tools that insert stents into ailing heart patients. Moreover, "[Ninety-nine percent] of pharmaceutical feedstocks and reagents are derived from petrochemicals." As for the N95 face masks that became the emblem of the coronavirus epidemic, they are made with petrochemicals.[6]

Petrochemical demand rises faster than GDP growth, sometimes twice as fast, and that means rising demand in that sector will offset the slack elsewhere.

SO WHEN WILL OIL DEMAND REACH ITS PEAK? IHS MARKIT'S Rivalry scenario, which is the planning case, points to the mid-2030s. In the alternative Autonomy scenario, the peak comes much earlier, as a result of strong government policies, a more rapid switch to electric cars, and the economic wounds of the 2020 coronavirus crisis. The actual answer will be determined by a concatenation of many forces— from what national governments and cities do in terms of regulation and incentives, to economic growth, to the availability of minerals, to the legal liability around autonomous vehicles and the security of the cyber systems controlling them, to the values and lifestyles of millennials, to social media, to the increase in air travel and petrochemicals, to geopolitical conflicts and social instability, to start-ups that have not yet started and new scientific and engineering breakthroughs, and so on. And of critical importance will be the long-term changes in behavior in commuting and travel wrought by the coronavirus shutdown. In short, the list is long. Is the "peak" in demand to be followed by a "plummet," as in collapsing demand? More likely is a gradual decline on a downward-sloping plateau. In terms of numbers, the planning scenario posits that the 100 million barrels of consumption prior to the coronavirus is around 113 million in 2050. That's certainly not the end of oil. Even in scenarios in which climate policies become much more aggressive, oil consumption falls to 60 to 80 million barrels per day in 2050.

What, then, is the future of the $5 trillion global oil and gas industry that supplies almost 60 percent of world energy? The industry will continue to need to find and develop another three to five billion barrels a year just to make up for the natural decline in oil fields, which happens after a field has been in production for some time. The International Energy Agency estimates that over $20 trillion of investment in oil and gas development will be required over the next two decades.

"Oil and gas companies" are adapting to the "After Paris" world. The large international companies have generally endorsed some form of carbon price. Some are now describing themselves as "gas and oil companies," owing to increasing emphasis on natural gas as an abundant, lower-carbon fuel. With gas, the industry will increasingly be competing to supply electric generation, meaning that its competitors will increasingly be both coal and renewables. World natural gas consumption is projected to grow at twice the rate of oil. The LNG segment of the business, which is knitting together a single global gas market, will grow faster. By 2050, natural gas demand is estimated to be 60 percent higher than it is today.

Some firms are laying out the ambition to become "energy companies," broadening into electric power, energy services, and new technologies. Whatever the name, the larger companies are increasing investment in new technologies, start-ups, and "low-carbon energy" and are bolstering internal R&D. The aims are multiple: to be more efficient, to meet environmental pressures and investor and regulatory requirements, to "solve" carbon, to participate in renewables and new technologies, to develop economic carbon capture, to play in the future of transportation, to be part of the digital economy, to ensure optionality, and to preserve their "license to operate." They are invest-ing with the "energy transition" in mind—in batteries, in fast charging for electric vehicles, in hydrogen, in wind farms and solar developers, even in fusion. There is a new emphasis on carbon capture. Some have adopted a target of net zero carbon for 2050, which, in addition to the preceding will, among other things, involve greater energy efficiency, biofuels, and reforestation.

Over the next several years, a world trying to regain a $90 trillion GDP and eventually be back on track to $100 trillion, but which is striving to stay below 2 degrees or 1.5 degrees, will still need a lot of energy. Achieving these targets and shifting the supply sources will re-quire the development of major new systems. Many of these will require scale and engineering, and technical and project management skills—all attributes that the oil and gas industry can bring to the table.

A prime example is hydrogen, which, as observed earlier, could po-tentially meet 10 percent or more of total energy requirements, and is becoming a focus for the oil and gas industry. Some companies are al-ready players in wind; and some, long accustomed to building and man-aging large complex offshore oil and gas platforms, are now entering the offshore wind business.

If the future is increasing electrification, to what degree will oil and gas companies be moving into electric power? Some already are. Financial returns will be a question. Power and renewable projects—"lower-carbon generation and distribution"—generally operate in highly-regulated markets and deliver lower rates of return than those traditionally of oil and gas projects. How will they square the circle with demands for returns from investors—which have to meet the re-tirement and pension needs of their fund-holders—and yet deliver an increasingly "green" portfolio for activist shareholders and millennial investors interested in "impact"? At the same time, the electricity busi-ness enables companies to participate more broadly in the changing energy value chain, provides more predictability in revenues, and off-sets volatility in oil and gas markets, especially in light of what happened in 2020.

With all these pressures around climate, companies will have to concentrate on being technology- and innovation-focused and, at the same time, relentlessly competitive, which means constant focus on costs and efficiency. There will be greater competition across a broad front—to attract talent, acquire low-cost oil and gas plays, develop projects, find low-carbon solutions, and innovate. Ultimately, this wide-ranging competition will determine whether the major energy provider of today will continue to play the same roles tomorrow—whatever the forms of energy to be provided.

Shale, with its growth over the last decade, has become a major segment in the overall U.S. economy. It has been an important market for manufacturing industries. Low-cost gas has benefited consumers and businesses, stimulating several hundred billion dollars of new investment in the United States. It has been a major factor in the development of a competitive global natural gas market. And, of course, shale oil has proved to be the most dynamic element in the world oil market in recent years.

The United States will continue to have an abundance of natural gas, but the hectic growth days of shale oil appear to be over. The United States will remain a major producer and will likely regain some of the output level lost from the coronavirus crisis; but it will not return to that high point of thirteen million barrels per day hit in February 2020, unless circumstances change significantly. The shale industry was already maturing before the coronavirus crisis, and companies were reshaping their businesses to deliver returns to investors. That would have taken time, but the crisis disrupted that process, and access to capital and rebuilding the relationship with investors will be a key challenge.

What about consumers? They are the ones, after all, who use the products. As one energy executive put it, if his company stopped producing oil tomorrow, that would not change consumption patterns. People would still be driving their cars, and another company would step in to fill all their gas tanks. In the absence of a carbon tax or significant incentives or higher gasoline taxes, how many consumers will willingly pay more for greener energy—such as buying an EV or a fuel cell vehicle, or choosing a greener yet pricier energy plan? Some will, some won't. In countries around the world, less economically advantaged communities could face higher energy prices, putting the goals of greener energy at odds with those of equity.

WHAT DO THE CHANGING WORLD ENERGY MARKETS MEAN for oil-exporting countries? Markets will go in cycles. They always have, and oil exporters will face volatility, although what happened in 2020 was never anticipated. They may well have to live with periods of lower revenues, which will mean austerity and lower economic growth, with greater risk of turmoil and political instability. This emphasizes the need for these countries to address their over-reliance on oil.

The overweening scale of the domestic oil business crowds out entrepreneurship and other sectors in many oil-exporting countries; it can promote rent-seeking and corruption. It also overvalues the exchange rate, hurting non-oil businesses. In the future, even with a rebound in prices, countries will need to manage oil revenues more prudently, with an eye on the longer term. That means more restrained budgeting and building up a sovereign wealth fund, which can invest outside the country and develop non-oil streams of revenues, helping to diversify the economy and hedge against lower oil and gas prices.

Petroleum-exporting countries will also find themselves competing with other exporting countries for new investment by companies that will be cost-conscious, selective, and focused on "capital disci-pline." That will push countries to shape fiscal and regulatory regimes that are competitive, attractive, stable, predictable, and transparent.

Experience proves how hard it is to diversify away from overdepen-dence. It requires a wide range of changes—in laws and regulations for small-and medium-sized companies, in the educational system, in ac-cess to investment capital, in labor markets, in the society's values and culture. These are not changes that can be accomplished in a short time. In the meantime, the flow of oil revenues creates a powerful countercurrent that favors the status quo.

AS THEY GROW, WIND AND SOLAR AND EVs WILL NEED "BIG shovels" to meet their increasing call on mined minerals and land itself. It is estimated that an onshore wind turbine requires fifteen hundred tons of iron, twenty-five hundred tons of concrete, and forty-five tons of plastic. About half a million pounds of raw materials have to be mined and processed to make a battery for an electric car.

The growth of renewables creates large economic opportunity for mineral-exporting countries—many located in the global South. These nations will face issues similar to those of oil-exporting countries. They will need to ensure the right regulatory frameworks, operating condi-tions, and business practices. The growth in demand for minerals will also shine a more intense light on the environmental aspects and labor conditions for mining and processing minerals. And as the demand grows, so will concerns about what might be called mineral security—that is, assuring reliable supply chains from mine to consumer.[7]

In a world of great power competition, the fragmenting of global-ization, and the rethinking of supply chains, geopolitics will become part of the new energy mix, as it continues to be in the current energy mix.

Conclusion

THE DISRUPTED FUTURE

W here does this new map of energy and geopolitics lead? The collapse of Soviet communism, the transformation of China, and India's move to open its economy—these together brought more than two-and-a-half-billion people into the world economy, creating connections and opportunities that would not have been imagined previously. The result was momentum toward a more collaborative world order that rested on an increasingly connected global economy, one facilitated by the internet and ever-cheaper communication, advances in transportation, and the flows of capital, skills and knowledge—and people. All this was captured in the term "globalization." And it had all been fueled by energy.[1]

But the momentum is now going in reverse. The world has become more fractured, with a resurgence of nationalism and populism and distrust, great power competition, and with a rising politics of suspicion and resentment. Globalization doesn't go away. But it becomes more fragmented, and more contentious, adding to the troubles along the already-troubled path to economic growth.

Before the coronavirus crisis, the $90 trillion global economy was well on its way to $100 trillion within the next five years. But the world economy is now tormented by lives upended and tragedy, unemployment, small businesses fighting for survival, companies under severe pressure, countries impoverished, hope vanquished for many, governments stretched to the extreme by debt, and enormous loss of economic output. It will likely take two to three years for the global economy just to work its way back to $90 trillion, and $100 trillion could be as much as a decade away—and this assumes therapeutics and vaccines arrive in a reasonable time.

Behaviors will be changed by the crisis. At least for a time, there will be an aversion to proximity to large groups, which will affect travel, events, and the way education and businesses operate. When it comes to transportation, people may revert to preferring to "own" their mobility—their personal car—rather than buying mobility when they need it, and, at least for a few years, opt to drive rather than fly when there is a choice. They will also be more cautious about using public transportation. The trend toward digitalization broadly-defined—new ways of working enabled by digital technologies, trading the physical world for the virtual world—has suddenly moved into hyper-gear. Work need not be concentrated in offices, companies can be run from homes, newspapers can be put out with almost no one in the newsroom; time spent commuting can be reduced; business meetings can be replaced by digital connecting. These impacts will last after lockdowns are well in the past. It took three years after 9/11 and more than seven years after the 2008 financial crisis for air travel in the United States to recover to the previous levels. The acceleration of innovation, especially in terms of artificial intelligence and machine learning and automation, will bring change for all kinds of work.

Oil's role will be challenged by these shifts in behavior, work, and daily life. It will, however, take a few years, post-vaccine, to understand the lasting impact on business and leisure travel, on education, on commuting, and on whether the "office of the future" will also now be at home.

The crisis will affect geopolitics as well, reinforcing trends already unfolding. In the face of nationalism and protectionism, the clash among nations will become sharper, international collaboration more difficult, and borders higher. International institutions will struggle to find their footing in a divided global community. The container ships will still set sail, but the global network of supply chains will be under pressure, as governments and companies reevaluate their dependence on those chains—more complex than many realized—and instead put more emphasis on security and resilience and "localization"—and creating local jobs. "Just in time" manufacturing and inventory management will make room for "just be sure." Automation and 3D manufacturing will facilitate this rebalancing in the world economy.

Nowhere will these divisions show up more clearly than in the divide between the two countries upon which, more than any other, world order depends. The United States and China are not decoupled. Despite their growing differences, extensive ligaments continue to tie them together; they share commonalities and mutual interests, including in a growing global economy and the avoidance of conflict. But they are increasingly at odds. The links are under ever-greater strain, and the divide grows deeper. The result could be, to paraphrase Deng Xiaoping, "one planet, two systems" when it comes to technology, the internet, finance, and commercial relations. The "WTO consensus" has given way to "great power competition" and increasing distrust, and to "strategic rivalry" and a high-tech arms race. All this is adding up to a new cold war. This polarization—and the risks that go with it, including the Thucydides Trap—will be a fundamental factor in world politics in the years ahead. The more entrenched the overall positions, the more difficult to resolve specific issues. This clash will hamper the workings of the global economy and, indeed, will contribute to its fragmentation.

The clash is creating growing quandaries for many other countries, which are so connected to both the United States and China but will feel increasing pressure to align to one side or the other. In the Soviet-American cold war, the Soviet Union was a minor player in the global economy. China, by contrast, is deeply embedded and indeed is one of the linchpins of today's world economy. In the summer of 2020, as tensions mounted between the United States and China, an alarmed Lee Hsien Loong, Singapore's prime minister, warned that Asia-Pacific nations "must avoid being caught in the middle or forced into invidious choices." As a senior official in one of the G20 countries put it, "When the United States and China go at it, everybody else in the world suffers.[2]

Energy—particularly oil and gas—will continue to be an integral part of the new geopolitics in the post-coronavirus world. The shale revolution has changed both the American economy and America's position in the world. The new oil order is dominated, owing to their sheer scale, by the Big Three—the United States, Russia, and Saudi Arabia. In the spring of 2020, a market collapse like none other brought them together. But their interests will likely diverge again, as markets and their own positions change, and as climate returns to center stage.

For Russia, oil and gas will remain fundamental to its quest to assert itself as a great power, its relations with Europe, the struggle over Ukraine, and its alignment with China. The Chinese economy will not grow as fast as in the past, but it will be growing off a much bigger economic base, and increasing quantities of energy will be required to assure that growth. This is why energy is a key element for China both in the South China Sea—which some see as the "accident" waiting to happen—and in the Belt and Road and the drive to rebalance the world economy. Oil, and more recently natural gas, will obviously remain central to the future of the Middle East—its economic prospects, the rivalries for regional predominance, governance, demographics, stability, and the region's relations with the rest of the world. Yet, ironically, this very centrality—and dependence—creates an imperative to make oil and gas less central for the future of the region.

WHILE THE PERENNIAL AND SOMETIMES UNEXPECTED GEO-political risks affecting oil will remain, they will be tempered by several factors. Even if the number of vehicles so far is small, the emergence of electricity as a competitor in transportation and the possibility of Auto-Tech provide an alternative to oil-based transportation and the unchallenged dominance of oil. The impact will be enhanced as automakers seek to make good on their promises to electrify their new car fleets, bolstered by governments promoting green recovery. The abundance unlocked by the North American shale revolution, backed up by Canadian oil sands and new production elsewhere, provides a significant security cushion against disruptions of supply. For the most part, wind and solar compete with natural gas and coal to generate electricity, not with oil for transportation. Yet the dramatic fall in the costs of wind and solar—along with their rapidly growing scale—change the balance in the overall energy mix as, at the same time, the world becomes more electric. The coronavirus crisis demonstrated the degree to which digitalization has become a competitor with transportation, using electrons to connect people rather than molecules to move them.

All the above is, in fact, part of the next energy transition—the effort to back away from oil and gas and coal, the products of organic material buried many millions of years ago. The main driver today is not energy security, as in past decades, but climate and the mobilization around it, particularly among younger people. For China and India, the drivers also include pollution and the dependence on oil and gas imports. At the same time, however, for those two countries—today the second and third largest energy consumers in the world—securing energy supplies, including oil and natural gas, is essential for fueling the economic growth they need to lift the incomes of their populations and to reduce pollution.

Will the COVID-19 crisis speed an energy transition or slow it? Some argue for "a green recovery," with government spending skewed to "climate-friendly infrastructure" and greater financial support for renewables and electric vehicles, as well as increasing restrictions on internal combustion engines and government-mandated "reallocation of capital." For local governments, "green recovery" and cleaner air become the rationale for restrictions on internal combustion engine cars and autos of all kinds, the closing off of roads to autos, and the multiplication of bike paths and pedestrian walkways.

Yet the notion of a fast track to a wholesale energy transition runs up against major obstacles—the sheer scale of the energy system that supports the world economy, the need for reliability, the demand for mineral resources for renewables, and the disruptions and conflicts that would result from speed. On top of all of that is the high cost of a fast transition and the question of who pays for it—especially given the staggering amounts of debt that governments took on in 2020 to fight the health and economic consequences of the coronavirus. In the spring of 2020, estimates based on OECD analysis indicated that its members, the developed countries, had already accumulated an additional $17 trillion dollars of debt to deal with the COVID-19 crisis.[3] Environmental ministers may seek to push aggressively ahead, but they will have to contend with finance ministers, who are worrying about budgets and deficits and the primary need to heal the economic wounds, promote recovery, and get people back to work. In short, for the next few decades, the world's energy supplies will come from a mixed system, one of rivalry and competition among energy choices.

In this system, oil will maintain a preeminent position as a global commodity, still the primary fuel that makes the world go round. Some will simply not want to hear that. But it is based on the reality of all the investment already made, lead times for new investment and innovation, supply chains, its central role in transportation, the need for plastics from building blocks of the modern world to hospital operating rooms, and the way the physical world is organized. As a result, oil—along with natural gas, which now is also a global commodity—will not only continue to play a large role in the world economy, but will also be central in the debates over the environment and climate, and certainly in the strategies of nations and in the contention among them.

How fast the mix changes will be determined, of course, not only by politics and policies, but by technology and innovation, which have been the ingredients of energy transitions since Abraham Darby lit up his furnace in 1709. That means the ability to move from idea and invention to technologies and innovation and finally into the marketplace. This is not something that necessarily happens fast—energy is not software. After all, the lithium battery was invented in the middle 1970s but took more than three decades before beginning to power cars on the road. The modern solar photovoltaics and wind industries began in the early 1970s but did not begin to attain scale until after 2010. Yet the pace of innovation is accelerating, as is the focus, owing in part to the climate agenda and government support, in part to decisions by investors, in part to the collaboration of different kinds of companies and innovators, and in part to the convergence of technologies and capabilities—from digital to new materials to artificial intelligence and machine learning to business models and more. The timing of what eventuates will also depend on the talent engaged, the financial resources that support that work, commitment, sheer grit, and the well of creativity upon which to draw. These will lead to the new technologies, disruptive and otherwise, that will shape the new map of energy and geopolitics.

But the map hardly assures us a straight line, for disruptions will with some frequency inevitably redirect the path. The shale revolution was not anticipated, nor was the financial crisis of 2008, nor the Arab Spring and the nuclear accident at Fukushima in 2011, nor the rebirth of the electric car, nor the plummeting in the costs of solar, nor an incredibly-transmissible bat virus that would lead to a pandemic and an economic dark age, nor massive protests in 2020 in the United States that would rock American politics.

Yet there are some disruptions we can anticipate, indeed clearly see, even if we cannot sketch out the precise routes by which they will take us from here. The struggles over climate will be one. But so also, in this era of rising tensions and a fragmenting global order, will be the clash of nations.

Epilogue

——— ——— ———

NET ZERO

Joe Biden wasted no time. It had only been a few hours since his swearing in as president of the United States on January 20, 2021. Sitting for the first time at the desk in the Oval Office, he signed an executive order that would make good, he said, on "the commitment I made that we will rejoin the Paris climate accord as of today." With that, there could be no doubt that "climate" would be, as he pledged, one of the top priorities of his administration. Former Secretary of State John Kerry was appointed U.S. Special Presidential Envoy for Climate, with the job of negotiating the next stage of a global compact; Gina McCarthy, former head of the Environmental Protection Agency, was designated to lead domestic climate policy from the White House; and every cabinet secretary was assigned a climate mandate.

The new administration made its goals clear—to decarbonize electricity by 2035 and then achieve by 2050 the big target for the entire United States—net zero carbon, otherwise known as "carbon neutrality." Other countries were already on board. France, Britain, and the

European Commission had all in 2019 adopted the goal of carbon neutrality, with Prime Minister Boris Johnson subsequently promising to make Britain "the Saudi Arabia of wind." In Asia, both South Korea and Japan adopted 2050 targets. But the most important commitment of all, owing to the scale of its emissions, was that of China. At the United Nations in September, 2020, President Xi Jinping announced that China too was taking "decisive steps" to net zero carbon. But he had a caveat—reaching the goal by 2060. China, which produces about 30 percent of anthropogenic CO_2, would need more time to shift away from its coal-based economy.

The pledges added up. By spring of 2021, over 70 percent of the world's total CO_2 emissions—and 80 percent of world GDP—were under the net zero umbrella.

To accelerate decarbonization—and mark the United States' return to the Paris accords—in April 2021 Biden hosted a virtual Leaders Climate Summit, at which he pledged to reduce U.S. emissions by at least 50 percent by 2030 against a baseline of 2005. This would be a very big leap from 2019, when emissions had been down 13 percent, largely owing to natural gas replacing coal in electric generation. Achieving such a goal would require, as the New York Times put it, "a drastic overhaul of American society," which could not be achieved by "market forces" alone.[1] It would require extensive government intervention across the entire economy through regulation, subsidies, incentives, penalties, directing of private investment—and very large amounts of government money. As one example, the administration's proposed $2.3 trillion infrastructure package apportioned 50 percent more money to electric vehicle promotion than to roads, bridges, ports, and waterways. While China's president Xi reiterated his 2060 target at the summit, he pushed back against criticism of China's emissions, saying that developed countries had a responsibility to "increase climate ambition and action," which would compensate for their emissions during their own industrializations.[2]

"Net zero carbon" has come to be the global target for meeting the Paris objective of holding temperature increases to no more than 2 degrees centigrade (or better, 1.5 degrees) above pre-industrial levels. (In Fahrenheit, that's either a 3.6 degree or a 2.7 degree increase). In the years since the 2015 agreement, "alignment with Paris" is no longer only the purview of governments. Financial firms, representing many tens of trillions of dollars of assets, have added "climate risk" to the screens through which they make investment and lending decisions. Over thirty central banks have elevated "climate" into their mandates. "Climate disclosure"—aiming to demonstrate how company strategies accord with the Paris goal—has become a requirement of company reporting. All of this increasingly focused on "Glasgow"—COP 26, the UN climate conference that is the successor to Paris 2015. The overriding objective of Glasgow is to translate aspirations into "climate action"—laws, regulations, and investments—to accelerate emission reductions—to get to carbon neutrality and to codify 1.5 degrees as the benchmark. Not 2 degrees.

To hasten global carbon neutrality and protect its own industries, the European Union developed a "carbon border adjustment mechanism." Essentially, that means carbon-related tariffs on goods coming from countries that do not have the same "ambition" as Europe—that is, that do not have the same tough standards that the EU is imposing on its own carbon-emitting companies. It is expected that this will be contentious with Europe's trading partners and challenged in the World Trade Organization.[3]

One striking difference between Europe and the United States is in the degree of consensus. In most of Europe, strong climate policies are embraced across the political spectrum. By contrast, divisions remain in the United States about policies, science, and government spending.[4]

By now the "What" has become clear in terms of energy transition— net zero carbon. But what remains uncertain is the "How"—how to get all the way to carbon neutrality in a global economy that currently relies on fossil fuels for 80 percent of its energy.[5] It is the great question that will be on the global agenda for years to come.

The question is further complicated by the different definitions of energy transition. For some, it means no carbon-based energy. For others, it means carbon capture, abatement, and offsets. A rich country in northern Europe has much more flexibility than a developing country—and the wealth—to pursue its ambitions. India, with its very large and young population, will increasingly become one of the main engines of the global economy. But, with hundreds of millions of people in poverty, the country is by necessity pursuing multiple paths of energy transition—very ambitious goals for wind and solar, but also commercial cooking fuels for people in villages, as an alternative to wood and waste, and an expanded natural gas system to reduce pollution. As Prime Minister Narendra Modi put it, the need is "to think both logically and ecologically."[6]

THE VERY FACT THAT SO MANY NATIONS HAVE VOLUNTARILY embraced something so fundamental and so challenging as carbon neutrality is remarkable. What makes it even more remarkable is that much of this was done during COVID-19 time, when lockdowns became ubiquitous and economic activity, suppressed. But then something else happened that also was remarkable. The time needed for vaccine development shrank from the traditional five to ten years to less than a year. "If the threat was big enough, speed was all that mattered," said Noubar Afeyan, cofounder of Moderna, which produced one of the leading vaccines in record time. The pandemic, he added, gave the necessary "permission to leap" technologically over the standard ways of innovating.[7] As vaccinations were rolled out into people's arms, despite variants in the coronavirus, the world was poised for a strong economic recovery that would propel global GDP back to and then above the pre-COVID year of 2019.

As economies opened up again, the unprecedented collapse in oil demand of the spring of 2020, which had driven petroleum prices down into negative territory and pushed the global oil industry toward collapse, came to an end. The OPEC-Plus deal that resulted, and the massive cutbacks in supply, had brought prices back from the abyss. As the prospect of vaccines became reality and as economic activity rebounded, oil demand rose, lifting oil prices back to a level that would permit new investment in oil and gas projects.

The U.S. shale industry had been particularly hard hit. But with prices recovering, the shale industry stabilized, and companies turned to mergers and scale as ways to bring down costs and ensure competitiveness. The second shale revolution also continued, as companies sought to deliver on the pledge that they would stay within cash flow and return money to investors. U.S. oil output, which in February 2020 had peaked at 13 million barrels per day—two-and-a-half times what it had been in 2008—fell to around 11 million barrels a day by the end of that year. Yet that still left the United States as the world's largest oil producer.

Moreover, the year ended with something of great significance, though hardly noticed. For the first time in seventy-two years—since that tanker with Middle East oil had sailed to the United States in 1948—the United States on a net basis was energy independent. In 2008, the United States spent almost $400 billion on oil imports. In 2020, that number was zero, meaning that hundreds of billions of dollars remained inside the U.S. economy each year.

But with the Biden administration so focused on climate, shale was not regarded as such a national asset. Indeed, Biden had been one of the few candidates in the Democratic primaries who had not wanted to "ban fracking." During the fall 2020 campaign, he had gone out of his way to say, "I will not ban fracking. Let me repeat, I will not ban fracking." He said it in Pennsylvania, a state that was benefitting from the shale revolution. Unstated was the reality that a ban on fracking—even if the federal government had the power to impose such—would really be an "import oil" policy because, whatever the source, almost all of the 280 million cars in the United States would need gasoline if they were to run. Moreover, in previous years, as a former member of the Senate Foreign Relations Committee and its sometime chairman, Biden had warned against "the costs to our foreign policy," of "our dependence" on imported oil.[8] Shale provided an answer to his concern.

Still, once in office, Biden imposed freezes on oil and gas leasing on federal lands, pending "review" of climate impacts. The federal government owns a lot of the United States—46 percent of the territory of the 11 western states and 61 percent of Alaska, but most of it is not prospective oil and gas. The main impact of the freeze was on the part of the Permian that is in New Mexico (about a third of the state's government revenues come from oil and gas), in Alaska, and in offshore waters—the last of which produces almost 2 million barrels per day, which is equivalent to Nigeria's output.

THE EFFECTS OF THE PANDEMIC-INDUCED 2020 COLLAPSE will continue to be felt by the global industry. The price collapse had forced a slashing of both budgets and investment in new supplies. Moreover, at the same time, the global oil and gas industry is having to adapt to a world that is aiming at net zero. Some features are common across most large companies. They are focusing on reducing emissions of CO_2 and methane from their operations and are putting increased emphasis on innovation, digital applications, new technologies, and corporate venture capital investing. They generally remain committed to expanding LNG as a lower-carbon fuel that can replace coal in electric generation, primarily in Asia. (Many of the major national oil companies are moving in the same directions.)

But they differ in crucial aspects of their overall strategies. Some are doubling down on efficiency in their operations and developing low-carbon and carbon management businesses. Others are making more dramatic changes. In response to government regulation, investor pressure, and the demands of "society," these companies are seeking to implement transitions in their own corporate identities. This is the shift from being IOCs—"international oil companies"—to becoming IECs—"integrated energy companies." Primarily, this means moving into the electricity business in multiple ways—as generators and traders and even retail providers, as operators of wind and solar, as owners of solar and battery companies, and as deployers of charging stations for electric vehicles. Some are going into offshore wind, leveraging their capabilities from their offshore oil and gas operations. Some are saying that they are aiming to take "responsibility" not only for their own emissions, but also for those of their customers—for instance, the emissions coming out of auto tailpipes and power plants—what are called "Scope 3" emissions. Though they vary in their projected levels of future investment in oil and gas production, the "integrated energy companies" recognize that they will need revenues from oil and gas to fund their large-scale migration into electricity and new technologies.

One who looks with a particular perspective on this move by oil companies into electricity is Ignacio Galán. When Galán became CEO of Iberdrola in 2001, the company was a Spanish electric utility, based on coal, nuclear, and hydro. He pushed the company toward renewables, specifically wind, at a time when wind was still relatively expensive and generally seen as marginal. Owing to his commitment to wind, Galán himself was sometimes regarded in some energy circles as a bit of a modern-day Don Quixote who was tilting at windmills. "Not many believed that what we were doing made any sense," he recalled. "We had a lot of people opposed." Today, however, Galán can look with a certain bemusement at the move by oil companies into electricity. For Iberdrola is a renewable energy major, with operations on four continents. It is the largest wind-generating company in the world outside China in terms of capacity, and, through its American subsidiaries, the third largest wind generator in the United States. "We used to regard oil companies as our enemy," he said. "Now they are becoming our competitors—and sometimes our partners."[9]

The falling costs of solar and wind are one of the bulwarks of net zero. They have come down dramatically—solar as much as 80 percent over the last decade; wind power costs by over 50 percent since 2010. These costs will likely continue to fall further, driven by economies of scale, increasing efficiencies, and technological evolution. For many utilities, solar and wind have become very competitive options; in some markets, the most competitive. They still, however, face the challenge of intermittency—the big swings in generation when the sun is or is not shining and the wind is or is not blowing—as well as of being available at peak times for consumption—which means, for homes, often around dinnertime. The new frontier for wind is offshore, where stronger and steadier winds could make wind energy more like "baseload"—that is, most consistent. However, costs still have to come down substantially for offshore to get to scale globally.

NET ZERO HAS GIVEN NEW MOMENTUM—AND URGENCY—TO the development of innovations across a wide front of technologies. But, at least for now, three technology areas stand out in the march to net zero.

One is carbon capture. As discussed in chapter 44, there is widespread recognition that carbon neutrality requires, as the UN Intergovernmental Panel on Climate Change puts it, "negative emissions"—that is, "use of carbon dioxide removal." The projected overall supply-demand numbers for the decades ahead don't work without it. Under any realistic scenario, oil and gas will remain significant components of the energy mix three decades from now. A billion cars with internal combustion engines are still likely to be on the road in 2050—and if not a billion, then 750 million. The world will still be using steel, cement, and fertilizer, the production of all of which releases emissions.

One way to remove carbon is by augmenting the "lungs of the

world" that are part of planet earth's natural carbon balance through the "nature-based solutions" discussed earlier in this book. At the top of that list for many is reforestation. Another way is through engineering—carbon capture and storage (CCS)—gathering the CO_2 emitted by industrial facilities and then piping it to a location where it can be sequestered underground. In Texas, a facility for "direct air capture" (DAC) is being developed to grab CO_2 directly from the air. And work continues on putting captured carbon to "use" in, for instance, the manufacture of cement.

The second area is hydrogen, now front and center. It may be the most abundant element in the universe, as well as the lightest, but when it comes to energy, despite previous bouts of hype (as noted in chapter 44), it has remained on the sidelines. That has changed. It has now become the new star of the energy transition. At least, the potential star.

Until now, hydrogen has had an important but limited market in industrial applications. Hydrogen has the highest energy density of any common fuel, which is why hydrogen—compressed from its gaseous state into a liquid at minus 423 degrees Fahrenheit—is also one of the main fuels used to blast rockets into outer space. But it is now coming to be seen by many as a fuel that will be crucial for getting to net zero. Hydrogen or hydrogen-derived fuels could provide the power for heavy trucks and shipping. But a much bigger role would be to replace natural gas as an industrial fuel and for heating. Hydrogen could also provide storage for electricity: renewable electricity could be used to make hydrogen, which could then be used as fuel when wind and solar output diminishes. Some European Union studies project up to a quarter of the energy needs of the EU being met by hydrogen by 2050. Achieving anything close to that would be a huge undertaking in terms of sheer scale.

Two big obstacles stand in the way of realizing the hydrogen hope—the cost and the pathways. And that gets to the matter of color, and here we must turn to the palette. "Gray hydrogen" is the hydrogen conventionally produced today as an industrial fuel, most commonly from natural gas ("brown hydrogen" if it is made from coal in China). But conventional hydrogen production results in carbon dioxide emissions. If something is to be done with the emissions, that requires carbon capture. With the addition of carbon capture, gray hydrogen is transmuted into "blue hydrogen." The greatest excitement is around "green hydrogen," which would be produced by using electricity to split a water molecule into hydrogen and oxygen molecules. What would make it green is the use of electricity generated by wind and solar. But currently the cost is higher than that of blue hydrogen—and much higher than the cost of conventional gray hydrogen. There's also a variant—hydrogen produced with nuclear energy, variously known as "yellow hydrogen" or "pink hydrogen," and other colors generated by other production processes.

The European Union is leading the global charge for hydrogen as a replacement fuel, and hydrogen strategies and initiatives are proliferating across the Continent. Across the world, countries and companies with existing or potential low-cost and abundant electricity are beginning to jostle to become hydrogen exporters; and some see the potential for a global trade in hydrogen.

At this point, the expectations for hydrogen are much greater than the capacity to deliver it in large volumes. Because of the great scale involved and the engineering capabilities required, and because the energy industry has the familiarity of working with it, hydrogen has become a focus for energy companies around the world. It's also of great interest to governments that recognize that they need a scale resource that is not intermittent. As a result, mounting efforts will go into hydrogen, with the hope of elevating, by 2050, the world's lightest element into a really heavy-duty player in the world's energy supply.

The third technology area is batteries and electricity storage. The move toward electric vehicles and the rapid growth of intermittent renewable power generation are making improving batteries, along with scaling up supply chains and manufacturing, a strategic priority.

The largest part of the growth will come from cars, as automakers pivot to electric. China, pushed by the government, is the world's largest electric car market. California has banned the selling of gasoline cars in the state after 2035, and Britain is aiming for 2030. A week after President Biden signed the executive order on climate change, General Motors CEO Mary Barra announced the company's ambition to give up on gasoline and produce only electric cars from 2035 onward. GM already has the Bolt and an electric Cadillac coming. Ford has brought out an electric version of the F-150, the best-selling pickup truck for almost half a century. Other companies are coming out with their own targets. Volvo pledges to go all-electric by 2030. The switch, said Volvo CEO Håkan Samuelsson, is based "on the expectation that legislation as well as a rapid expansion of accessible high-quality charging infrastructure will accelerate consumer acceptance of fully electric cars." One of his senior colleagues added, "There are some leaps of faith and some bets that need to be made for this road to be viable."[10]

The cost of lithium-ion (Li-ion) batteries has fallen dramatically in recent years as manufacturing has scaled up. But continued cost and technology improvements are crucial for electric cars to compete with gasoline-powered cars in terms of convenience and cost without subsidies. That is also the case for what is known as the "hybrid" combination of renewables and longer-duration battery storage to compete with power from conventional electric generation. This will involve not only improvements in today's batteries, but also developing new and different battery technologies.

One other big battery issue is emerging—disposal. What to do with the growing number of batteries, with their toxic materials, that will reach the end of their lives in the years ahead? To address this, a rapid build-out of recycling facilities will need to be paired with innovation to improve recovery rates and costs; and policy and regulation will emerge to help establish a circular supply chain.

———

GROWTH ON THIS SCALE IS RAISING GROWING CONCERNS and alarm about battery supply chains and, more broadly, critical minerals and the geopolitics around them. Demand for copper and for key metals used in Li-ion batteries, in addition to lithium—such as, nickel, manganese, and cobalt—will rise sharply. Copper demand will also rise sharply with the modernization and expansion of electric power transmission systems. Can the capacity to mine and refine the metals keep up? There is also concern about mining and work conditions in some countries, as well as about whether battery manufacturing will compete with other industries for supplies. And, seeing these as critical materials, will countries increasingly be competing with each other?

One country is already ahead. China moved early to take a leading position in batteries. It is home to over 80 percent of global battery cell manufacturing capacity. But the combination of high growth and dependence on imports from China has prompted the Biden administration to focus in on "risks in the supply chain for high-capacity batteries, including electric-vehicle batteries" and to promote the building of manufacturing capacity in the United States as a strategic priority. The familiar call for "energy security" has been replaced by the call for "battery security." The European Union is moving on the same path with its European Battery Alliance. In 2011, the EU identified fourteen "critical raw materials." By 2020, it had raised that number to thirty, in the process warning that "access to resources is a strategic security question for Europe's ambition to deliver the Green Deal." It noted that 95 percent of Europe's rare earths, needed both for electric cars and wind turbines, comes from China. While 60 percent of Europe's cobalt originates from mines in the Democratic Republic of the Congo, over 80 percent of what Europe actually imports is refined in China. "The era of a conciliatory or naive Europe that relies on others to look over it is interests is over," said Thierry Breton, Europe's industry commissioner.[11]

After fighting broke out on the border between India and China in the Galwan river valley high in the Himalayas in June, 2020, Prime Minister Narendra Modi's government stepped up efforts to reduce India's reliance on Chinese supply chains. He went further, calling for international companies to reorient their supply chains toward India. "Global supply chains should not only be based on cost," he said. "They should also be based on trust."[12]

For its part, China is also moving to reengineer its own supply chains as part of its "dual circulation" economic strategy, launching a campaign in particular to reduce its dependence on high tech and other key imports from the United States and instead to become more self-sufficient.

When it comes to minerals required for the energy transition, the mega challenge has been identified by the International Energy Agency (IEA)—that the 2050 net zero target will "supercharge demand for critical minerals" as the world moves from a "fuel intensive to a mineral intensive energy system." The result will likely be bottlenecks, shortages, and price spikes. An EV uses six times more minerals than a conventional car; a wind turbine, nine times more minerals than a gas-powered plant. Demand for minerals will skyrocket—lithium by as much as 4300 percent, cobalt and nickel as much as 2500 percent. This, says the IEA, points to potentially large shortfalls—a mine typically takes more than sixteen years from initial discovery to first production. Moreover, the concentration in terms of producing countries is much higher for minerals than for oil. The top three oil producers in the world are responsible for about 30 percent of total liquids production. For lithium, the top three control over 80 percent of supply; China controls 60 percent of rare earths output needed for wind towers; the Democratic Republic of the Congo, 70 percent of the cobalt required for EV batteries.[13]

This great expansion in the demand for minerals will entail a very large growth in mining, which, observes the IEA, will greatly increase emissions as well as pose environmental and social challenges. The latter will include a surge in demand for scarce water, threats to biodiversity, impacts on land use and local communities, and mounting waste and waste products from the mining process. Moreover, more than a million children work in mining, many in what are called "artisanal" mines, which, despite the "artisanal" name, means subsistence mines in developing countries, hand-dug and depending on hard hand labor.

Overall, the drive for net zero carbon by 2050 has been subject to something of a whiplash. At the end of 2020, the IEA warned that investment in oil and gas, including new exploration, had fallen too much, was inadequate, and needed to step up in order to avoid disruption and a "supply crunch." Yet less than six month later, in May 2021, the IEA reversed itself. It called for an end to all new oil and gas investment "as of today" in order to achieve the 2050 goals.

This was a stunning change from what had been its established position. In the same report, the IEA also called for the end of coal-fired electricity generation by 2040, some nineteen years away, which would have big impact on China, India, and other Asian countries. That, however, received much less attention, eclipsed by the focus on the 180 degree turn on oil and gas. Australia and Norway both quickly dissented from the new IEA position, as did Japan, traditionally a stalwart of the IEA. A senior Japanese official put it diplomatically. Such a policy, he said, is "not necessarily in line with the Japanese government's policy."[14]

The Biden administration embraced the IEA's new stance. Biden almost immediately issued a new executive order proposing to empower federal agencies to direct the flow of most energy investment in the United States. "Climate-related financial risk," the order said, would become a major regulatory criterion for judging the appropriateness of lending and investment, whether by banks and other financial institutions or by pension funds. Government agencies would be further charged to ferret out "hidden" energy investment risks. The overall aim would be to assure that all lending and investment complied with "achieving net-zero greenhouse gas emissions." "The touchstone," explained a top administration official, "is that climate risk is financial risk."[15]

———

FOR MANY DECADES, OF COURSE, THE SUPPLY CHAINS FOR global oil and, at times, for natural gas have been deeply emmeshed with geopolitics. But those were rather special cases, and generally not true for the massively dense weave of supply chains that came to characterize the global trading system that had emerged with the end of the Cold War and the subsequent full-throttle entry of China and India into the world economy in the 1990s and 2000s. The supply chain issues were largely premised around efficiency, costs and economics, logistics, trade barriers, intellectual property, technology, and sheer coordination. But that has changed with the supplanting of the era of the "WTO consensus" by the new era of great power competition and strategic rivalry. Geopolitical clash has now become entangled with a broad range of supply chains and, more broadly, trade—from computer chips and telecommunications equipment to food and medical supplies. And that is also proving true for the big, new supply chains that are developing for the net zero carbon future.

The Biden administration is seeking to bring greater order and stability to its policies toward China. But, while recognizing the extensive interdependence between the two countries, its policies are rooted in what has now become the embedded view that China is much more a rival and even an adversary than a partner, a usurper of the international order rather than a major stakeholder.

This is the message of the Interim 2021 National Security Strategic Guidance from the Biden administration, which describes China as "the only competitor potentially capable of combining its economic, diplomatic, military, and technological power to mount a sustained challenge to a stable and open international system." It also warns that China, along with Russia, has "invested heavily in efforts meant to check U.S. strengths and prevent us from defending our interests and allies around the world." U.S. Secretary of State Tony Blinken reinforced that message, declaring that China "poses the most significant challenge of any nation-state in the world to the United States."[16]

The 2021 National Security Strategic Guidance differs markedly from the Obama administration's 2015 version—just six years earlier—when Joe Biden was vice president and Tony Blinken deputy secretary of state. "The scope of our cooperation is unprecedented," said the 2015 document. "The United States welcomes the rise of a stable, peaceful, and prosperous China" and the prospect of a "constructive relationship with China."[17] But those optimistic words turned out to be one of the last gasps of the WTO consensus, inscribed as they were in the immediate afterglow of the 2014 Obama-Xi Beijing deal on climate and in anticipation of the 2015 Paris Climate Conference. The difference between the two strategy documents demonstrates how abruptly the terrain of international relations has changed in just six years.

Many factors have so dramatically reshaped Washington's view. On the economic side, they have been trade practices, intellectual property, and technology. Strategically, they have been the South China Sea, pressure on Taiwan, and China's military buildup. Politically, they have been the absorption of Hong Kong, the treatment of the Uighurs, and the tighter controls that have come from the reassertion of the primacy of the communist party.

In order to develop a common approach to addressing China, Washington is seeking to restore the transatlantic alliance, battered during the Trump years. It is also bolstering "the Quad"—the security dialogue among the United States, India, Japan, and Australia—that has also included such activities as joint naval exercises in the Bay of Bengal.

The Chinese have responded in kind. Both officials and strategists decry Washington's efforts to form an Indo-Pacific coalition that would stretch from Japan to India—what they call an "encirclement" strategy, an effort to hold China back, prevent its economic development, and stymie its regaining its appropriate place in the world—altogether hearkening back, as they put it, to the "Century of Humiliation." They convey that the United States' political and economic position in the world is eroding and vehemently push back on what they call interference in China's internal affairs and assaults on its sovereignty. Foreign minister Wang Yi denounced what he called "the U.S. attempt to suppress China and start a new Cold War."

President Xi, at the World Economic Forum, warned, in language clearly directed to the United States, against "unilateralism" and attempting "to impose hierarchy on human civilization, or to force one's own history, culture, and social system upon others" and that a new "Cold War" or "trade war or tech war" or "decoupling, supply disruption, or sanctions" will "push the world into division or even confrontation." He also pointedly reminded other countries of what is economically at stake for them should they get on Beijing's wrong side—access to "the huge China market and enormous domestic demand." A couple of months later, speaking to the Boao Forum, regarded as China's Asian version of Davos, he added that "we must not let . . . unilateralism pursued by certain countries to set the pace for the whole world." When it came to an internal audience in northwest China, he was more explicit. "The biggest source of chaos in the present-day world is the United States," he is reported to have said. "The United States is the biggest threat to our country's development and security."[18]

The tensions are reflected across the entire range of relationships—from restrictions on students and expulsion of media by both countries to military maneuvers. In 2021, as the USS *Theodore Roosevelt* carrier strike force sailed on a "freedom of navigation operation" into the South China Sea that China claims as its own, Chinese bombers and jets practiced a mock "attack" on the carrier group, going so far as to simulate launching anti-ship missiles. And it was not something that the Chinese went out of their way to disguise.

Here is where the geopolitical and energy maps overlap. The great power rivalry will create challenges for the world economy, including intensified competition for resources and additional pressures on what will become the increasingly stressed supply chains for net zero carbon.

AS SUMMER WAS COMING TO THE MIDDLE EAST IN 2020, A Boeing 787 Dreamliner landed at Israel's Ben Gurion Airport carrying COVID-19 medical supplies for Palestinians. But it also carried a mysterious message, for its markings were those never before seen in Israel—that of Etihad, the national airline of Abu Dhabi, the largest of the emirates that compose the United Arab Emirates (UAE). This meant that the almost-unthinkable had just happened—a plane from one of the major Arab states of the Persian Gulf had landed in Israel.

The mystery was abruptly solved several weeks later with a stunning announcement—Israel and the UAE were about to sign a peace treaty and open diplomatic relations. This became known as the Abraham Accords, named for the biblical patriarch of the three great monotheistic faiths of the Middle East. The UAE would be just the third country in the region to normalize relations with Israel, joining Egypt and Jordan. This overturned a decades-long expectation that any opening between Israel and most of the Arab world would have to wait on a resolution of the Israeli-Palestinian issue. This new agreement did address that issue to some degree; the deal forestalled the possibility of Israel's formally annexing part of the West Bank.

The peace treaty represented a dramatic redrawing of the political map of the Middle East. On the diplomatic front, there had been nothing that epochal since President Anwar Sadat of Egypt went to Jerusalem in 1977 and made peace with Israel two years later. Bahrain, Morocco, and Sudan quickly followed the UAE. While the UAE-Israel treaty had been facilitated by the Trump administration, it had been several years in the making. The way had been paved partly by the warming of relations between Egypt and Israel brought on by offshore natural gas discoveries and the formation of the Eastern Mediterranean Gas Forum, and partly by Israel's export of inexpensive natural gas to Jordan, helping to relieve pressure on that country's finances.

What certainly brought the UAE and Israel together was their common opposition to Iran: its perpetual hostility, and its increasing threat in the Middle East, both directly, through the growing might of its missile and drone programs, and indirectly, through its militias and allies. For Israel, the issue was existential, as Iran rejected its right to exist; for the UAE, what they saw was the implementation of a strategy of encirclement by Iran and its proxies. The key date for the new relationship, however, was 2015 and the nuclear deal with Iran, which lifted sanctions and was interpreted as a tilt by Washington toward Tehran. Following that, the two countries stepped up their cooperation, including on intelligence.

But Iran was not the only reason for this opening. Israel and the UAE also converged on their concerns about Turkey's drive to assert its influence and Ottoman vocation across the Middle East, and what appeared to be its alignment with the Muslim Brotherhood. Moreover, the relationship with Israel would open the door for the UAE to collaboration on high-tech military and cyber defenses—that need brought home by the Houthi and Iranian proxy drone attacks on Saudi Arabia.

The economic dimension loomed large as well. Opening relations would connect Israel's innovative, high-tech, start-up economy with Abu Dhabi, which is moving in the same direction, and with Dubai, already a global financial center. Business took off almost overnight between what the UAE economics minister called "two economic powerhouses." In less than a year, fifteen nonstop flights a day were carrying tourists from Israel to Dubai, deals were flowing, and a proposed investment by an Abu Dhabi sovereign wealth in Israel's offshore gas industry was announced.

There was one further factor—the United States itself. Better to be aligned and thus prepared for a prospective new Democratic administration that would likely resume nuclear negotiations with Iran, which in turn would lead to an easing of sanctions, ending of Iran's isolation and strengthening its economic and strategic position. An even deeper issue concerned the United States itself—the rise of shale oil. A United States that was energy independent, no longer an importer of oil, could over time become less interested in the Middle East and the security of the region. Better for two of the countries in the region that were among the strongest, both economically and militarily, to build their own security architecture as a bulwark against the day when the U.S. might stand back—or as a prominent Democratic senator proposed "right size its commitment to the Middle East.

In the days just before Joe Biden's inauguration, one more step was taken in the region. Saudi Arabia ended its blockade of Qatar, and Qatar pledged to align itself more closely with the other members of the Gulf Cooperation Council. Saudi airspace opened up to Qatar Airways, which meant that its planes no longer needed to arch over Iran on takeoffs and landings. Saudi Arabia also allowed Israeli jets carrying tourists to Dubai to overfly its territory—perhaps a harbinger of more change to come.

But some things were not subject to change. Seven years after the last such battle, a fierce eleven-day air war erupted in the spring of 2021 between Hamas and Israel. Hamas, with support from Iran, launched four thousand rockets on Israel, the majority of which, though not all, were intercepted by Israel's Iron Dome air defense system. Israel responded with air strikes aimed at knocking out the launching points and the large network of underground tunnels that were Hamas's military stronghold. The conflict, which ended with yet another cease-fire, underscored the persistence of instability in the region.

GEOPHYSICAL MAPS CHANGE VERY SLOWLY. BUT POLITICAL, technical, and economic maps can change quickly, revealing new topographies that present multiple challenges and need to be traversed with care and thought. We are on such terrain today. The drive for net zero carbon in a matter of just a few decades will mean remaking the global economy—and doing so in a remarkably short time. It will require huge investment, bring dislocations, add to financial burdens on governments, and impose heavy costs on some parts of the economy. At the same time, it will create major new economic opportunities, open new frontiers for technology and innovation, and stimulate entrepreneurship and creativity. While it will present new avenues for cooperation, it will also create new risks for conflict. It will create new tensions between developed and developing nations. It will change the balance among nations and reorder the competitive playing field among companies, including some that don't yet exist. There will certainly be surprises and unexpected developments. How the new map unfolds will be central to economies and how people live, the future of the international order, and relations among nations in what has become a wary age of growing great power competition. Of one thing we can be sure. The emergence of climate change as one of the defining features of the new map is opening a new era in the relationship between energy and nations.

Appendix

THE FOUR GHOSTS WHO HAUNT THE SOUTH CHINA SEA

This essay first appeared in The Atlantic

The South China Sea is the most important body of water for the world economy—through it passes at least one-third of global trade. It is also the most dangerous body of water in the world, the place where the militaries of the United States and China could most easily collide.

Chinese and American warships have just barely averted several incidents there over the past few years, and the Chinese military has warned off U.S. jets flying above it. The two nations carry out competing naval exercises in those waters. Given what is called the growing "strategic rivalry" between Washington and Beijing, the specter of an "accident" that in turn triggers a larger military confrontation preoccupies strategists in both capitals.

These tensions grow out of a disagreement between the two countries as to whether the South China Sea is or is not Chinese territory, a quarrel that speaks to a deeper dispute about maritime sovereignty, how it is decided upon, and the fundamental rights of movement in those waters.

The standoff over the South China Sea thus has many levels of complexity. It is not simply about one body of water, or a single boundary. As Tommy Koh, a senior Singaporean diplomat who led negotiations to create the United Nations Convention on the Law of the Sea, told me, "the South China Sea is about law, power, and resources, and about history."

That history is haunted in particular by four ghosts, long-departed men from centuries past whose shadows fall across the South China Sea, their legacies shaping the deepening rivalry in the region; historical figures whose lives and work have framed the disputes about sovereignty and freedom of navigation, the competition of navies, as well as war and its costs.

When I was beginning *The New Map*, I participated in a conference on the challenges of globalization and international commerce at the U.S. Naval War College in Newport, Rhode Island. When it came my turn to talk, the commanders of virtually all the world's navies were there in the auditorium, a galaxy of admirals, all resplendent in their dress uniforms. Among them was Admiral Wu Shengli, the head of China's navy at the time and the man who was driving its expansion to compete with the American navy. By then the South China Sea had already become a center of contention. Wu sat in the middle of the audience, in the fifth or sixth row, his gaze unwavering throughout.

That was when I started seeing ghosts: that of China's greatest seafarer, a predecessor to Wu; of the Dutch lawyer who penned the legal brief that now underpins the American argument against China's claims; of the American admiral whose writings offered a foundation for both the U.S. Navy's earlier and Chinese navy's current maritime development; and of the British journalist who argued that the costs of conflict were too high, even for those who would be victorious.

For modern China, claims to the South China Sea center around what is called the "9-Dash Line"—literal dashes that, on the Chinese map, hug the coasts of other nations and encompass 80 to 90 percent of the waters of the South China Sea. Derived from a map drawn by a Chinese cartographer in 1936 in response to what Beijing calls the "Century of Humiliation," the 9-Dash Line is, according to Shan Zhiqiang, the former editor of *Chinese National Geography* magazine, "now deeply engraved in the hearts and minds of the Chinese people." Chinese schoolchildren have for decades been taught that their country's border extends more than a thousand miles to the coast of Malaysia. Beijing's claims are bolstered by military bases that it has built in recent years on tiny islands and on 3,200 acres of reclaimed land scattered in the middle of the sea.

Beijing bases its claim of "indisputable sovereignty" upon history—that, as an official position paper put it, "Chinese activities in the South China Sea date back to over two thousand years ago." These "historic claims," in the words of a Chinese government think tank, have "a foundation in international law, including the customary law of discovery, occupation, and historic title."

The U.S. replies that, under international law, the South China Sea is an open water—what is often called "Asia's maritime commons"—for all nations, a view shared by the countries that border its waters, as well as by Australia, Britain, and Japan. As such, says the U.S. State Department, China "has no legal grounds" for its South China Sea claims and "no coherent legal basis" for the 9-Dash Line. "China's maritime claims," a U.S. government policy paper argued, "pose the greatest threat to the freedom of the seas in modern times."

And this brings us back to the four ghosts.

IN 1381, DURING A BATTLE IN SOUTHWEST CHINA, A MUSLIM boy was captured by soldiers of the Ming dynasty, castrated, and sent to work in the royal household of Prince Zhu Di. As time went on, the boy—renamed Zheng He—grew up to become a confidant of the prince and, eventually, one of his most able military leaders.

When Zhu became emperor, determined that China must be a great maritime power, he ordered a frenetic shipbuilding campaign that launched huge fleets carrying up to 30,000 personnel. They transported both a wide range of Chinese goods and the most advanced ordnance of the day—guns, cannonballs, and rockets. The biggest boats were enormous treasure ships that were as much as ten times larger in size than those Christopher Columbus would captain to the New World almost a century later. These Chinese voyages would take two or three years, with eunuchs in command of each of the fleets. But the commander in chief, above all others, was Zheng. He eventually became known as the Three-Jeweled Eunuch, in honor of the "three jewels" central to the dominant Buddhist faith of Zhu's reign.

Admiral Zheng's first voyage, in 1405, put to sea with an armada of 255 ships, of which 62 were treasure ships. Altogether Zheng commanded seven voyages, some sailing as far as the east coast of Africa, to modern Kenya. Along the way, his fleet would trade Chinese goods and products with the locals while projecting the power and majesty of China—in Zheng's words, "making manifest the transforming power of imperial virtue." One can imagine the impact on those ashore when they caught sight of the approaching giant fleets, and especially the huge treasure ships, with their tail sails filling the skies and the fierce dragon eyes painted on their prows, bearing down on the shore.

On the return voyages home, Zheng's fleets carried not only a wide variety of products and novelties—including precious stones, spices, camels, and ostriches—but also rulers and ambassadors, who would pay homage and tribute before the emperor. Zheng's armadas, as the historian John Keay has written, also "demonstrated maritime mastery of the entire Indian Ocean."

In 1433, on a final voyage homeward across the Indian Ocean—nine years after the death of his patron, Zhu—Zheng died. The great navy he had built did not long survive him. Eventually, on the orders of the new emperor, China's fleet, which had numbered as many as

3,500 ships, was burned. Bureaucrats argued that the navy was wasting money needed to resist encroaching Mongols in the north (though of course, they also saw the navy as the power base for their great rivals, the eunuchs). The legacy of the Three-Jeweled Eunuch was to be expunged from history, the memory of his seaborne exploits almost obliterated.

As China once again has turned to the sea in the twenty-first century, Zheng has been resuscitated as the symbol of the country's traditional engagement and trading relationship with Southeast and South Asia—and as "the most towering maritime figure" in the nation's history. The admiral was celebrated with a widely watched series on Chinese television, and in 2005, on the 600th anniversary of his first voyage, a $50 million museum dedicated to him was opened in Nanjing. A nineteen-year-old girl from an island off Kenya, distinguished by her seemingly Asian features, was invited to the museum's opening as a putative descendant of Chinese who had sailed with Zheng, ostensibly living proof of how far-reaching and "manifest" was the Three-Jeweled Eunuch's seagoing prowess. Today, Zheng and his voyages are the great embodiment of "Chinese activities in the South China Sea," and the claims of history based upon it, his legacy enshrined centuries later in the 9-Dash Line.

IF ZHENG PROVIDED THE NARRATIVE OF CHINA'S HISTORIC maritime rights, then the Dutch lawyer Hugo Grotius did the opposite, laying the foundations for the concept of free passage through the world's oceans, and embodying the "rule of law" as opposed to the legacy of history.

Though of worldwide import, Grotius's arguments arose, ironically, from a specific event at one corner of the South China Sea. In 1603, after the burning of China's fleet and the erasing of the memory of Zheng, Dutch ships attacked a Portuguese vessel in the South China Sea in revenge for Portuguese attacks on Dutch shipping. This marked the beginning of a global struggle between Portugal and the Netherlands for control of colonies and, in Southeast Asia, the spice trade. The Portuguese ship was a tempting prize, laden with silk, gold, porcelain, spices, and many other goods.

But when the booty got back to the Netherlands, the Dutch needed legal ammunition to justify the seizure and secure their profit. They turned to Grotius who, although just twenty-one, was already known as a dazzling prodigy—he had entered Leiden University at eleven.

In his legal brief, Grotius pulverized the Portuguese argument that the South China Sea was theirs because they had "discovered" the sailing routes to it, as though Zheng He and all of the other eunuch captains, along with the Arab and Southeast Asian merchants before them, had never existed. Instead, Grotius argued for freedom of the seas and of commerce and asserted that these rights were universal in their application. Thus, he insisted, the Dutch seizure was wholly justified in retaliation for Portuguese interference with Dutch shipping. Part of the brief was published in what became his great work, *Mare Liberum*, or *The Freedom of the Seas*. The water was, like the air and the sky, the common property of humanity, Grotius wrote. No nation could own them or prevent another from sailing through them. "Every nation," he declared, "is free to travel to every other nation, and to trade with it."

Grotius went on to occupy several distinguished legal and civic positions. But then, caught on the wrong side in a religious battle in the Netherlands, he was sentenced to life imprisonment. Smuggled out of jail in a book chest, he managed to make his way to Paris, where he wrote another landmark book, *On the Law of War and Peace*, which outlined both the basis of a "just war" and the rules for the conduct of warfare. Adam Smith, author of *The Wealth of Nations*, later said that

"Grotius seems to have been the first who attempted to give the world anything like a regular system of natural jurisprudence."

Much admired by Sweden's king, Grotius was appointed Sweden's ambassador to France. On a trip back from Sweden in 1645, he was tossed for three days in the Baltic Sea by a violent storm, wrecking the ship, and Grotius eventually washed up on a beach in northern Germany. There, "the father of the law of the sea," as he would later be called, died from a calamity at sea. His legacy lived on, however: The United Nations Convention on the Law of the Sea, the defining international document governing maritime rules, can be directly traced back to his work.

IN 1897, THEODORE ROOSEVELT, THEN THE ASSISTANT SEC- retary of the Navy, traveled up to the U.S. Naval War College. In his lecture there, Roosevelt propounded the argument for a much stronger U.S. Navy—"a first-class fleet of first-class battleships"—as the best guarantor of peace. The speech brought him national attention.

Roosevelt visited the War College for a second purpose as well: to meet with a faculty member, Admiral Alfred Thayer Mahan, who would have more influence on him in regard to naval power than any other single person, and whose spirit pervades today's disputes over the South China Sea and the collision of U.S. and Chinese naval power.

Mahan—despite the objections of his father, a professor at the Army's West Point military academy—had attended the U.S. Naval Academy. But when he served at sea, his commanders judged him to be deficient in practical command. He did not disagree. "I have known myself too long not to know that I am the man of thought, not the man of action," Mahan wrote to Roosevelt. But he was determined, as he put it, "to be of some use to a navy, despite adverse reports." And he would be. Beginning with *The Influence of Sea Power Upon History*, his many books and articles would make him the world's most influential theorist of naval strategy.

Sea power, Mahan wrote, was essential to protect a nation's commerce, its security, and its position, and it rested on "three pillars"— overseas commerce, naval and merchant fleets, and bases along maritime lanes. The great objective was to assure "command of the sea" and "the overbearing power that can only be exercised by great navies," which meant the ability to dominate naval passages and the "sea lines of communication."

His influence on the United States was clear and direct. Roosevelt went on to become vice president and then, in September 1901, after the assassination of William McKinley, ascended to the presidency. With that, as one scholar has written, "Mahan's philosophy of sea power entered the White House." Roosevelt was relentless in his commitment to a modern navy, culminating with his launch of the Great White Fleet on a round-the-world voyage, which announced America's new role as a global power.

Mahan's impact was also global. The Japanese translation of *The Influence of Sea Power Upon History* sold several thousand copies in a matter of days, and he was offered a teaching post at Japan's Naval Staff College. On a visit to Britain, he received honorary degrees from Cambridge and Oxford and dined with Queen Victoria. Yet no nation took Mahan more to heart than Germany. "I am just now not reading but devouring . . . Mahan's book and am trying to learn it by heart," Kaiser Wilhelm II of Germany wrote. "It is on board all my ships and constantly quoted by my captains and officers."

When Mahan died in 1914, Roosevelt wrote, "There was no one else in his class, or anywhere near it." Decades later, one strategist observed, "Few persons leave so deep an imprint on world events as that left by Mahan." That imprint is clear today in China, and particularly as it relates to the South China Sea.

Beijing maintains as a "core interest" that Taiwan is an integral part of China. In 1996, Beijing, fearing that the lead candidate in Taiwan's presidential election might move toward official independence, launched missile tests and live fire in waters very near the island, effectively blockading its western ports. The United States responded by dispatching an aircraft-carrier group into the Taiwan Strait, ostensibly to avoid "bad weather." The crisis subsided, but the Mahanian lesson for Beijing was clear: The ability to deploy and demonstrate sea power was of paramount importance.

There are many other strands in Chinese military debates, but Mahan's focus on maritime power and "command of the seas" provides a framework for understanding Chinese naval strategy. More than a century after his death, he is much quoted and cited by Chinese thinkers and continues to shape their views. As the strategist Robert Kaplan wrote: "The Chinese are the Mahanians now."

On a clear Sunday morning in August 2014, Chinese naval personnel gathered in the northern port of Weihai. They were there not to mark a victory, the usual reason for such gatherings, but to mark a defeat—China's loss to the Japanese in the First Sino-Japanese War of 1894–95, which had been sealed by the destruction of the Chinese fleet at Weihai. As a result, Japan gained control over Korea and Taiwan, and Weihai passed under British control, altogether a particularly humiliating chapter in China's Century of Humiliation.

At the 2014 ceremony, white chrysanthemums and red roses were scattered over the waters to mourn the Chinese losses. The most prominent speaker that day was Admiral Wu Shengli. In his remarks, one could hear echoes of Mahan.

"The rise of great nations is also the rise of great maritime powers," Wu said. "History reminds us that a country will not prosper without maritime power." The Century of Humiliation, he argued, was the result of insufficient naval strength, which the Weihai defeat had demonstrated. But today, "the sea is no obstacle; the history of national humiliation is gone, never to return."

Mahan was writing amid the first age of globalization in the late nineteenth and early twentieth centuries, when the world was being knit together by technology—steamships, railways, the telegraph—and by flows of investment and trade. He provided the intellectual rationale in that age for what became a global race to build up navies.

In looking for analogies for the wider risks that might be unleashed by the U.S.-China naval competition in the South China Sea, analysts are drawn again and again to that vivid example of strategic rivalry from more than a century ago: the Anglo-German naval race that helped set the stage for World War I. So worrying is it that in his book *On China*, Henry Kissinger concludes with an epilogue titled "Does History Repeat Itself?" entirely devoted to this military buildup. Yet Kissinger goes on to say with some uneasiness, "Historical analogies are by nature inexact."

The Anglo-German naval race was the defining strategic competition of the time. It was also a significant part of a fever that convinced people that war between Britain and Germany was inevitable. That was the conclusion that Winston Churchill, the first lord of the admiralty, came to in 1911. From then on, as he later wrote, he prepared "for an attack by Germany as if it might come the next day."

Yet there were some who disagreed with that assessment, and vigorously so. One of them is the fourth ghost who haunts the South China Sea.

AMONG THE VOICES AT THE BEGINNING OF THE TWENTIETH century arguing that war between Germany and Britain need not be inevitable, none was more powerful than that of a slight, frail-looking sometime-journalist named Norman Angell. He would have enormous influence in convincing people that war had become irrational. He would even receive the Nobel Peace Prize for making the case that "war is a quite inadequate method for solving international disputes." (That the award was made in 1934 prompted him to remark, with a certain dryness, "It would have been more logical to have awarded it at the earlier date.") Angell emphasized the benefits of a connected world economy and the costs of conflict, a particularly relevant message for a United States and a China that are so economically interdependent with each other and so embedded in a wider global economy on which their respective prosperities rely.

Angell came to his calling by a rather circuitous and incongruous route. At sixteen, he went to work as a newspaper reporter, first in his native Britain, then in the United States. He ended up northeast of Los Angeles, in sparsely populated Bakersfield, where he worked as a ranch hand and a mail carrier, homesteaded outside the city, speculated unsuccessfully in land, searched for gold, and tried his hand at oil exploration, all to no avail. Having failed to find his fortune, he left and eventually ended up in Paris, where he worked for English-language newspapers.

By then he had become obsessed with the rise of mass media and alarmed about what he saw as the emergence of mass psychology and the rising temper of virulent nationalism and intolerance in Europe. In 1903 he published his first book, *Patriotism Under Three Flags*, arguing that "emotionalism," or extreme jingoism, worked against the interests of the polity.

Angell then landed a job as the publisher of the European edition of the *Daily Mail*, at the time the largest-circulation newspaper in the world. Prompted by the Anglo-German naval race, Angell hurriedly wrote a new book, *Europe's Optical Illusion*, in which he insisted that he was no pacifist and was not opposed to Britain's military spending, but that, owing to how much more interconnected the world economy had become and the dense ligaments of trade and investment that by then joined nations, the costs of war would far outweigh the gains even for the victor. To Angell's chagrin, he could not find a publisher, and ended up self-publishing and distributing the book himself.

Despite its inauspicious start, the book caught on. A top British diplomat said it had "set my brain in a whirl." One newspaper called it "the most discussed book of recent years." No greater commendation could have been imagined than what the Foreign Secretary himself, Sir Edward Grey, told a banquet, "Just lately there has been published a very interesting little book, *Europe's Optical Illusion*, the moral of which is to try to impress upon nations that commercially and financially their interests are so bound up together that even the victor in a quarrel between them is bound to lose a good deal more than he can possibly gain."

The book became a best seller, and Norman Angell was launched. So, too, was "Norman Angell": Up to this time, he had written under his real name, Ralph Lane, adopting "Norman Angell" to separate his books from his work for the *Daily Mail*. In subsequent editions, Angell rechristened the book *The Great Illusion*.

There were critics, among them Admiral Mahan, who dismissed Angell's argument that growing interdependence made war irrational. "Nationality will not be discarded in face of the remapping of the world," the admiral wrote in words that have some echo today.

Critics notwithstanding, Angell was only gaining in influence. Even Kaiser Wilhelm was reported to have read the book "with keen interest and discussed it a good deal." The Anglo-German naval race continued under full steam; yet the two powers had demonstrated restraint during a Balkans crisis in 1912. This Angell took as a sign of rationality over emotion. On a trip to the United States in February 1914, he told a reporter, "There will never be another war between European powers." In June 1914, the British fleet made a weeklong friendship visit to the German port of Kiel, giving strength to his claim. While it was there, some 800 miles to the south in Sarajevo, Franz Ferdinand, the archduke of the Austro-Hungarian empire, was assassinated. Five weeks later, World War I began.

Angell has been ridiculed over the years since for allegedly saying that the powerful economic links of the first modern age of globalization made war impossible. But, although a man of many words, sometimes too many, that actually is not what he said. His thesis was "not that war is impossible, but that it is futile." Given the grim decades that followed the First World War, who can say that he was wrong?

Indeed, the war's aftermath would prove Angell right: The lasting costs far outweighed whatever might have been gained. It is a message that haunts today's rising tensions between the United States and China.

HISTORY VERSUS INTERNATIONAL LAW, NATIONALISM AND military power versus interdependence and common interests—these define the contention over the South China Sea.

And so when you hear *historic claims*, think Admiral Zheng He. When it's *freedom of the seas*, it's Hugo Grotius. When it's *the U.S.-China arms race*, then it's the other admiral, Alfred Thayer Mahan. And with the growing rift between Washington and Beijing, think of Norman Angell and the costs of confrontation between two nations that are so economically entwined in such a highly-connected world economy.

These are the four ghosts who haunt those troubled waters.

Acknowledgments

Great appreciation goes first to my editor, Ann Godoff, publisher of Penguin Press, who, with her great sense of story and for the moment, encouraged this project from the beginning, helped me think it through and shape it, and from whose guidance I much benefited. Will Heyward at Penguin has been deeply engaged in editing and the thought process and brought welcome perspectives. Also at Penguin, my thanks to Elisabeth Calamari for the opportunity to work with her again. And thank you to everyone else at Penguin for all their efforts in a time of great challenge.

Deep thanks to Stuart Proffitt at Penguin in London, insightful and wise, who is incisive in finding the point and doing so with precision. And thanks to Penelope Vogler and her colleagues in London.

Suzanne Gluck, my agent at WME, has been a comrade and sage counselor long before this book. And appreciation to her colleague Andrea Blatt.

This book greatly benefitted from the commitment of Elena Pravettoni over the entire project. A researcher and analyst of considerable talent and acuity, she brought judgment and knowledge to every step of this story.

I had the good fortune to work again with Ruth Mandel, a superb and creative and sometimes magical photo editor, and with Virginia Mason, expert cartographer, who ensured that the maps tell the story of a book called The New Map. Matthew Luckwitz did the same for the graphics, and Michael Blea applied his creative design to the photo gallery. I am very grateful to Freda Amar and Christa Temple, who supported the endeavor over the entire time and deciphered my occasional indecipherableness.

Steven Weisman, life friend and writer and editor of uncanny skill, adeptly and incisively and with good judgment read this manuscript as he has so many other works of mine. Jamey Rosenfield, with whom I've collaborated for many years, brought, as always, his rigor, perspective, questioning, and sense of structure to our discussions on this book.

I'm also very deeply grateful for their careful reading of the entire work and thoughtful critiques and continuing advice and insights—and commitment of time—all along the way to Atul Arya, Bhushan Bahree, Jason Bordoff, Jim Burkhard, Carlos Pascual, Jeff Marn, and Sue Lena Thompson, who also collaborated on the photo gallery.

I am very appreciative to the people who took the time to read and comment on parts of the book. I have benefited from their expertise and advice. For their insight on the Middle East and markets: Frances Cook, Kristin Smith Diwan, Roger Diwan, Martin Indyk, and Meghan O'Sullivan. For their perspectives on Asia: James Clad, Jin-Yong Cai, Bonnie Glaser, Gauri Jauhar, and Xizhou Zhou. For their knowledge on Russia, Central Asia, and European gas: Simon Blakey, Thane Gustafson, Laurent Ruseckas, Matt Sagers, Shankari Srinivasan, and Michael Stoppard. And for their perspectives on North American energy, technology, and renewables: Kevin Birn, Raoul LeBlanc, Charles Leykum, Jeff Meyer, Anna Mosby, Steven Koonin, Nina Sovich, and Edurne Zoco.

I am very fortunate to be part of IHS Markit with highly knowledgeable colleagues across the entire canvas of energy and the economy, and I thank the many to whom I turned, including Aaron Brady, Vera de Ladoucette, Andrew Ellis, Mark Eramo, Karim Fawaz, Judson Jacobs, Dewey Johnson, Amy Kipp, Steven Knell, Alejandra Leon, Dylan Mair, Eduard Sala de Vedruna, Nirmal Shani, Zbyszko Tabernacki, Linda Toyias, John Webb, Stanislav Yazynin, and Irina Zamarina. And my great appreciation to the other colleagues who helped on specific points in very timely ways.

IHS Markit, with its people around the world, provides a global perspective on the world economy, and I acknowledge and thank Lance Uggla, chairman and CEO, and the senior management team of IHS Markit for their support and leadership—Shane Akeroyd, Brian Crotty, Jonathan Gear, Sari Granat, Adam Kansler, Will Meldrum, Sally Moore, Yaacov Mutnikas, Edouard Tavernier, and Ronnie West, as well as Todd Hyatt and our director, Lord Browne of Madingley.

I appreciate the dialogue over many years with MIT professor and former U.S. energy secretary Ernest Moniz and the collaboration with his Energy Future Initiative, as indicated in chapter 44. John Harper, corporate historian at Chevron, graciously provided guidance to the company's archives and to vivid discoveries about "discovery." Marsha Salisbury mined the rich archives of *Journal of Commerce (JOC)*, an extraordinary resource for the development of the global economy. I thank her and Peter Tirschwell, who heads *JOC*, now part of IHS Markit, and thanks to John Heimlich for his help on airlines. Special thanks for his insights to Dr. James LeDuc, director of the Galveston National Laboratory and former director of the division of viral diseases at the Centers for Disease Control. And I certainly want to express my appreciation to the members of the Brookings Institution's Energy Security Roundtable, which I have the privilege to chair, for their stimulating discussions over these several years.

And last, and most important, deep appreciation to my family for their patience and encouragement and advice—Rebecca, Alex, and Jessica. And to my toughest but favorite critic, my wife, Angela Stent, who while I was doing this book was doing her own work on the end of the old cold war and the start of new ones, but who supported and engaged with this project. I owe her deep thanks for her partnership all across the years. Thank you!

And one last word, of course. I am solely responsible for the content and perspectives in all the pages that precede.

Notes

Data sources for this book include the U.S. Energy Information Administration and its *Monthly Energy Review*; the International Energy Agency; World Bank; International Monetary Fund; IHS Markit energy, economic, automotive, and maritime databases; *BP Statistical Review* and *BP Statistical Review All Data*; OPEC; International Renewable Energy Agency; Eurostat; the OECD; Peterson Institute for International Economics; World Trade Organization; ASEAN; United Nations; International Institute for Strategic Studies; Jane's; and Johns Hopkins University Coronavirus Resource Center. Note: Titles of media articles have been shortened for space.

AMERICA'S NEW MAP

Chapter 1: The Gas Man

1. Loren C. Steffy, *George P. Mitchell: Fracking, Sustainability, and an Unorthodox Quest to Save the Planet* (College Station: Texas A&M University Press, 2019), p. 174 ("livable forest"), p. 23; interview with Dan Steward; Gregory Zuckerman, *The Frackers: The Outrageous Inside Story of the New Billionaire Wildcatters* (New York: Portfolio/Penguin, 2013), p. 21; Dan Steward, *The Barnett Shale Play: Phoenix of the Fort Worth Basin* (Fort Worth: Fort Worth Geological Society, 2007); Russell Gold, *The Boom: How Fracking Ignited the American Energy Revolution and Changed the World* (New York: Simon & Schuster, 2014).

2. Interview with Dan Steward; Steward, *The Barnett Shale Play*; Gold, *The Boom*; Steffy, *George P. Mitchell*, p. 23 ("sad"); Roger Galatas, "Why George Mitchell Sold the Woodlands," *The Woodlands History*, December 2011 ("hated").

3. Interview with Dan Steward; Steffy, *George P. Mitchell*, p. 254 (quarter billion).

4. Interviews with Dan Steward, Nick Steinsberger, and Larry Nichols.

5. *Balancing Natural Gas Policy: Fueling the Demands of a Growing Economy*, vol. 1 (Washington, D.C.: National Petroleum Council, September 2003), p. 16; National Petroleum Council, Committee on Natural Gas, September 3, 2003, transcript, pp. 30–31 ("hell of a big difference").

6. Interview with Larry Nichols.

7. Russell Gold, "Natural Gas Costs Hurt US Firms," *Wall Street Journal*, February 17, 2004.

8. *Potential Supply of Natural Gas in the United States*, Report of the Potential Gas Committee (2011, 2019). President Obama, Remarks on America's Energy Security, Georgetown University, March 30, 2011.

Chapter 2: The "Discovery" of Shale Oil

1. Interview with Mark Papa; Lawrence C. Strauss, "The Accidental Oil Man," *Barron's*, October 22, 2011.

2. Interview with Mark Papa; EOG Presentation, 2010.

3. Interview with John Hess; *Time*, December 1, 1952.

4. Interviews with Harold Hamm and John Hess; North Dakota Industrial Commission, Department of Mineral Resources, Annual Oil Production Statistics.

5. Cornell Lab of Ornithology, "All About Birds: Say's Phoebe"; Scott R. Loss, Tom Will, and Peter P. Marra, "The Impact of Free-Ranging Domestic Cats on Wildlife of the United States," *Nature Communications* 4 (2013); Christopher Helman, "Judge Throws Out Criminal Case Against Oil Companies for Killing Birds at Drilling Sites," *Forbes*, January 18, 2012; *United States v. Brigham Oil & Gas, L.P.*, 840 F. Supp. 2d 1202 (D.N.D. 2012); "Federal Court Holds That the Migratory Bird Treaty Act Does Not Apply to Lawful Activities That Result in the Incidental Taking of Protected Birds," Stoel Rives LLP, January 30, 2012.

6. Edgar Wesley Owen, *Trek of the Oil Finders: A History of Exploration for Petroleum* (Tulsa: American Association of Petroleum Geologists, 1975), pp. 886, 890.

7. Jon Meacham, *Destiny and Power: The American Odyssey of George Herbert Walker Bush* (New York: Random House, 2016), p. 92.

8. Richard Nehring, "Hubbert's Unreliability," *Oil and Gas Journal*, April 17, 2006; Leta Smith, Sang-Won Kim, Pete Stark, and Rick Chamberlain, "The Shale Gale Goes Oily," IHS CERA, 2011.

9. Interview with Scott Sheffield; Pioneer Natural Resources, "Wolfcamp Horizontal Play," Board Presentation, November 16, 2011; *Pioneering Independent* (Irving: Pioneer Natural Resources, 2018), chapters 8 and 9.

Chapter 3: "If You Had Told Me Ten Years Ago": The Manufacturing Renaissance

1. David Mitchell, "Change Is Coming," *New Orleans Advocate*, September 23, 2015.

2. Jason Thomas, "Commodities, and the Global Slowdown," Carlyle Group, January 2020; Mohsen Bonakdarpour, IHS Markit Economics; PricewaterhouseCoopers, *Impacts of the Natural Gas, Oil, and Petrochemical Industry on the U.S. Economy in 2018* (PWC, 2020).

3. Daniel Raimi, *The Fracking Debate: The Risks, Benefits and Uncertainties of the Shale Revolution* (New York: Columbia University Press, 2018); International Energy Agency, *Tracking Fuel Supply* (Paris: November 2019); Oil and Gas Climate Initiative, "Keeping the Accelerator on Methane Reduction," Blog, April 2020, https://oilandgasclimateinitiative.com/keeping-the-accelerator-on-methane-reduction/.

4. IHS, *America's New Energy Future: The Unconventional Oil and Gas Revolution and the United States Economy*, vol. 1 (October 2012); vol. 2 (December 2012); vol. 3 (September 2013); Jeff Meyer, "Trade Savings: How the Shale Revolution Helped Moderate the Trade Deficit," IHS Markit Report, July 2020.

5. American Chemistry Council, "Notes on Shale Gas, Manufacturing, and the Chemical Industry," February 2020.

6. Daniel Fisher, "Shale Gas and Buffett's Billions Fuel Turnaround at Dow Chemical," *Forbes*, October 15, 2014 ("pivoted"); Zain Shauk, "Cheap Natural Gas Feeds Chemical Industry Boom," *Houston Chronicle*, April 19, 2012; Alex MacDonald, "Voestalpine Bets Big on U.S. Shale-Gas Boom," *Market Watch*, May 24, 2013.

Chapter 4: The New Gas Exporter

1. Interview with Charif Souki.

2. *Journal of Commerce*, February 3, 1959.

3. Interview with Michael Smith.

4. Interview with Charif Souki; Gregory Zuckerman, *The Frackers: The Outrageous Inside Story of the New Billionaire Wildcatters* (New York: Portfolio/Penguin 2013), pp. 316–17.

5. Interview with Michael Smith.

6. Eliza Notides, "The US Department of Energy Speaks," *IHS CERA Alert*, May 21, 2013; Department of Energy, "Energy Department Authorizes Second Proposed Facility to Export Liquefied Natural Gas," May 17, 2013; Christopher Smith, Testimony on the Effects of LNG Exports on US Foreign Policy, Committee on Oversight and Government Reform, U.S. House of Representatives, April 30, 2014.

7. Steve Holland and David Brunnstrom, "Trump Urges India's Modi to Fix Deficit," Reuters, June 26, 2017; Remarks by President Trump, President Moon, Commerce Secretary Ross, and NEC Director Cohn in Bilateral Meeting, June 30, 2017.

8. Interview with oil company CEO.

Chapter 5: Closing and Opening: Mexico and Brazil

1. Interview.

2. Elizabeth Malkin, "To Halt Energy Slide," *New York Times*, April 11, 2019; Jude Webster and Michael Stott, "Mexico: Lopez Obrador Makes a Big Bet," *Financial Times*, October 3, 2019 ("technocrats" and "sovereignty"); Sergio Chapa, "Mexico's New President Takes Nationalist Tone," *Houston Chronicle*, March 21, 2019 ("transformation").

3. Andres Schipani and Bryan Harris, "Can Brazil's Pension Reform Kick-Start the Economy?," *Financial Times*, October 22, 2019.

Chapter 6: Pipeline Battles

1. Remarks by President Obama on American-Made Energy, Cushing, Oklahoma, March 22, 2012; Jane Mayer, "Taking It to the Streets," *New Yorker*, November 20, 2011 ("game over"); Kevin Birn and Cathy Crawford, "The GHG Intensity of Canadian Oil Sands Production: A New Analysis," IHS Markit Canadian Oil Sands Dialogue, June, 2020. On permitting pipelines, National Petroleum Council, *Dynamic Delivery: America's Evolving Oil and Natural Gas Transportation Infrastructure* (Washington, D.C., 2019).

2. Interview with Carlos Pascual.

3. "Keystone XL Pipeline Permit Determination," Press Statement, John Kerry, Secretary of State, Washington, D.C., November 6, 2015, https://2009-2017.state.gov /secretary/remarks/2015/11/249249.htm; "Background Briefing on the Keystone XL Pipeline," Special Briefing, Office of the Spokesperson, via teleconference, November 6, 2015, https://2009-2017.state.gov/r/pa/prs/ps/2015/11/249266.htm.

4. U.S. Army Corps of Engineers, *Dakota Access Pipeline Environmental Statement*, December 9, 2015; Dakota Access Pipeline, "Addressing Misconceptions About the Dakota Access Pipeline," https://daplpipelinefacts.com/The-Facts.html; Daryl Own, "The Untold Story of the Dakota Access Pipeline," *LSU Journal of Energy and Resources* 6 (Spring 2018).

5. Charlotte Alter, "Inside Alexandria Ocasio-Cortez's Unlikely Rise," *Time*, March 21, 2019; Earthjustice, "The Dakota Access Pipeline," https://earthjustice.org /cases/2016/the-dakota-access-pipeline; "Paying for Standing Rock," *Wall Street Journal*, September 29, 2017 ("bison"); Blake Nicholson, "State, Feds Address Cleanup at Oil Pipeline Protest Camp," AP News, February 15, 2017.

Chapter 7: The Shale Era

1. *Journal of Commerce*, July 8, 1946; March 9, May 12, 1948.

2. Daniel Yergin, *The Prize: The Epic Quest for Oil, Money & Power* (New York: Simon & Schuster, 2009), chapters 28–31; IHS Energy, *US Crude Oil Export Decision: Assessing the Impact of the Export Ban and Free Trade on the US Economy* (Houston: 2014).

3. Lisa Murkowski, "Opening Remarks," CERAWeek 2015; Lisa Murkowski, John McCain, and Bob Corker, "The U.S. Needs to End Its Ban on Crude Oil Exports,"

Foreign Policy, April 14, 2015; Maroš Šefčovič, "A 21st Century Transatlantic Energy Compact," Peterson Institute for International Economics, Washington, D.C., October 22, 2015.

4. Daniel Yergin and Joseph Stanislaw, *The Commanding Heights: The Battle for the World Economy* (New York: Touchstone, 2002).

5. Maroš Šefčovič, "A 21st Century Transatlantic Energy Compact."

Chapter 8: The Rebalancing of Geopolitics

1. St. Petersburg International Economic Forum, June 2013; Thomas Donilon, "Energy and American Power: Farewell to Declinism," *Foreign Affairs*, June 2013; Michael R. Pompeo, Speech at CERAWeek: "U.S. Foreign Policy in the New Age of Discovery," March 12, 2019. On the impact on U.S. foreign relations, see Meghan O'Sullivan, *Windfall: How the New Energy Abundance Upends Global Politics and Strengthens America's Power* (New York: Simon & Schuster, 2017), chapters 5 and 6.

2. Bruce Jones, David Steve, and Emily O'Brien, *Fueling a New Order? The New Geopolitical and Security Consequences of Energy* (Washington, D.C.: Brookings Institution, 2014), p. 12.

3. *Foreign Relations of the United States, 1950, vol. 5,* (Washington, D.C.: United States Government Printing Office, 1978) p. 1191.

4. Trade Partnership Worldwide, "Trade and American Jobs: The Impact of Trade on U.S. and State-Level Employment," February 2019.

5. Carlos Pascual, "The New Geopolitics of Energy," Columbia Center on Global Economic Policy, September 2015, pp. 14–15; Javier Blas, "Iran Sanctions Push Oil Price," *Financial Times*, January 23, 2012; "Iran: EU Oil Sanctions," BBC, January 23, 2012.

6. Christopher Alessi and Sarah McFarlane, "Europe Opens Up to Other Gas Suppliers," *Wall Street Journal*, September 27, 2018.

7. Interview (Korea).

8. Interview.

9. Raoul LeBlanc, "The Impact of Financial Discipline on Crude Supply," IHS Markit, January 23, 2020.

RUSSIA'S MAP

Chapter 9: Putin's Great Project

1. Angela Stent, *The Limits of Partnership: U.S.-Russian Relations in the Twenty-First Century* (Princeton, NJ: Princeton University Press, 2014), p. 191 ("energy superpower"); Oliver Stone, *The Putin Interviews: Oliver Stone Interviews Vladimir Putin* (New York: Hot Books, 2017), p. 149 ("important elements").

2. Vagit Alekperov, *Oil of Russia: Past, Present and Future* (Minneapolis: East View Press, 2011), p. 98 ("breadbasket"), p. 128.

3. Alekperov, *Oil of Russia*, pp. 173–74 ("fuel crisis").

4. Alekperov, *Oil of Russia*, p. 252.

5. Yegor Gaidar, "The Soviet Collapse: Grain and Oil," American Enterprise Institute, April 2007 ("timeline").

6. Interview with Mikhail Gorbachev; Yegor Gaidar, *Collapse of an Empire: Lessons for Modern Russia* (Washington, DC: Brookings Institution Press, 2007), p. 165 ("no national economy").

7. Serhii Plokhy, *The Last Empire: The Final Days of the Soviet Union* (London: One World, 2014), pp. 303–10, 367, 377.

8. Interview with Yegor Gaidar ("*chudo*"). On the "Wild 90's" and the Russian oil industry, see Thane Gustafson, *Wheel of Fortune: The Battle for Oil and Power in Russia* (Cambridge: Harvard University Press, 2012), chapters 2–5.

9. Fiona Hill and Clifford G. Gaddy, *Mr. Putin: Operative in the Kremlin* (Washington, D.C.: Brookings Institution Press, 2015), pp. 107–15, 181–83; Angela Stent, *Putin's World: Russia Against the West and with the Rest* (New York: Twelve, 2019), p. 4 ("judoist"), pp. 83–86; Stent, *Limits of Partnership*, pp. 4, 85–86.

10. Interview.

Chapter 10: Crises over Gas

1. Paul R. Magosci, *A History of Ukraine: The Land and Its Peoples* (Toronto: University of Toronto Press, 2010), pp. 72–73; Serhii Plokhy, *The Gates of Europe: A History of Ukraine* (New York: Basic Books, 2015), p. 167.

2. Serhii Plokhy and M. E. Sarotte, "The Shoals of Illusion," *Foreign Affairs*, November 22, 2019 ("born nuclear"); Steven Pifer, "The Budapest Memorandum and U.S. Obligations," Brookings Institution, December 4, 2014.

3. Thane Gustafson, *The Bridge: Natural Gas in a Redivided Europe* (Cambridge, MA: Harvard University Press, 2020), chapter 9.

4. Serhii Plokhy, *The Gates of Europe*, pp. 333–34; Angela Stent, *The Limits of Partnership: US-Russian Relations in the Twenty-First Century* (Princeton: Princeton University Press, 2014), chapter 5; Samuel Charap and Timothy J. Colton, *Everyone Loses: The Ukraine Crisis and the Ruinous Contest for Post-Soviet Eurasia* (New York: Routledge, 2017), p. 89 ("direct threat").

5. Jonathan Stern, "The Russian-Ukrainian Gas Crisis of January 2006," *Oxford Institute for Energy Studies*, January 16, 2006 ("one-third"); "Beginning of Meeting on Economic Issues," Kremlin, December 8, 2005; http://en.kremlin.ru/events/president/transcripts/23321; Thane Gustafson and Matthew Sagers, "Gas Transit Through Ukraine: The Struggle for Ukraine's Crown Jewels," CERA, 2003.

6. Condoleezza Rice, Remarks at the State Department Correspondents Breakfast, January 5, 2006; Jonathan Stern, "The Russian-Ukrainian Gas Crisis."

7. Simon Blakey and Thane Gustafson, "Russian-Ukrainian Gas: Why It's Different This Time," CERA, 2009 (Putin).

Chapter 11: Clash over Energy Security

1. Thane Gustafson and Simon Blakey, "It's Not Over Till It's Over: The Russian-Ukrainian Gas Crisis in Perspective," CERA, 2009.

2. "Official Launch of Construction of Nord Stream," press release, April 9, 2010 ("long-term goals"); Thane Gustafson, *The Bridge: Natural Gas in a Redivided Europe* (Cambridge, MA: Harvard University Press, 2020), p. 372 ("priority"); Gerrit Wiesmann, "Russia-EU Gas Pipeline," *Financial Times*, November 8, 2011 ("temptation" and "exclusive"); Isabel Gorst and Neil Buckley, "Russia Opens $10bn Nord Stream," *Financial Times*, September 6, 2011 ("civilized").

3. European Commission, "Third Energy Package," https://ec.europa.eu/energy/en/topics/markets-and-consumers/market-legislation/third-energy-package.

4. German Federal Court of Auditors, "2016 Report—Implementation of Energy Transition," December 21, 2016, www.bundesrechnungshof.de/en/veroeffentlichungen/products/beratungsberichte/sammlung/2016-report-implementation-of-energy-transition.

5. Angela Stent, *From Embargo to Ostpolitik: The Political Economy of West German–Soviet Relations, 1955–1980* (Cambridge: Cambridge University Press, 1981), p. 69 ("drown us"); Edwin L. Dale, "Soviet Oil Feeds Dispute in the West," *New York Times*, June 5, 1961; Raymond P. Brandt, "Oil a New Soviet Weapon in Economic and Political Offensive Against West," *St. Louis Post-Dispatch*, June 4, 1961.

6. Charles Moore, *Margaret Thatcher: The Authorized Biography*, vol. 1, *From Grantham to the Falklands* (New York: Knopf, 2013), pp. 578–79 (Reagan).

7. Moore, *Margaret Thatcher*, p. 584; Gustafson, *The Bridge*, p. 160; George Shultz, *Turmoil and Triumph: Diplomacy, Power, and the Victory of the American Ideal* (New York: Charles Scribner's Sons, 1993), p. 140. ("wasting asset"), p. 89.

Chapter 12: Ukraine and New Sanctions

1. Edward J. Epstein, *How America Lost Its Secrets* (New York: Vintage, 2017), pp. 143–44 (Putin); Angela Stent, *The Limits of Partnership: US-Russian Relations in the Twenty-First Century* (Princeton: Princeton University Press, 2014) p. 270.

2. Stent, *The Limits of Partnership*, pp. 219, 270 ("slouch"), 293 ("regional power").

3. Jeffrey Goldberg, "The Obama Doctrine," *Atlantic*, April 2016 ("core"); James Marson, "Putin to the West," *Time*, May 25, 2009 ("Big Russia").

4. Oliver Stone, *The Putin Interviews: Oliver Stone Interviews Vladimir Putin* (New York: Hot Books, 2017) p. 65 ("signal").

5. Mark Kramer, "Why Did Russia Give Away Crimea," May 19, 2014, Cold War International History Project e-Dossier No. 47; William Taubman, *Khrushchev: The Man and His Era* (New York: Norton, 2003), chapters 9 and 10.

6. U.S. Treasury Secretary Jacob J. Lew, "Evolution of Sanctions and Lessons for the Future," Carnegie Endowment for International Peace, Washington, D.C., March 30, 2016; Ministry of Justice and Security, Government of the Netherlands, "Decision on Prosecution MH17," Letter to Parliament, June 19, 2019.

7. Lew, "Evolution of Sanctions"; Jacob Lew and Richard Nephew, "The Use and Misuse of Economic Statecraft," *Foreign Affairs*, November/December 2018.

8. U.S. Geological Survey, "Circum-Arctic Resource Appraisal: Estimates of Undiscovered Oil and Gas North of the Arctic Circle," 2008; "Russia Plants Flag Under N Pole," BBC News, August 2, 2007.

9. U.S. Energy Information Administration, *Technically Recoverable Shale Oil and Shale Gas Resources: An Assessment of 137 Shale Formations in 41 Countries Outside the United States*, June 2013, table 4, p. 8; Timothy Gardner, "Exxon Winds Down Drilling," Reuters, September 19, 2014; Jack Farchy, "Gazprom Neft Strives to Go It Alone," *Financial Times*, January 3, 2017.

Chapter 13: Oil and the State

1. Interview with Alexei Kudrin.

2. International Monetary Fund, "Russian Federation: 2019 Article IV Consultation Staff Report," August 2019.

Chapter 14: Pushback

1. Donald Tusk, "A United Europe Can End Russia's Energy Stranglehold," *Financial Times*, April 21, 2014; Maroš Šefčovič. "Nord Stream II—Energy Union at the Crossroads," Brussels, April 6, 2016; "Germany to Back Nord Stream 2 Despite Ukraine Tensions," EURACTIV.com with Reuters, December 4, 2018 (commercial project); Thane Gustafson, *The Bridge: Natural Gas in a Redivided Europe*, (Cambridge: Harvard University Press, 2020), chapter 12.

2. "Sally Yates and James Clapper Testify on Russian Election Interference," *Washington Post*, May 8, 2017; Ben Smith's Blog, "Hillary: Putin 'Doesn't Have a Soul,'" *Politico*, January 6, 2008; Michael Crowley and Julia Ioffe, "Why Putin Hates Hillary," *Politico*, July 25, 2016.

3. Polina Nikolskaya, "Russia's Largest Oil Producer Says New US Sanctions Are Going to Backfire," *Business Insider*, August 3, 2017.

4. Statement by President Donald J. Trump on Signing the "Countering America's Adversaries Through Sanctions Act," August 2, 2017.

5. Gustafson, *The Bridge*, p. 411.

6. Elena Mazneva, Patrick Donahue, and Anna Shiryaevskaya, "Germany, Austria Tell U.S. Not to Interfere," *Bloomberg*, June 15, 2017 ("energy supply" and "try to favor"); Alissa de Carbonnel and Vera Eckert, "EU Stalls Russian Gas Pipeline," Reuters, March 21, 2017.

7. Wolfgang Ischinger, "Why Europeans Oppose the Russian Sanctions Bill," *Wall Street Journal*, July 17, 2017; Andrew Rettman, "Merkel: Nord Stream 2 Is 'Political,'" *EUobserver*, April 11, 2018; United States Senate Letter to Secretary of the Treasury Steven Mnuchin and Deputy Secretary of State John J. Sullivan, March 15, 2018; CNBC, July 11, 2018 ("totally controlled"); Steven Erlanger and Julie Hirschfeld Davis, "Trump vs. Merkel," *New York Times*, July 11, 2018; Tom Di-Christopher, "Behind Nord Stream 2," CNBC, July 11, 2018; "Trump and Juncker Agree," CNBC, July 25, 2018.

8. Roman Olearchyk and Henry Foy, "Zelensky Stands Firm," *Financial Times*, December 10, 2019; Helen Maguire and Ulf Mauder, Putin-Merkel press conference, January 11, 2020, http://en.kremlin.ru/events/president/transcripts /62565.

9. European Commission Fact Sheet, "Liquefied Natural Gas and Gas Storage Will Boost EU's Energy Security," February 16, 2016; Groningen, Field Summary Report, IHS Markit, February 2020.

10. Andrius Sytas and Nerijus Adomaitis, "Lithuania Installs LNG Terminal to End Dependence on Russian Gas," *Daily Mail*, October 27, 2014.

11. Association of Gas Producers of Ukraine, http://agpu.org.ua/en/.

12. "Ceremony of First Tanker Loading," Kremlin, December 8, 2017, http://en.kremlin .ru/events/president/transcripts/56338; Henry Foy, "Russia's LNG Ambitions," *Financial Times*, December 27, 2017; Matthew Sagers, Dena Sholk, Anna Galtsova, and Thane Gustafson, "Russia's New LNG Strategy," IHS Markit, April 2018.

13. Gregory Meyer, Ed Crooks, David Sheppard, and Andrew Ward, "Gas from Russian Arctic to Warm Homes in Boston," *Financial Times*, January 22, 2018; www .markey.senate.gov/news/press-releases/senator-markey-calls-for-greater -scrutiny-of-lng-shipments-from-russia-to-the-united-states.

14. Henry Foy, "Russia's Novatek Shows Resilience," *Financial Times*, August 1, 2018.

15. John Webb and Stanislav Yazynin, "Russia's Northern Sea Route," IHS Energy, April 2015.

16. Sagers et al., "Russia's New LNG Strategy"; Foy, "Russia's LNG Ambitions" ("confirmation").

Chapter 15: Pivoting to the East

1. "Merkel Sees No End to EU Sanctions," Reuters, August 19, 2016; James Henderson and Tatiana Mitrova, "Energy Relations Between Russia and China: Playing Chess with the Dragon," Oxford Institute for Energy Studies, August 2016; Sergei Karaganov, "Global Tendencies and Russian Policies," *Russia in Global Affairs*, February 13, 2016 ("rosy dreams"); "Press Statement Following Russian-Chinese Talks," Kremlin, May 20, 2014, http://en.kremlin.ru/events/president/transcripts /21047; Angela Stent, *Putin's World: Russia Against the West and with the Rest* (New York: Twelve, 2019), p. 103 ("own world"); Angela Stent, *The Limits of Partnership: US-Russian Relations in the Twenty-First Century* (Princeton: Princeton University Press, 2014).

2. Bobo Lo, *Russia and the New World Disorder* (London: Chatham House, 2015), p. 133 ("pivot to the east"); transcript of Putin conversation with Lionel Barber and Henry Foy, *Financial Times*, June 27, 2019, https://www.ft.com/content /878d2344-98f0-11e9-9573-ee5cbb98ed36 ("eggs").

3. Brian Spegele, Wayne Ma, and Gregory L. White, "Russia and China Agree," *Wall Street Journal*, May 21, 2014; "Replies to Journalists' Questions Following a Visit to China," Kremlin, May 21, 2014, http://en.kremlin.ru/events/president/news/21064.

4. Erica Downs, "China's Quest for Oil Self-Sufficiency in the 1960s," unpublished manuscript, 2001 ("scabs").

5. "Press Statements Following Russian-Chinese Talks," Kremlin, May 8, 2015, http://en.kremlin.ru/events/president/transcripts/49433 ("key partner"); Cheang Ming, "'Best Time in History' for China-Russia Relationship," CNBC, July 4, 2017; "Putin: Outside Interference in South China Sea Dispute Will Do Only Harm," Reuters, September 5, 2016; Nikolai Novichkov and James Hardy, "Russia Ready to Supply 'Standard' Su-3s," *Jane's*, November 25, 2014; Alexander Gabuev, "China and Russia: Friends with Strategic Benefits," Carnegie, April 7, 2017; "Russia-China Military Ties 'at All-Time High,'" *RT*, November 23, 2016.

6. Oliver Stone, *The Putin Interviews: Oliver Stone Interviews Vladimir Putin* (New York: Hot Books, 2017), p. 90 ("global leadership").

7. Dragoș Tîrnoveanu, "Russia, China and the Far East Question," *The Diplomat*, January 20, 2016.

8. "Pancake Diplomacy," *The Week*, September 12, 2018.

Chapter 16: The Heartland

1. Halford Mackinder, "The Geographical Pivot of History," *Geographical Journal* 23, no. 4 (April 1904), pp. 421–37.

2. Dan Morgan and David B. Ottaway, "Azerbaijan's Riches Alter the Chessboard," *Washington Post*, October 4, 1998 "major goal" and "real country."

3. Daniel Yergin, *The Quest: Energy, Security, and the Remaking of the Modern World* (New York: Penguin, 2011), chapters 2 and 3; Amy Myers Jaffe and Robert A. Manning, "The Myth of the Caspian 'Great Game': The Real Geopolitics of Energy," *Survival* 40, no. 4 (Winter 1998–99), p. 116 (Richardson).

4. Georgi Gotev, "Kazakhstan's Transition," EURACTIV.com, April 24, 2019.

5. "Plenary Session of St Petersburg International Economic Forum," Kremlin, June 7, 2019, http://en.kremlin.ru/events/president/news/60707; Radina Gigova, "Xi Gets Russian Ice Cream from 'Best and Bosom Friend'," CNN, June 17, 2019.

6. Power of Siberia pipeline ceremony, https://www.youtube.com/watch?v =vzafjfHkxPk Anna Galtsova, Jenny Yang, and Sofia Galas, "Power of Siberia: Upcoming Debut to the China Gas Market," IHS Markit, November 25, 2019; Putin speech March 10, 2020: http://en.kremlin.ru/events/president/news/62964; *The Economist*, March 10, 2020.

CHINA'S MAP

Chapter 17: The "G2"

1. Congressional Research Service, "U.S-China Investment Ties," August 28, 2019, and "U.S.-China Trade and Economic Relations," January 29, 2020; NAFSA: Association of International Educators (Chinese university students).

2. Bill Clinton, *My Life* (New York: Vintage, 2005), p. 922; Bill Clinton, Speech, Johns Hokins SAIS, Washington,D.C., March 9, 2000.

3. Jacqueline Varas, "Trade Policy Under President Trump," American Action Forum, Insight, December 13, 2016.

4. Graham Allison, *Destined for War: Can America and China Escape Thucydides's Trap?* (Boston: Houghton Mifflin Harcourt, 2017); "Xi Jinping's Speech on China-U.S. Relations in Seattle," *Beijing Review*, September 23, 2015.

5. The author of *The China Dream*, Colonel Liu Mingfu, proposes "a hundred year marathon" in which China will become preeminent. That became the title for Michael Pillsbury's influential book *The Hundred-Year Marathon: China's Secret Strategy to Replace America as the Global Superpower* (New York: St. Martin's Griffin, 2016. Also, Michael Fabey, *Crashback: The Power Clash Between the U.S. and China in the Pacific* (New York: Scribner, 2017), p. 59; Minghao Zhao, "Is a New Cold War Inevitable? Chinese Perspectives on US-China Strategic Competition," *Chinese Journal of International Politics* 12, no. 3 (2019), p. 374 ("capitalist world" and "co-equal").

6. Allison, *Destined for War*, p. 13.

7. IHS Markit AutoInsight.

8. Nicholas Eberstadt, "With Great Demographics Comes Great Power," *Foreign Affairs*, July/August 2019; Nicholas Eberstadt, "China's Demographic Outlook to 2040 and Its Implications," American Enterprise Institute, January 2019.

9. "Xi Wants High-Tech Fighting Force," *China Daily*, May 1, 2018; "Military expenditure by country," SIPRI, 2019; "The U.S.-China Military Scorecard," RAND Corporation, 2015; Elbridge Colby, "How to Win America's Next War," *Foreign Policy*, May 5, 2019 ("wide variety"); *Jane's*, "China's Advanced Weapons Systems," May 12, 2018. That report quotes a Chinese work, *The Science of Military Strategy*: "The war we need to prepare for . . . is a large-scale, and highly intensive local war from the sea" (p. 8).

10. Michael Gordon, "Marines Plan to Retool to Meet China Threat," *Wall Street Journal*, March 22, 2020; David H. Berger, *Force Design*, United States Marine Corps, March 2020 ("great power/peer level"), p. 2.

11. Evan Osnos, "Making China Great Again," *New Yorker*, January 1, 2018.

12. "President Xi Jinping's Keynote Speech," *China Daily*, May 14, 2017.

13. Interview with James Stavridis; Allison, *Destined for War*, p. 150 ("tension"); Robert D. Kaplan, scholar of geography and geopolitics, describes the South China Sea as "now a principal node of global power politics, critical to the preservation of the worldwide balance of power." Robert D. Kaplan, *Asia's Cauldron: The South China Sea and the End of a Stable Pacific* (New York: Random House, 2015), p. 49.

Chapter 18: "Dangerous Ground"

1. Centre des Archives Diplomatiques du Ministère de l'Europe et des Affaires Étrangères, 32 CPCOM/79, ASIE 1930–1940, CHINE, E 513-0 sd/e 749.

2. Monique Chemillier-Gendreau, *Sovereignty over the Paracel and Spratly Islands* (Cambridge, MA: Kluwer Law International, 2000), pp. 226–27 ("quirks of geology"); Clive Schofield and Ian Storey, *The South China Sea Dispute: Increasing*

Stakes and Rising Tension (Washington, D.C.: Jamestown Foundation: 2009), p. 11 ("Dangerous Ground"); David Hancox and Victor Prescott, *A Geographical Description of the Spratly Islands and an Account of Hydrographic Surveys Amongst Those Islands*, International Boundaries Research Unit, Maritime Briefing, vol. 1, no. 6 (1995), p. 38 ("mariner's . . . guarantee"); Stein Tonnesson, "The South China Sea in the Age of European Decline," *Modern Asian Studies* 40 (2006), pp. 1–57, 3 ("Europe-dominated waters").

3. Centre des Archives Diplomatiques du Ministère de l'Europe et des Affaires Étrangères, 32 CPCOM/79, ASIE 1930–1940, CHINE, E 513-0 sd/e 749.

4. "How Much Trade Transits the South China Sea?," China Power, Project CSIS, October 10, 2019 (world trade); Schofield and Storey, *South China Sea Dispute*, p. 9 (tuna); Alan Dupont, "Maritime Disputes in the South China Sea: ASEAN's Dilemma," *Perspectives on the South China Sea*, ed. Murray Hiebert, Phuong Nguyen, and Gregory Poling (Washington, D.C.: CSIS, 2014), p. 46 ("strategic commodity"); Ian Storey, "Disputes in the South China Sea: Southeast Asia's Troubled Waters," *Politique Étrangère* 79, no. 3 (2014), p. 11 ("entire world").

5. Chemillier-Gendreau, *Sovereignty over the Paracel and Spratly*, pp. 205–6, 210–11 ("megalomania"); Bill Hayton, *The South China Sea: The Struggle for Power in Asia* (New Haven Yale University Press, 2014), pp. 54–55 ("navy is weak").

6. State Council, White Paper on South China Sea, July 2016, english.www.gov.cn /state_council/ministries/2016/07/13/content_281475392503075.htm ("9 Islets"); Henry Kissinger, *On China* (New York: Penguin, 2011); Hayton, *The South China Sea*, p. 55 ("mapmakers"). On the "unequal treaties," Denis Twitchett and John K. Fairbank, *The Cambridge History of China*, vol. 10, *Late Ch'ing, 1800–1911, Part 1* (Cambridge: Cambridge University Press, 1978), chapter 5.

7. State Council, White Paper on South China Sea, http://english.www.gov.cn/state _council/ministries/2016/07/13/content_281475392503075.htm ("reviewed and approved"); Peter A. Dutton, "Through a Chinese Lens," *Proceedings of the U.S. Naval Institute* 136, no. 4 (April 2010), p. 25; Zou Keyuan, "The Chinese Traditional Maritime Boundary Line in the South China Sea and Its Legal Consequences for the Resolution of the Dispute over the Spratly Islands," *International Journal of Maritime and Coastal Law* 14, no. 1 (1999), p. 33; Hayton, *The South China Sea*, p. 56 ("Loving the nation"); William Callahan, "The Cartography of National Humiliation and the Emergence of China's Geobody," *Public Culture* vol. 21, issue 1, Winter 2009, p. 154 ("common people"); "Spratlys—Spratly Islands (Nansha Islands) of China," www.spratlys.org/islands-names/1935.htm (chronology of maps). On Bai Meichu and the *Journal of Geographical Studies*, see Pei-yin Lin and Weipin Tsai, *Print, Profit, and Perception: Ideas, Information and Knowledge in Chinese Societies, 1895–1949* (Leiden: Brill, 2014), pp. 105–11, and Brian Moloughney and Peter Zarrow, eds., *Transforming Society: The Making of a Modern Academic Discipline in Twentieth Century China* (Hong Kong: Chinese University of Hong Kong Press, 2011), pp. 310–17, 331–32 (more patriotic).

8. Tonnesson, "The South China Sea in the Age of European Decline," p. 29.

9. Philip Short, *Mao: A Life* (New York: Henry Holt, 2000), pp. 418–420 ("stood up"); Zheng Wang, "The Nine-Dashed Line: 'Engraved in our Hearts'," *The Diplomat*, August 25, 2014.

Chapter 19: The Three Questions

1. Position Paper of the Government of the People's Republic of China on the Matter of Jurisdiction in the South Sea Arbitration Initiated by the Republic of the Philippines, December 7, 2014.

2. Zhiguo Gao and Bing Bing Jia, "The Nine-Dash Line in the South China Sea: History, Status, and Implications," *American Journal of International Law* 107, no. 1 (January 2013), p. 101; Bill Hayton, *The South China Sea: The Struggle for Power in Asia* (New Haven Yale University Press, 2014), pp. 24–26; Florian Dupuy and Pierre-Marie Dupuy, "A Legal Analysis of China's Historic Rights Claim in the South China," *American Journal of International Law* 107, no. 1 (January 2013), pp. 124, 136–41; Robert Beckman, "The UN Convention of the Law of the Sea and the Maritime Disputes in the South China Sea," *American Journal of International Law* 107, no. 1 (January 2013), p. 143; U.S. Department of State, "Limits in the Seas: Maritime Claims in the South China Sea," December 5, 2014, p. 9.

3. The Asia Maritime Transparency Initiative at CSIS in Washington for information and maps on the South China Sea: https://amti.csis.org and https://amti.csis.org/island-tracker/china/; David E. Sanger and Rick Gladstone, "Piling Sand in a Disputed Sea," *New York Times*, April 8, 2015 ("facts on the water").

4. U.S. Department of State, "Limits in the Seas," pp. 16, 19; Zhiguo Gao and Bing Bing Jia, "The Nine-Dash Line in the South China Sea," p. 102; Dupuy and Dupuy, "A Legal Analysis of China's Historic Rights Claim," p. 139.

5. Interview with Robert Beckman.

6. Jeff Himmelman, "A Game of Shark and Minnow," *New York Times Magazine*, October 27, 2013; Ben Blanchard, "Duterte Aligns Philippines," Reuters, October 20, 2016. China's historical chronology in State Council, People's Republic of China, "China Adheres to the Position of Settling Through Negotiation the Relevant Disputes Between China and the Philippines in the South China Sea," July 2016; State Council, White Paper on South China Sea, July 2016, http://english.www.gov.cn/state_council/ministries/2016/07/13/content_281475392503075.htm (illegal claim).

7. Interviews with Robert Beckman and Antonio Carpio.

8. Shih Hsiu-Chuan, "Ma Addresses Nation's Role in S China Sea," *Taipei Times*, September 2, 2014; "Joining the Dashes," *The Economist*, October 4, 2014.

Chapter 20: "Count on the Wisdom of Following Generations"

1. Ezra Vogel, *Deng Xiaoping and the Transformation of China* (Cambridge: Harvard University Press, 2013); Richard Baum, *Burying Mao: Chinese Politics in the Age of Deng Xiaoping* (Princeton: Princeton University Press, 1994); Daniel Yergin and Joseph Stanislaw, *The Commanding Heights: The Battle for the World Economy* (New York: Touchstone, 2002), p. 197.

2. Bill Hayton, *The South China Sea: The Struggle for Power in Asia* (New Haven: Yale University Press, 2014), pp. 28, 121 ("not wise enough").

3. Carlyle A. Thayer, "Recent Developments in the South China Sea: Implications for Peace, Stability, and Cooperation in the Region," *South China Sea Studies*, March 24, 2011, p. 3; U.S. Department of State, "Limits in the Seas"; Tran Truong Thuy and Le Thuy Trang, *Power, Law, and Maritime Order in the South China Sea* (Lanham: Lexington Books, 2015), pp. 103–15.

4. Interviews; Hillary Rodham Clinton, *Hard Choices* (New York: Simon & Schuster, 2014), p. 79; Edward Wong, "Chinese Military Seeks to Extend Its Naval Power," *New York Times*, July 23, 2010.

Chapter 21: The Role of History

1. Interview with Tommy Koh.

2. Louise Levathes, *When China Ruled the Seas: The Treasure Fleet of the Dragon Throne, 1405–1433* (New York: Oxford University Press, 1994), pp. 170, 88, 20; Geoff Wade, "The Zheng He Voyages: A Reassessment," *Journal of the Malaysian Branch of the Royal Asiatic Society* 78, no. 1 (2005), pp. 37–58; Valerie Hansen, *The Open Empire: A History of China to 1800* (New York: Norton, 2015), pp. 352–60.

3. Michael Fabey, *Crashback: The Power Clash Between the U.S. and China in the Pacific* (New York: Scribner, 2017), p. 65; Joseph Kahn, "China Has an Ancient Mariner to Tell You About," *New York Times*, July 20, 2005.

4. Kurt Campbell, *The Pivot: The Future of American Statecraft in Asia* (New York: Twelve, 2016), pp. 185–87; Patrick E. Tyler, "China Warns U.S. to Keep Away from Taiwan Strait," *New York Times*, March 18, 1996.

5. Peng Yining, "Sea Change," *China Daily*, September 24, 2014; Jeremy Page, "As China Expands Its Navy," *Wall Street Journal*, March 30, 2015.

6. Henry Kissinger, *On China* (New York: Penguin, 2011), pp. 514–27.

Chapter 22: Oil and Water?

1. H. C. Ling, *The Petroleum Industry of the People's Republic of China* (Palo Alto, CA: Hoover Institution Press, 1975).

2. *Far Eastern Economic Review*, February 2004 ("Malacca Dilemma").

3. Xizhou Zhou, "Battle for the Blue Skies: How China's Anti-Smog Campaign Triggered a Natural Gas Crisis and a Switch to 'Clean Coal,'" IHS Markit, March 15, 2018; Jenny Yang and Xiaomin Liu, "The Battle for the Blue Sky Continues," IHS Markit, July 9, 2018.

4. "In High Seas, China Moves Unilaterally," *New York Times*, May 8, 2014; John Ruwitch and Nguyen Phuong Ling, "Chinese Oil Rig Moved Away," Reuters, July 16, 2014 ("bowed"); "China, Vietnam to Address Maritime Disputes Without Using 'Megaphone Diplomacy': Xinhua," Reuters, December 27, 2014; Khanh Vu, "Chinese Ship Leaves Vietnam's Waters," Reuters, October 24, 2019; Carl Thayer, "A Difficult Summer in the South China Sea," *The Diplomat*, November 2019.

5. Dylan Mair and Rachel Calvert, "Energy Drivers for Offshore Cooperation," revised 2019, IHS Markit, unpublished; U.S. Energy Information Administration, "South China Sea," February 7, 2013; U.S. EIA, "Contested Areas of South China Sea

Likely Have Few Conventional Oil and Gas Resources," April 3, 2013; Lee Hsien Loong, "The Endangered Asian Century," *Foreign Affairs*, July-August 2020, p. 59.

Chapter 23: China's New Treasure Ships

1. Arthur Donovan and Joseph Bonney, *The Box That Changed the World: Fifty Years of Container Shipping* (East Windsor: Commonwealth Business Media, 2006), p. 46 (most consequential); Marc Levinson, *The Box: How the Shipping Container Made the World Smaller and the World Economy Bigger* (Princeton: Princeton University Press, 2006), p. 15 ("no engine"); Peter Drucker, *Innovation and Entrepreneurship*, www.academia.edu/38623791/Innovation_and_entrepreneurship_ -_Peter_F_Drucker, p. 31.

2. *Journal of Commerce*, June 3, 2001; Donovan and Bonney, *The Box*, pp. 51–52 ("We are convinced"), p. 97 (thievery); Levinson, *The Box*, p. 165.

3. IHS Markit PIERS and IHS World Trade Service (container ports); *Journal of Commerce*, June 3, 2001; Donovan and Bonney, *The Box*, p. 177.

Chapter 24: The Test of Prudence

1. Sarah Raine and Christian Le Mière, *Regional Disorder: The South China Sea Dispute* (New York: Routledge, 2013), p. 167 ("power politics").

2. Interview with Bilahari Kausikan.

3. Xi Jinping, "China-US Ties," *China Daily*, September 24, 2015. Xi Jinping, *The Governance of China* (Beijing: Foreign Language Press, 2014), pp. 479–80.

4. Xi Jinping, "Achieving Rejuvenation Is the Dream of the Chinese People," *Governance of China*, pp. 37–39; "Xi Pledges Great Renewal of Chinese Nation," *People's Daily*, November 30, 2012; Edward Wong, "Signals of a More Open Economy in China," *New York Times*, December 9, 2012.

5. Elizabeth C. Economy, *The Third Revolution: Xi Jinping and the New Chinese State* (New York: Oxford University Press, 2018); Xi Jinping, Speech, at the National People's Congress, *China Daily*, March 22, 2018; "Xi Jinping Promises More Assertive Chinese Foreign Policy," *Financial Times*, March 20, 2018 ("east wind").

6. "'Leave Immediately': US Navy Plane Warned over South China Sea," CNN, August 24, 2018, www.stltoday.com/news/world/leave-immediately-us-navy-plane -warned-over-south-china-sea/article_fcb06c65-9775-5f07-b3d9-bfa4cea0a28d .html.

7. Michael R. Pompeo, "America's Indo-Pacific Economic Vision," Remarks at the U.S. Chamber of Commerce, Washington, D.C., July 30, 2018; Joe Gould, "U.S. Senate Panel Sets Sights," *Defense News*, June 6, 2018.

8. Interview with Tommy Koh.

9. Interview with Chan Heng Chee.

10. Interview; Adam P. Liff and G. John Ickenberry, "Racing Toward Tragedy?: China's Rise, Military Competition in the Asia Pacific, and the Security Dilemma," *International Security* 39, no. 2 (Fall 2014), p. 78.

11. Liff and Ickenberry, "Racing Toward Tragedy?," p. 72 (Rudd); interview with Peter Ho ("distracted"); Council on Foreign Relations, "Conflict in the South China Sea: Contingency Planning Memorandum Update"; Interview (hostage).

12. *Defusing the South China Sea Disputes: A Regional Blueprint* (Washington, D.C.: CSIS, 2018); Raine and Le Mière, *Regional Disorder*, pp. 179–214; Bonnie Glaser, "A Step Forward in U.S.-China Military Ties: Two CBM Agreements," Asia Maritime Transparency Initiative, November 11, 2014.

13. Interview with Yoriko Kawaguchi; Yoji Koda, "Japan's Perspectives on U.S. Policy Toward the South China Sea," in *Perspectives on the South China Sea*, ed. Murray Hiebert, Phuong Nguyen, and Gregory Poling (Washington, D.C.: CSIS, 2014), pp. 85–87; Ian Storey and Lin Cheng-yi, *The South China Sea Dispute: Navigating Diplomatic and Strategic Tensions* (Singapore: ISEAS, 2016), p. 5 (Abe).

14. Jacqueline Varas, "Trade Policy Under President Trump," American Action Forum, Insight, December 13, 2016; Thomas Franck, "Trump Doubles Down: 'Trade Wars Are Good, and Easy to Win,'" CNBC, March 2, 2018.

15. Kurt M. Campbell and Jake Sulivan, "Competition without Catastrophe: How America Can Both Challenge and Coexist with China," *Foreign Affairs*, September/October 2019; Michael D. Swaine, "Chinese Views on the U.S. National Security and National Defense Strategies," Carnegie Endowment for International Peace, May 1, 2018 ("unprecedented"); "China in U.S. National Security Strategy Reports, 1987–2017," USC U.S.-China Institute, December 18, 2017, https://china.usc.edu/china-us-national-security-strategy-reports-1987-2017 (previous presidents); Susan Thornton, "Is American Diplomacy with China Dead?," *Foreign Service Journal*, July-August 2019; David Shambaugh, "U.S.-China Rivalry in Southeast Asia," *International Security*, Spring 2018, p. 88.

16. *National Security Strategy of the United States*, December 2017.

17. *Summary of the 2018 National Defense Strategy of the United States of America*.

18. "Remarks by Secretary Esper in a Joint Press Conference with Senior Afghan Officials and Resolute Support Mission Commander," https://www.defense.gov/Newsroom/Transcripts/Transcript/Article/1994448/remarks-by-secretary-esper-in-a-joint-press-conference-with-senior-afghan-offic/; Discussion ('asleep').

19. "Vice President Pence on the Administration's Policy Toward China," Hudson Institute, Washington, D.C., October 4, 2018.

20. State Council, People's Republic of China, "China's National Defense in the New Era," July 2019, english.www.gov.cn/archive/whitepaper/201907/24/content_WS5d3941ddc6d08408f502283d.html; "China Warns of War in Case of Move Towards Taiwan Independence," CNBC, July 23, 2019 (Wu Qian).

21. Anthony H. Cordesman, "China's New 2019 Defense White Paper," CSIS, July 24, 2019; interview. interview; National Security Council, "White House Strategic Approach to the People's Republic," May 26, 2020.

22. Elsa B. Kania, "Made in China 2025, Explained," *The Diplomat*, February 1, 2019; Pence, "Administration's Policy Toward China."

23. "Huawei Is at the Centre of Political Controversy," *The Economist*, April 27, 2019; Robert Zoellick, "Can America and China Be Stakeholders?," December 4, 2019 (Carnegie Endowment for International Peace); also, Robert Zoellick, "The China Challenge," *The National Interest*, February 2020.

Chapter 25: Belt and Road Building

1. Xi Jinping, "Work Together to Build the Silk Road Economic Belt," in Xi Jinping, *The Governance of China* (Beijing: Foreign Language Press, 2014), pp. 315–19.

2. Craig Benjamin, *Empires of Ancient Eurasia: The First Silk Roads Era, 100 BCE–250 CE* (Cambridge: Cambridge University Press, 2018), p. 144; Valerie Hansen, *The Silk Road: A New History* (Oxford: Oxford University Press, 2012), pp. 5–8, plate 2 (von Richthofen and his map).

3. Hansen, *The Silk Road*, pp. 14–16 (wrapping).

4. Xi Jinping, Speech to Indonesian Parliament, Jakarta, October 2, 2013.

5. Nadège Rolland, "China's Eurasian Century," National Bureau of Asian Research, May 23, 2017, p. 51.

6. Ministry of Foreign Affairs, "Foreign Ministry Spokesperson Hua Chunying's Remarks on the U.S. House of Representatives Passing the Uyghur Human Rights Policy Act of 2019," December 4, 2019; Jonathan D. Pollack and Jeffrey A. Bader, *Looking Before We Leap: Weighing the Risks of US-China Disengagement*, Brookings Foreign Policy Brief, July 2019; Lindsay Maizland; Lucy Hornsby, "Chinese Official Defends Mass Incarceration of Uighurs," *Financial Times*, March 12, 2019; Steven Lee Myers, "China Defends Crackdown on Muslims, and Criticizes *Times* Article," November 18, 2019 ("fakery"); Sheena Chestnut Greitens, Myunhee Lee, and Emir Yazici, "Counterterrorism and Preventative Repression: China's Changing Strategy on Xinjiang," *International Security*, Winter 2019/2020, pp. 9–47.

7. Evan Osnos, "Making China Great Again," *New Yorker*, January 8, 2018 (Marshall Plan).

8. Interview with Bilahari Kausikan; Kurt Campbell, *The Pivot: The Future of American Statecraft in Asia* (New York: Twelve, 2016), chapter 1.

9. David F. Gordon, Haoyu Tong, and Tabatha Anderson, "Beyond the Myths—Towards a Realistic Assessment of China's Belt and Road Initiative: The Development-Finance Dimension," International Institute of Strategic Studies, March 2020, p. 7; Liff and Ickenberry, "Racing Toward Tragedy?," p. 58("ill-thought out"), p. 53 (predominance).

10. Rolland, "China's Eurasian Century," p. 96 ("strategic hinterland"), p. 46 ("western values"); interview.

11. Erich Schwartzel, "China's Hollywood Challenge," *Wall Street Journal*, December 28, 2017; Rolland, "China's Eurasian Century," p. 191; Asian Infrastructure Investment Bank, www.aiib.org/en/index.html.

12. Interview; "Panama President Cheers China's 'Belt and Road' Initiative," AP News, April 2, 2019; IHS Markit, "Which Countries Are in China's Belt and Road Initiative?," May 2019.

13. Rolland, "China's Eurasian Century," pp. 84, 82, 87; Stefan Reidy, "The New Silk Road: What Should Shippers of Goods Expect from the New Era of Trans-Eurasian Freight Forwarding," Arviem, October 3, 2017.

14. Interview ("financially sound").

15. "China's Super Link to Gwadar Port," *South China Morning Post*, https://multimedia.scmp.com/news/china/article/One-Belt-One-Road/pakistan.html.

16. Jeremy Page and Saeed Shah, "China's Global Building Spree," *Wall Street Journal*, July 22, 2018; Saeed Shah and Bill Spindle, "U.S. Seeks to Avoid a Pakistan Bailout," *Wall Street Journal*, July 31, 2018 ("bail out"); International Monetary Fund, "IMF Executive Board Approves US$6 Billion 39-Month EFF Arrangement for Pakistan," July 3, 2019.

17. Christine Lagarde, "Belt and Road Initiative: Strategies to Deliver in the Next Phase," IMF-PBC Conference, Beijing, April 17, 2018; Maria Abi-Habib, "How China Got Sri Lanka to Cough Up a Port," *New York Times*, June 25, 2018; Matt Ferchen and Anarkale Perera, "Why Unsustainable Chinese Deals Are a Two-Way Street," Carnegie-Tsinghua Center for Global Policy, July 23, 2019.

18. Amanda Erickson, "Malaysia Cancels Two Big Chinese Projects," *Washington Post*, August 21, 2018; Lucy Horby, "Mahathir Mohamad Warns Against 'New Colonialism'," *Financial Times*, August 20, 2018.

19. Ministry of Finance, People's Republic of China, "Debt Sustainability Framework for Participating Countries of the Belt and Road Initiative," April 25, 2019; IHS Markit, "Changing Role of Chinese Lending," August 2019; Gordon, Tong, and Anderson, "Realistic Assessment of China's Belt and Road," pp. 11–15.

20. Asia Maritime Transparency Initiative, "Ports and Partnerships: Delhi Invests in Indian Ocean Leadership," https://amti.csis.org/ports-and-partnerships-delhi -invests-in-indian-ocean-leadership/; Harsh V. Pant, "India Challenges China's Belt-Road Intentions," Yale Global Online, June 22, 2017; Dhruva Jaishankar, "Acting East: India in the Indo-Pacific," Brookings India, October 2019, pp. 4, 11 ("deep distrust").

MAPS OF THE MIDDLE EAST

Chapter 26: Lines in the Sand

1. Missy Ryan, "Islamic State Threat," Reuters, August 21, 2014.

2. Roger Adelson, *Mark Sykes: Portrait of an Amateur* (London: Jonathan Cape, 1975); Christopher Simon Sykes, *The Man Who Created the Middle East: A Story of Empire, Conflict, and the Sykes-Picot Agreement* (London: William Collins, 2017); James Barr, *A Line in the Sand: Britain, France, and the Struggle That Shaped the Middle East* (New York: Norton, 2012), pp. 7 ("Mad Mullah"), p. 15 ("young").

3. David Fromkin, *A Peace to End All Peace: The Fall of the Ottoman Empire and the Creation of the Modern Middle East* (New York: Henry Holt, 2009), p. 149 ("cease to be"); Barr, *A Line in the Sand*, p. 13 ("French endeavours"), p. 67 ("Crusades").

4. Fromkin, *A Peace to End All Peace*, p. 141 ("hornet's nest"), pp. 103, 192; Sean McMeekin argues that it should be called the Sazanov-Sykes-Picot agreement because of the role of Russian foreign minister Sergei Sazanov in defining Russia's imperial objectives for the partition of the Ottoman Empire. *The Ottoman Endgame: War, Revolution, and the Making of the Modern Middle East 1908–1923* (New York: Penguin, 2016), pp. 286–89.

5. Bruce Masters, *The Arabs of the Ottoman Empire, 1516–1918: A Social and Cultural History* (New York: Cambridge University Press, 2013), p. 181 ("ethnic lines").

6. Scott Anderson, *Lawrence in Arabia: War, Deceit, Imperial Folly and the Making of the Modern Middle East* (London: Anchor Books, 2014), chapter 7; Barr, *A Line in the Sand*, p. 18 (Mecca 1).

7. Balfour Declaration, November 2, 1917; Mark Tessler, *A History of the Israeli-Palestinian Conflict*, 2nd ed. (Bloomington: Indiana University Press, 2009), p. 153 ("Handsome as a picture"); Isaiah Friedman, *Palestine, a Twice-Promised Land* (Abingdon: Routledge, 2017), p. 220 ("cousins by blood"); Chaim Weizmann, *Trial and Error: Autobiography* (New York: Harper, 1949).

8. Barr, *A Line in the Sand*, p. 57 (oil report); V. H. Rothwell, "Mesopotamia in British War Aims, 1914–1918," *The Historical Journal* 13, no. 2 (1970), pp. 289–90.

9. Adelson, *Mark Sykes*, pp. 294–95; Sykes, *The Man Who Created the Middle East*, pp. 324–30.

10. Margaret Macmillan, *Paris 1919: Six Months That Changed the World* (New York: Random House, 2003), p. 396 ("dogfight," "greatest oil-field"), chapters 8 and 27.

11. Macmillan, *Paris 1919*, pp. 397–98; Lady Bell, ed., *The Letters of Gertrude Bell* (London: Ernest Benn, 1927), p. 620 (anthem).

12. Ali A. Allawi, *Faisal of Iraq* (New Haven: Yale University Press, 2014), p. 538.

13. Gamal Abdel Nasser, *Egypt's Liberation: The Philosophy of the Revolution* (Cairo, 1958); Said K. Aburish, *Nasser: The Last Arab* (London: Duckworth, 2004), p. 25 (Arab defeat); Anas Alahmed, "Voice of the Arabs Radio: Its Effects and Political Power During the Nasser Era (1953–1967)," March 12, 2011, https://papers.ssrn.com/sol3/papers.cfm?abstract_id=2047212.

14. Michael Oren, *Six Days of War: June 1967 and the Making of the Modern Middle East* (New York: Ballantine, 2003), p. 93; Jesse Ferris, *Nasser's Gamble: How Intervention in Yemen Caused the Six-Day War and the Decline of Egyptian Power* (Princeton Princeton University Press, 2013).

Chapter 27: Iran's Revolution

1. Ray Takeyh, *Guardians of the Revolution: Iran and the World in the Age of the Ayatollahs* (Oxford: Oxford University Press, 2009), p. 133; "Iran's Supreme Leader Calls the Saudi Leaders 'Idiots,'" DW, May 28, 2017, www.dw.com/en/irans-supreme-leader-calls-the-saudi-leaders-idiots/a-39013367; Imam Khamenei's Hajj Message—2016.

2. Jeffrey Goldberg, "Saudi Crown Prince: Iran's Supreme Leader," *Atlantic*, April 2, 2018; Mohammed bin Salman interview with *Time*, April 5, 2018.

3. Albert Hourani, *A History of the Arab Peoples* (London: Faber & Faber, 2005), pp. 30–37; Ira M. Lapidus, *A History of Islamic Societies* (Cambridge: Cambridge University Press, 2014), pp. 66–69; Pew Research Center, Forum on Religion & Public Life, "Mapping the Global Muslim Population," October 2009, "Muslims," December 18, 2012.

4. William Shawcross, *The Shah's Last Ride* (New York: Simon & Schuster, 1988), p. 179 ("megalomaniac").

5. Richard Falk, "Trusting Khomeini," *New York Times*, February 16, 1979 ("humane governance"); Ervand Abrahamian, *A History of Modern Iran* (Cambridge: Cambridge University Press, 2008), pp. 163–64 ("powers unimagined"); Kim Ghattas, *Black Wave: Saudi Arabia, Iran, and the Forty-Five Year Rivalry That Unraveled Culture, Religion, and Collective Memory in the Middle East* (New York: Henry Holt, 2020), chapters 1, 2, and 4 (on how Khomeini made his way to power).

6. Pierre Razoux, *The Iran-Iraq War* (Cambridge: Harvard University Press, 2015), p. 2; Suzanne Maloney, *Iran's Political Economy Since the Revolution* (Cambridge: Cambridge University Press, 2015), chapter 4.

Chapter 28: Wars in the Gulf

1. Steve Coll, *Ghost Wars: The Secret History of the CIA, Afghanistan, and bin Laden, from the Soviet Invasion to September 10, 2001* (New York: Penguin, 2005), chapter 2.

2. Coll, *Ghost Wars*, chapter 2; Stuart Eizenstat, *President Carter: The White House Years* (New York: St. Martin's Press, 2018), pp. 653–54 ("sermon" and doctrine).

3. "Saddam Hussein Talks to the FBI: Interviews and Conversations with 'High Value Detainee # 1,' Session 3, in 2004," National Security Archive Electronic Briefing Book No. 279, Posted July 1, 2009; *Foreign Relations of the United States 1977–80*, Volume XVIII, no. 139; Pierre Razoux, *The Iran-Iraq War* (Cambridge: Harvard University Press, 2015), pp. 7–8 ("One moment").

4. Williamson Murray and Kevin M. Woods, *The Iran-Iraq War* (Cambridge: Cambridge University Press, 2014), p. 36 ("satanic turbans"), p. 41 ("blasphemy"), p. 45 ("chaos"), p. 65 ("disintegration"); Pierre Razoux, *The Iran-Iraq War*, p. 3 ("little Satan").

5. Murray and Woods, *The Iran-Iraq War*, p. 85 ("their noses").

6. Rodger W. Claire, *Raid on the Sun: Inside Israel's Secret Campaign That Denied Saddam the Bomb* (New York: Broadway Books, 2005); Razoux, *The Iran-Iraq War*, pp. 465–66 ("cup of poison").

7. Adel Al Toraifi, "Understanding the Role of State Identity in Foreign Policy Decision-making: The Rise of Saudi-Iranian Rapprochement (1997–2009)," PhD thesis (London School of Economics, 2012), p. 149 ("harmed relations"), p. 154 ("geographic reality"); Bruce Riedel, *Kings and Presidents: Saudi Arabia and the United States Since FDR* (Washington, D.C.: Brookings Institution Press, 2018), p. 97.

8. Kevin M. Woods, *The Mother of All Battles: Saddam Hussein's Strategic Plan for the Persian Gulf Wars* (Annapolis: Naval Institute Press, 2008), p. 96 ("Iraqization"), p. 197 ("really harshly").

9. George H. W. Bush and Brent Scowcroft, *A World Transformed* (New York: Vintage, 1999), pp. 319–20.

10. Woods, *Mother of All Battles*, p. 186 ("camel").

11. Bush and Scowcroft, *A World Transformed*, p. 489; Woods, *The Mother of All Battles*, pp. 243–45 ("strongest . . . powers").

12. Micah Sifry and Christopher Cerf, *The Iraq War Reader: History, Documents, Opinions* (New York: Simon & Schuster, 2003), p. 618.

13. Sifry and Cerf, *The Iraq War Reader*, p. 269.

14. Strategic Studies Institute and U.S Army War College Press, *The U.S. Army in the Iraq War: Surge and Withdrawal 2003–2006*, vol. 1, pp. 140–44; Thomas E. Ricks, *Fiasco: The American Military Adventure in Iraq* (New York: Penguin, 2007), p. 163 ("snatched defeat"); Jim Mattis and Bing West, *Call Sign Chaos: Learning to Lead* (New York: Random House, 2019); Interview with Meghan O'Sullivan (payment to soldiers).

15. Susan Sachs and Kirk Semple, "Ex-Leader, Found Hiding in Hole," *New York Times*, December 14, 2003.

16. "Interviewing Saddam: FBI Agent Gets to the Truth," January 8, 2008, https://archives.fbi.gov/archives/news/stories/2008/january/piro012808.

Chapter 29: A Regional Cold War

1. Kenneth M. Pollack, *The Persian Puzzle: The Conflict Between Iran and America* (New York: Random House, 2004), pp. 282, 324 (masterminded), chapter 11; Bruce Riedel, *Kings and Presidents: Saudi Arabia and the United States Since FDR* (Washington, D.C.: Brookings Institution Press, 2018), pp. 122–25.

2. Karim Sadjadpour, *Reading Khamenei: The World View of Iran's Most Powerful Leader* (Washington, D.C.: Carnegie Endowment for International Peace, 2009).

3. Adel Al Toraifi, "Understanding the Role of State Identity in Foreign Policy Decision-Making," PhD thesis, (London School of Economics, 2012), pp. 200, 218 ("our security"), 215 ("at the disposal").

4. "U.S. Embassy Cables: Saudi King's Advice for Barack Obama," *Guardian*, November 28, 2010; Jay Solomon, *The Iran Wars* (New York: Random House, 2016).

5. Ray Takeyh, *Guardians of the Revolution: Iran and the World in the Age of the Ayatollahs* (Oxford: Oxford University Press, 2009), p. 248 ("500 kilometers"), p. 161 ("preeminent power"); Joshua Teitelbaum, *What Iranian Leaders Really Say About Doing Away with Israel: A Refutation of the Campaign to Excuse Ahmadinejad's Incitement to Genocide* (Jerusalem: Jerusalem Center for Public Affairs, 2008), p. 66.

6. Ross Colvin, "'Cut Off Head'," Reuters, November 29, 2010.

7. Interview with Carlos Pascual; Carlos Pascual, "The New Geopolitics of Energy," Center for Global Energy Policy, Columbia University, 2015; Richard Nephew, *The Art of Sanctions: A View from the Field* (New York: Columbia University Press, 2018), chapter 7.

8. Solomon, *Iran Wars*, p. 246.

9. "Iran Nuclear Crisis: What Are the Sanctions?," BBC News, March 30, 2015 ("wheels of industry"); Suzanne Maloney, *Iran's Political Economy Since the Revolution* (Cambridge: Cambridge University Press, 2015), p. 493 ("worst conditions").

10. William Burns, *The Back Channel: A Memoir of American Diplomacy and the Case for Its Renewal* (New York: Random House, 2019), chapter 9; Wendy R. Sherman, "How We Got the Iranian Deal," *Foreign Affairs*, September/October 2018.

11. Burns, *The Back Channel*, p. 375 ("good day"), pp. 377–81.

12. David E. Sanger and William J. Broad, "Now the Hardest Part: Making the Iran Deal Work," *New York Times*, October 17, 2015.

13. "Viewpoint: How U.S. Can Reach New Iran Deal—After Trump," BBC News, November 11, 2018; Remarks by President Trump on the Joint Comprehensive Plan of Action, May 8, 2018.

14. "Rouhani: Iran Will Export Crude Oil," AP, September 4, 2018.

15. Vali Nasr, *The Shia Revival: How Conflicts Within Islam Will Shape the Future* (New York: Norton, 2016), pp. 114–15, 143; Jeffrey Feitman, "Hezbollah: Revolutionary Iran's Most Successful Export," Brookings Institution, January 17, 2019; Kim Ghattas, *Black Wave: Saudi Arabia, Iran, and the Forty-Five Year Rivalry That Unraveled Culture, Religion, and Collective Memory in the Middle East* (New York: Henry Holt, 2020), chapters 7 and 16 (Iran and the rise of Hezbollah).

Chapter 30: The Struggle for Iraq

1. Mohamad Bazzi, "King Salman's War," *Politico*, January 25, 2015.

2. Joby Warrick, *Black Flags: The Rise of ISIS* (New York: Doubleday, 2015), p. 300 ("middle of the road").

3. Dexter Filkins, "The Shadow Commander," *New Yorker*, September 30, 2013; Ali Alfoneh, "Brigadier General Qassem Soleimani: A Biography," AEI, no. 1 (January 2011); Ali Soufan, "Qassem Soleimani and Iran's Unique Regional Strategy," *CTC Sentinel* 11, no. 10 (November 2018).

4. Filkins, "The Shadow Commander."

5. "Iraq Oil Rush," *New York Times*, June 22, 2008 ("suspicions"); "Iraq—Systematic Country Diagnostic," World Bank, February 2017.

6. Interview with Dr. Ashti Hawrami; IHS Markit, "KRG and Iraq Could Reach Oil Export and Revenue Deal in December, According to Turkey," December 4, 2013.

7. Renad Mansour and Faleh A. Jabar, "The Popular Mobilization Forces and Iraq's Future," Carnegie Endowment for International Peace, April 28, 2017.

8. Tim Arango et al., "The Iran Cables," *New York Times*, November 19, 2019 ("eyes closed," oil contracts); Strategic Studies Institute and U.S. Army War College Press, *The U.S. Army in the Iraq War*, vol. 2, *Surge and Withdrawal, 2007–2011* (2019), p. 639.

9. Mustafa Salim and Liz Sly, "Widespread Unrest Erupts in Southern Iraq Amid Acute Shortages of Water, Electricity," *Washington Post*, July 14, 2018.

Chapter 31: The Arc of Confrontation

1. Robert Gates, *Duty: Memoirs of a Secretary at War* (New York: Knopf, 2014), p. 505; George H. W. Bush and Brent Scowcroft, *A World Transformed* (New York: Knopf, 1998), p. 341 ("my wise friend").

2. Interview with Frank Wisner; Hillary Rodham Clinton, *Hard Choices* (New York: Simon & Schuster, 2014), p. 343 ("peaceful transition"); Gates, *Duty*, pp. 505–6 ("exactly," "now").

3. William Burns, *The Back Channel: A Memoir of American Diplomacy and the Case for Its Renewal* (New York: Random House, 2019), p. 466 ("expectations"); Gates, *Duty*, p. 507 ("our allies").

4. David D. Kirkpatrick and Steven Lee Myers, "Libya Attack Brings Challenges for U.S.," *New York Times*, September 12, 2012.

5. Allan R. Millett and Peter Maslowski, *For the Common Defense: A Military History of the United States from 1607 to 2012* (New York: Simon & Schuster, 1994), ("lost province"); Souad Mekhennet and Joby Warrick, "U.S. Increasingly Sees Iran's Hand in the Arming of Bahraini Militants," *Washington Post*, April 1, 2017 (Soleimani).

6. Michael Knights and Matthew Levitt, "The Evolution of Shia Insurgency in Bahrain," The Washington Institute, January 2018, p. 23.

7. "Interview with Syrian President Bashar al-Assad," *Wall Street Journal*, January 31, 2011, Jay Solomon, *The Iran Wars* (New York: Random House, 2016), pp. 104–108.

8. Patrick Seale, *Asad: The Struggle for the Middle East* (Berkeley: University of California Press, 1988), p. 352 (*fatwa*); Moshe Ma'oz, *Asad: The Sphinx of Damascus* (New York: Grove Weidenfeld, 1988).

9. Nasser Menhall, Petition to the Legal Adviser, U.S. Department of State, October 30, 2013.

10. James Bennet, "The Enigma of Damascus," *New York Times Magazine*, July 10, 2005.

11. Barack Obama, "The Future of Syria," White House Blog, August 18, 2011.

12. C. J. Chivers and Eric Schmitt, "Saudis Step Up Help for Rebels in Syria with Croatian Arms," *New York Times*, February 25, 2013.

13. Martin Dempsey in "Obama at War," *Frontline*, PBS, season 2015, episode 10, transcript, www.pbs.org/wgbh/frontline/film/obama-at-war/transcript/.

14. Susan Rice, *Tough Love: My Story of the Things Worth Fighting For* (New York: Simon & Schuster, 2019), pp. 362–69; Burns, *Back Channel*, pp. 328–30; Jeffrey Goldberg, "The Obama Doctrine," *Atlantic*, April 2016 ("playbook").

15. Greg Jaffe, "The Problem with Obama's Account of the Syrian Red Line Incident," *Washington Post*, October 4, 2016 ("stunning twist").

16. Mounir al-Rabih, "Hezbollah and Iran in Syria," The Washington Institute, November 13, 2017.

17. On the Syrian Kurds, articles by Robin Wright in the *New Yorker*, April 4, October 20, 2019; United Nations High Commissioner for Refugees (UNHCR)—Syria emergency, www.unhcr.org/en-us/syria-emergency.html.

18. Gregory D. Johnsen, *The Last Refuge: Yemen, Al-Qaeda, and America's War in Arabia* (New York: Norton, 2014), chapter 17; Robert F. Worth, *A Rage for Order: The Middle East in Turmoil from Tahrir Square to ISIS* (New York: Farrar, Straus, and Giroux, 2016), chapters 4 and 6.

19. Michael Knights, "The Houthi War Machine," *CTC Sentinel*, September 2018, vol. 11, issue 8; Nicholas Niarchos, "How the U.S. Is Making the War in Yemen Worse," *New Yorker*, January 15, 2018.

20. Gerald M. Feierstein, "Iran's Role in Yemen and Prospect for Peace," Middle East Institute, December 6, 2018 (air service).

21. Peter Salisbury, "Yemen and the Saudi-Iranian 'Cold War,'" Chatham House, February 2015, pp. 8–9 ("a victory"); Sayed Hassan Nasrallah, "The Resistance Axis Triumphs, Israel Panics," Speech, March 18, 2017 ("decisive victory").

22. Sami Aboudi and Stephanie Nebehay, "Saudi Oil Tanker Hit in Houthi Attack off Yemen—Coalition," Reuters, April 3, 2018.

23. "How—and Why—to the End the War in Yemen," *The Economist*, November 30, 2017.

24. Peter Salisbury, "Yemen: National Chaos, Local Order," Research Paper, Chatham House, December 2017, p. 45.

25. Salisbury in Niarchos, "How the U.S. Is Making the War in Yemen Worse."

26. Sam Dagher, "What Iran Is Really Up To in Syria," *The Atlantic*, February 14, 2018; Kambiz Foroohar and Ladane Nasseri, "Iran Wields Power from Syria to Gulf as Rise Alarms Sunni Rivals," *Bloomberg*, February 18, 2015.

Chapter 32: The Rise of the "Eastern Med"

1. IHS Markit, "Israel: Oil & Gas Risk Commentary," August 2019; IHS Markit, "Levantine Basin," Basin Insights Profile, October 19, 2016.

2. Interview with Eli Groner; Shoshanna Solomon, "Octogenarian Geologist," *Times of Israel*, April 18, 2018.

3. "Hezbollah Issues Fresh Threat," *The Times of Israel*, February 18, 2018 ("Cease operating"); Judah Ari Gross, "With High-Tech Warships," *The Times of Israel*, February 5, 2018.

4. Tarek el-Molla and Yuval Steinitz at CERAWeek 2019.

Chapter 33: "The Answer"

1. Richard P. Mitchell, *The Society of the Muslim Brothers* (New York: Oxford University Press, 1993), pp. 2, 8–9 ("We are brothers"); Gudrun Krämer, *Hasan al-Banna* (Oxford: Oneworld, 2010), chapter 2; Eric Trager, *Arab Fall: How the Muslim Brotherhood Won and Lost Egypt in 891 Days* (Washington, D.C.: Georgetown University Press, 2016), pp. 46–51.

2. Lawrence Wright, *The Looming Tower: Al Qaeda and the Road to 9/11* (New York: Vintage, 2011), pp. 9, 15.

3. Wright, *The Looming Tower*, pp. 34–35; Fawaz A. Gerges, *Making the Arab World: Nasser, Qutb, and the Clash That Shaped the Middle East* (Princeton: Princeton University Press, 2018), pp. 140–45, 223–32; John Calvert, *Sayyid Qutb and the Origins of Radical Islamism* (New York: Oxford, 2018), pp. 182–87.

4. Yaroslav Trofimov, *The Siege of Mecca: The Forgotten Uprising in Islam's Holiest Shrine and the Birth of al-Qaeda* (New York: Penguin, 2007), p. 7 ("modern times"); Robert Lacey, *Inside the Kingdom: Kings, Clerics, Modernists, Terrorists, and the Struggle for Saudi Arabia* (New York: Penguin, 2010), chapters 2 and 3.

5. Trofimov, *The Siege of Mecca*, p. 248.

6. Shadi Hamid, *The Temptations of Power: Islamists & Illiberal Democracy in the New Middle East* (Oxford: Oxford University Press, 2015), pp. 14–15; Shadi Hamid, "Islamists and the Brotherhood," in *The Arab Awakening: America and the Transformation of the Middle East* (Washington, D.C.: Brookings Institution Press, 2011), pp. 31–34; Shadi Hamid, *Islamic Exceptionalism: How the Struggle over Islam is*

Reshaping the World (New York: St. Martin's Press, 2016), p. 80 ("affiliates"); Eric Trager, *Arab Fall*, pp. 48–55; 94–95 ("a unified Islamic state, or a global Islamic union" and a "global Islamic state"); Robert Worth, *A Rage for Order* (New York: Farrar, Straus, and Giroux, 2016), chapter 5.

7. Wright, *The Looming Tower*, pp. 54–55, 211; Peter Bergen, *The Longest War: The Enduring Conflict Between America and Al-Qaeda* (New York: Free Press, 2011), chapter 2; Steve Coll, *The Bin Ladens: An Arabian Family in the American Century* (New York: Penguin Press, 2008).

8. Thomas Hegghammer, *Jihad in Saudi Arabia: Violence and Pan-Islamism Since 1979* (Cambridge: Cambridge University Press, 2010).

9. Lukáš Tichý and Jan Eichler, "Terrorist Attacks on the Energy Sector: The Case of Al Qaeda and the Islamic State," *Studies in Conflict & Terrorism* 41, no. 6 (2018), pp. 455, 465; Lisa Marshall, "Terrorism and Friendship in Algeria," *Colorado School of Mines Magazine* 103, no. 2 (Summer 2013).

10. Charles Lister, *The Syrian Jihad: Al-Qaeda, the Islamic State and the Evolution of an Insurgency* (New York: Oxford University Press, 2015), p. 192 ("find them").

11. Lister, *The Syrian Jihad*, p. 214; David Ignatius, "How Isis Spread in the Middle East," *Atlantic*, October 29, 2015.

12. Joby Warrick, *Black Flags: The Rise of ISIS* (New York: Doubleday, 2015), p. 299 ("Persians").

13. Interviews.

14. Alissa J. Rubin, "Militant Leader," *New York Times*, July 5, 2014; Warrick, *Black Flags*, pp. 304–305 ("Rome"); Liz Sly, "The Hidden Hand," *Washington Post*, April 4, 2015; Hugh Kennedy, *Caliphate: The History of an Idea* (New York: Basic Books, 2016), pp. 63, 70; Damien McElroy, "Rome Will Be Conquered Next Says Leader of Islamic State," *Telegraph*, July 1, 2014.

15. "Timeline: The Rise, Spread, and Fall of the Islamic State," Wilson Center, October 28, 2019.

16. "In Audio Recording, ISIS Leader Abu Bakr Al-Baghdadi Says Operations Underway, Urges 'Caliphate Soldiers' to Free Captives," MEMRI, September 16, 2019.

17. Department of Defense Briefing on Baghdadi raid, October 30, 2019, www.defense.gov/Newsroom/Transcripts/Transcript/Article/2004092/department-of-defense-press-briefing-by-assistant-to-the-secretary-of-defense-f/.

Chapter 34: Oil Shock

1. Neil Hume and Anjli Raval, "Iraq Violence Lights Fuse to Oil Price Spike," *Financial Times*, June 20, 2014.

2. Ali Al-Naimi, *Out of the Desert: My Journey from Nomadic Bedouin to the Heart of Global Oil* (New York: Portfolio/Penguin, 2016), pp. 286–88; Robert McNally, *Crude Volatility: The History and the Future of Boom-Bust Oil Prices* (New York: Columbia University Press, 2017), pp. 212–16; Raf Sanchez, "Barack Obama: Iran Could Be a 'Successful Power," *Telegraph*, December 29, 2014.

3. "Sheikhs v Shale," *The Economist*, December 12, 2014; Asjylyn Loder, "Shale Producers Clobbered," *Bloomberg*, September 10, 2015.

4. Interview ("Every study we saw").

5. Ali Al-Naimi, Speech, CERAWeek, February 23, 2016; Al-Naimi, *Out of the Desert*; Daniel Yergin, "The Global Battle for Oil Market Share," *Wall Street Journal*, December 15, 2015; Wael Mahdi, "Saudi Prince Affirms Oil Strategy as Market Seen as 'Excellent,'" *Bloomberg Business*, April 28, 2015.

6. "Meeting with Deputy Crown Prince of Saudi Arabia Mohammad bin Salman Al Saud," September 4, 2016, http://en.kremlin.ru/events/president/news/52825.

7. Rania El Gamal, Alex Lawler, and Vladimir Soldatkin, "OPEC Agrees Modest Oil Output Curbs in First Deal Since 2008," Reuters, September 28, 2016.

8. "OPEC's Sanusi Barkindo," Bloomberg TV, December 1, 2016.

9. Jim Burkhard, Bhushan Bahree, and Aaron Brady, "The New Math of Oil: The 'Inadvertent Swing Supplier'—the United States," IHS Markit Strategic Report, February 2015; Vladimir Soldatkin and Oksana Kobzeva, "Russia Offers to Sell Gas," Reuters, December 8, 2017 ("rivals into partners").

10. Donald J. Trump, tweet, April 20, 2018.

11. Donald J. Trump, tweet, June 22, 2018.

12. "Trump Says He Wants to Go Slower on Sanctions," Reuters, November 5, 2018.

13. Ana Vanessa Herrero, "After U.S. Backs Juan Guaidó as Venezuela's Leader, Maduro Cuts Ties," *New York Times*, January 23, 2019 ("usurper").

14. *The Economist*, May 11, 2019.

15. Rawan Shaif and Jeremy Binnie, "Attack on Saudi Oil Facilities Deepens Regional Malaise," *Jane's*, October 9, 2019.

16. Ali Soufan, "Qassem Soleimani and Iran's Unique Regional Strategy," *CTC Sentinel*, November 2018.

17. Peter Baker et al., "7 Days in January," *New York Times*, January 12, 2020; Parisa Hafezi, "Iran's Khamenei Stands by Guards," Reuters, January 17, 2020.

Chapter 35: Run for the Future

1. Stephen Kalin and Katie Paul, "Saudi Arabia Says It Has Seized over $100 Billion in Corruption Purge," Reuters, January 30, 2018.

2. "Saudi Arabia's Heir to the Throne Talks to 60 Minutes," CBS, March 9, 2018; "Full Transcript of Saudi Crown Prince's CBS Interview with Nora O'Donnell, Including Unaired Answers," Arabia5AM, October 1, 2019.

3. Samia Nakhoul, William Maclean, and Marwa Rashad, "Saudi Prince Unveils Sweeping Plans," Reuters, April 25, 2016 ("at the mercy"); John Kemp, "Saudi Arabia Will Struggle to Kick Its Addiction," Reuters, April 27, 2016.

4. W. H. Berg to R. W. Hanna, March 14, 1938, and W. J. Lenahan to L. N. Hamiton, "Declaration of Commercial Production," October 30, 1938, Chevron archives; Telegram to E.A. Skinner, March 4, 1938, Saudi Aramco archives.

5. "Mohammed bin Salman Talks to *Time*," *Time*, April 5, 2018 ("mud huts").

6. International Monetary Fund, "Saudi Arabia: Selected Issues," Country Report 19/291, September 2019, p. 4.

7. Ezzoubeir Jabrane, "Grand Mufti of Saudi Arabia: 'Concerts and Cinema Are Haram,'" *Morocco World News*, January 15, 2017; Kristin Smith Diwan, "Let Me Entertain You: Saudi Arabia's New Enthusiasm for Fun," Arab Gulf States Institute in Washington, March 9, 2018.

8. James Lemoyne, "Mideast Tensions; Ban on Driving by Women Reaffirmed by Saudis," *New York Times*, November 15, 1990; "Saudi Arabia—Fatwa on Women's Driving of Automobiles (Shaikh Abdel Aziz Bin Abdallah Bin Baz), 1990."

9. Kim Mackrael, Paul Vieira, and Donna Abdulaziz, "Saudi Students Fret over Future After Order to Leave Canada," *Wall Street Journal*, August 9, 2018.

10. *The Abu Dhabi Economic Vision for 2030* (Abu Dhabi Economic Council for Development, 2008); Abu Dhabi Chamber and IHS Markit, *Biannual Abu Dhabi Economic Report*, December, 2019; "Nearly 50 Years On, Adviser Remembers First Abu Dhabi Master Plan," *The National*, June 14, 2016 ("village"); "Sheikh Mohammed bin Zayed's Inspirational Vision," *The National*, February 10, 2015; interviews.

11. Vision 2030, my.gov.sa.

12. "Full Transcript of Saudi Crown Prince's CBS Interview," Arabia5AM, October 1, 2019 ("book a room").

13. International Monetary Fund, "Saudi Arabia: Selected Issues," Country Report 19/291, September 2019, p. 12 (jobs).

14. On East Asian "miracle," Daniel Yergin and Joseph Stanislaw, *The Commanding Heights: The Battle for the World Economy* (New York: Touchstone, 2002), chapters 6 and 7.

15. "Erdoğan Seeks to Expand Turkey's Influence in the Middle East Through Diplomacy—and Force," *The Conversation*, April 17, 2018; "Erdoğan: Turkey Is the Only Country That Can Lead the Muslim World," *Yeni Şafak*, October 15, 2018.

16. Adam Taylor, "As America Tries to End 'Endless Wars,' America's Biggest Mideast Base is Getting Bigger," *Washington Post*, August 21, 2019. 2019 ("in the center").

17. "Interview with Muhammad bin Salman," *The Economist*, January 6, 2016.

18. "Transcript: Muhammad bin Salman," *The Economist*, January 6, 2016.

19. "Saudi Crown Prince Discusses Trump, Aramco, Arrests: Transcript," *Bloomberg*, October 5, 2018 ("investment powerhouse"); International Monetary Fund, "Saudi Arabia: Selected Issues," Country Report 19/291, September 2019, p. 9.

Chapter 36: The Plague

1. Clifford Krauss, "Saudi Oil Price Cut," *New York Times*, March 9, 2020; Nelli Sharushkina and Amena Bakr, "Saudi-Russia Rift," *International Oil Daily*, March 9, 2020 ("wondering" and "more time"); Nelli Sharush, "Russia Braces for Price War," *International Oil Daily*, March 11, 2020 ("variants"); Alexander Novak Interview on Russia Channel 1, April 13, 2020.

2. "Sechin: Low Oil Prices," Interfax, March 20, 2020; Joshua Yaffa, "Russian-Saudi Oil War Went Awry," *The New Yorker*, April 15, 2020 ("strategic threat").

3. *Mapping the Global Future: Report of the National Intelligence Council's 2020 Project* (December 2004); Centers for Disease Control and Prevention; Bill Gates Ted Talk, April 3, 2015, https://www.youtube.com/watch?v=6Af6b_wyiwI.

4. Alexander Novak interview, Ekho Moskvy Radio, April 2, 2020.

5. Interview ("dire"); interview with Don Sullivan; March 13, 2020, letter to Crown Prince; March 25, 2020, letter to Honorable Mike Pompeo; Tucker Higgins, "Ted Cruz, Other Senators, Warn Saudis," CNBC, March 30, 2020 ("economic warfare"); Lutz Kilian, Michael D. Plante, and Xiaoqing Zhou, "How Falling Oil Prices in Early 2020 Weakened the U.S. Economy," Dallas Federal Reserve, May 2020.

6. Coronavirus Task Force Briefing, March 19, 2020 ("person driving the car"); Javier Blas, "Trump's Oil," *Bloomberg*, April 13, 2020 ("so low" and "wiped out"); Coronavirus Task Force Briefing, April 1, 2020; Donald J. Trump, Tweet, April 2, 2020 ("my friend").

7. Frank Kane, "Saudi Arabia Calls 'Urgent Meeting of Oil Producers'," *Arab News*, April 2, 2020; Meeting on the Situation in Global Energy Markets, April 3, 2020, Kremlin website.

8. Mohammad Barkindo, Remarks to OPEC+ Ministerial Meeting, April 9, 2020.

9. Coronavirus Task Force Briefing, April 8, 2020 ("hated OPEC"); Dan Brouillette, Remarks for G20 Extraordinary Energy Ministers Meeting, April 10, 2020; Benoit Faucon, Summer Said, and Tim Puko, "U.S., Saudi Arabia, Russia Lead Pact," *Wall Street Journal*, April 12, 2020; "The Largest Oil Supply Cut in History," *Oil Market Insight*, IHS Markit, April 12, 2020.

10. Anjli Raval, "Saudi Arabia Says Price War," *Financial Times*, April 14, 2020 ("departure" and "divorce lawyers"); *Petroleum Intelligence Weekly*, April 17, 2020.

11. Jim Burkhard, "Covid-19 Oil Prices," *Oil Market Briefing*, IHS Markit, April 20, 2020.

ROADMAP

Chapter 37: The Electric Charge

1. Interview with J. B. Straubel.

2. Interview with J. B. Straubel; IHS Markit, "Reinventing the Wheel: The Future of Cars, Oil, Chemicals, and Electric Power," Part I Report, June 2017, p. VI-3; David Keyton, "Three Win Nobel in Chemistry," Phys.org, October 9, 2019.

3. Interview with J. B. Straubel.

4. Ashlee Vance, *Elon Musk: Tesla, SpaceX, and the Quest for a Fantastic Future* (London: Virgin Books/Penguin, 2016), p. 161 ("sucked"); Levi Tillemann, *The Great Race: The Global Quest for the Car of the Future* (New York: Simon & Schuster, 2015), p. 152 (Lutz).

5. Tillemann, *The Great Race*, p. 241.

6. Interview with Daniel Akerson.

7. Interview with Mary Barra.

8. Interview with Mary Barra.

9. Virginia Gewin, "Turning Point: Daniel Carder," *Nature*, November 18, 2015; Gregory J. Thompson, "In Use Emissions Testing of Light-Duty Diesel Vehicles in the United States," Final Report, Center for Alternative Fuels, Engines and Emissions, West Virginia University, May 15, 2014; Philip E. Ross, "How Engineers at West Virginia University," *IEEE Spectrum*, September 22, 2015; David Morgan, "West Virginia Engineer," Reuters, September 23, 2015.

10. Chris Reiter and Elisabeth Behrmann, "VW's Diesel Woes Reach $30 Billion," *Bloomberg*, September 29, 2017; "VW Diesel Crisis: Timeline," Cars.com, December 7, 2017; "Is Diesel Dead in the EU and Will Electrification Take Its Place?," IHS Markit AutoIntelligence Strategic Report, November 27, 2017.

11. Jeff Meyer, "Will Houston Become More Like Oslo: Perspective on City Vehicle Restrictions," IHS Markit Mobility and Energy Future Strategic Report, May 2020; Elena Pravettoni, "Cars and the City of the Future: London Case Study," IHS Markit Mobility and Energy Future Strategic Report, August 1 2018; "Angela Merkel verteidigt Diesel-Autos," *Frankfurter Allgemeine*, March 23, 2017; Kylie MacLellan and Guy Faulconbridge, "Electric Cars Win?," Reuters, July 26, 2017.

12. Herbert Diess, Speeches: at the Global Board Meeting, January 16, 2020, November 4, 2019, September 9, 2019, Annual General Meeting, May 14, 2019.

13. Jack Ewing, "Volvo, Betting on Electric," *New York Times*, July 5, 2017.

14. Interview with Bill Ford; Akiko Fujita, "Toyota Chairman," CNBC, September 5, 2017 (Toyota); Elon Musk | Full Interview | Code Conference 2016, YouTube, June 2, 2016, www.youtube.com/watch?v=wsixsRI-Sz4.

15. Jim Burkhard, "Electric Geography: Nearly Half of U.S. EVs in One State—California," IHS Markit Mobility and Energy Future Strategic Report, October 22, 2018.

16. MacLellan and Faulconbridge, "Electric Cars Win?" ("Tavares").

17. Interview (Norway); Tillemann, *The Great Race*, p. 180 (Senate staffer).

18. California Air Resources Board, https://ww2.arb.ca.gov/our-work/programs/volkswagen-zero-emission-vehicle-zev-investment-commitment/about.

19. Tillemann, *The Great Race*, p. 99 (tractors), p. 115 ("day by day").

20. Premasish Das and Xiaonan Feng, "Electricity Takes the Bus," Strategic Report, IHS Markit, June 2018.

21. IHS Markit, "Reinventing the Wheel," p. IV-10; Jeff Meyer et al., "From 'Carrot' to 'Stick': How China's EV Policy Support Is Evolving," IHS Markit Mobility and Energy Future Insight, September 18, 2019.

22. *Pulse of Change: Update on LV and EV Sales*, "Global EV Sales in 2019 Edge Out Higher than in 2018—But Pace of Growth Sharply Decelerates," February 14, 2020, IHS Markit Mobility and Energy Future Insight.

23. "Nitin Gadkari Tells Carmakers: Move to Electric Cars," NDTV, September 7, 2017.

24. "Inside India's Messy Electric Vehicle Revolution," *New York Times*, August 22, 2019; Dharmendra Pradhan and Ajit Jindal, Remarks at India Energy Forum by CERAWeek, October 14–16, 2018.

25. Sintia Radu, "Toyota and Mazda Join Forces," *Washington Post*, August 4, 2017.

26. India Energy Forum 2017 by CERAWeek, October 8–10, 2017.

27. *Pulse of Change: Update on LV and EV Sales*, "Global EV Sales."

28. Interview with J. B. Straubel; MIT Energy Initiative, *Insights into Future Mobility* (Cambridge, MA: MIT Energy Initiative, 2019); Elena Pravettoni, Sam Huntington, and Youmin Rong, "How Fast Are Electric Vehicle Battery Costs Falling," IHS Markit Power and Renewables Strategic Report, October 2019.

29. IHS Markit Lithium and Battery Materials Service.

30. IHS Markit and Energy Futures Initiative, *Advancing the Landscape of Clean Energy Innovation*, Breakthrough Energy Coalition, February 2019.

31. Vaclav Smil, *Energy Transitions: Global and National Perspectives* (Santa Barbara: Praeger, 2017).

32. Interview with Bill Ford.

Chapter 38: Enter the Robot

1. "Radio-Driven Auto Runs Down Escort," *New York Times*, July 28, 1925; IHS Markit, *Reinventing the Wheel*, p. III-16.

2. "The DARPA Grand Challenge: Ten Years Later," DARPA, March 13, 2014.

3. Interview with Sebastian Thrun; Christine O'Toole, "What Drives Red Whittaker," *Pittsburgh Quarterly*, Winter 2018; "The Great Robot Race," *Nova*, PBS, March 28, 2006.

4. Interview with Sebastian Thrun; "The Great Robot Race."

5. Interview with Sebastian Thrun; Conor Dougherty, "How Larry Page's Obsession Became Google's Business," *New York Times*, January 22, 2016; Arjun Kharpal, "Google's Larry Page Disguised Himself," CNBC, May 11, 2017.

6. Interview with Lawrence Burns; Scott Corwin and Rob Norton, "The Thought-Leader Interview: Lawrence Burns," *strategy+business*, Autumn 2010; Lawrence Burns and Christopher Shulgin, *Autonomy: The Quest to Build the Driverless Car and How It Will Reshape Our World* (New York: HarperCollins, 2018).

7. Lindsay Chappell, "2007: The Moment Self-Driving Cars Became Real," *Automotive News*, December 19, 2016 (Whittaker).

8. Interview with Sebastian Thrun; Sebastian Thrun, "What We're Driving At," Google Blog, October 9, 2010.

9. Burkhard Bilger, "Auto Correct," *New Yorker*, November 18, 2013.

10. "Taxonomy, and Definitions for Terms Related to Driving Automation Systems for On-Road Motor Vehicles," SAE, June 6, 2015, www.sae.org/standards/content /j3016_201806/.

11. Rebecca Yergin, "IoT Update: Navigating the Course of Spectrum for Connected and Automated Vehicle Technologies," *National Law Review*, November 30, 2018.

12. National Science and Technology Council and the U.S. Department of Transportation, *Assuring America's Leadership in Automated Vehicles Technologies: Automated Vehicles 4.0* (Washington D.C., 2020); Rebecca Yergin, "NHTSA Continues to Ramp Up Exploration of Automated Driving Technologies," Covington & Burling, Blog, April 2020.

13. Marco della Cava, "Garage Startup Uses Deep Learning to Teach Cars to Drive," *USA Today*, August 30, 2016.

Chapter 39: Hailing the Future

1. Interview with Garrett Camp; "UberCab" pitch deck, December 2008.

2. Adam Lashinsky, *Wild Ride: Inside Uber's Quest for World Domination* (New York: Portfolio/Penguin, 2017), pp. 80–81, 91.

3. Megan Rose Dickey, "Lyft's Rides Are So Social," *Business Insider*, March 16, 2014; Travis Kalanick, Uber Policy White Paper, "Principled Innovation: Addressing the Regulatory Ambiguity," April 12, 2013 ("compete").

4. "Didi Chuxing's Founder Cheng Wei," *Times of India*, August 8, 2016.

5. Interview with Jean Liu.

6. Mike Isaacs, *Super Pumped: The Battle for Uber* (New York: W. W. Norton, 2019), chapters 27–30; Farhad Manjoo, "Uber's Lesson," *New York Times*, June 21, 2017; Anita Balakrishnan, "Here's the Full 13-Page Report of Recommendations for Uber," CNBC, June 13, 2017.

Chapter 40: Auto-Tech

1. Michael Sivak, "Younger Persons Are Still Less Likely to Have a Driver's License Than in the 1980s," January 6, 2020, https://www.greencarcongress.com/2020/01/20200106-sivak.html.

2. Adrienne LaFrance, "The High-Stakes Race to Rid the World of Human Drivers," *Atlantic*, December 1, 2015; IHS Markit, "Mobility and Energy Future: 2019 Update," July 2019.

3. Lawrence D. Burns with Christopher Shulgan, *Autonomy: The Quest to Build the Driverless Car—And How It Will Reshape Our World* (New York: HarperCollins, 2018).

4. Interview with Bill Ford.

5. Mike Colias, Tim Higgins, and William Boston, "Will Tech Leave Detroit in the Dust?," *Wall Street Journal*, October 20, 2018 (Toyoda).

6. Interview with Bill Ford.

7. Interview with Mary Barra.

8. Interview with Bill Ford.

CLIMATE MAP

Chapter 41: Energy Transition

1. On net zero carbon: David Victor, "Deep Decarbonization: A Realistic Way Forward on Climate Change," *Yale Environment* 360, January 28, 2020; John Deutch, "Is Net Zero Carbon by 2050 Possible?," June 2020, https://eartharxiv.org/bvf5c/. On coal and Darby: Peter Brimblecombe, "Attitudes and Responses Towards Air Pollution in Medieval England," *Journal of the Air Pollution Control*

Association 26, no. 10 (October 1976), pp. 941–45; R. A. Mott, "Abraham Darby (I and II) the Coal Iron Industry," *Transactions of the Newcomen Society* 31, no. 1 (1957); Nancy Cox, "Imagination and Innovation of an Industrial Pioneer: The First Abraham Darby," *The Industrial Archaeology Review* 12 (1990), pp. 127–44; "Shopshire: Industrial Heritage," BBC, http://www.bbc.co.uk/shropshire/content /articles/2009/02/12/abraham_darby_feature.shtml ("doubt me"); Barrie Trinder, *The Darbys of Coalbrookdale* (Chichester, UK: Phillimore, 1978), chapter 2; Barbara Freese, *Coal: A Human History* (New York: Penguin Books, 2004), pp. 24–32.

2. Vaclav Smil, *Energy Transitions: Global and National Perspectives* (Santa Barbara: Praeger, 2017).

3. Intergovernmental Panel on Climate Change, *Climate Change 2007: The Physical Science Basis* (New York: Cambridge University Press, 2007); Rajendra Pachauri, CERAWeek, February 11, 2008; Gayathri Vaidyanathan, "U.N. Climate Science Body Launches Search to Replace a Strong Leader," *E&E News*, February 25, 2015.

4. Intergovernmental Panel on Climate Change, *Climate Change 2014: Synthesis Report. Contribution of Working Groups I, II, and III to the Fifth Assessment Report of the Intergovernmental Panel on Climate Change* (Geneva: IPCC, 2014); Steven Koonin, "Climate Science Is Not Settled," *Wall Street Journal*, September 19, 2014.

5. Mark Landler and Helene Cooper, "After a Bitter Campaign, Forging an Alliance," *New York Times*, March 18, 2010 ("worst meeting"); Bruce Einhorn, "Why the US-China Emissions Pact Could Be a Climate Change Breakthrough," *Bloomberg*, November 12, 2014.

6. Fiona Harvey, "Paris Climate Change Agreement," *Guardian*, December 14, 2015; Suzanne Goldenberg, John Vidal, Lenore Taylor, Adam Vaughan, and Fiona Harvey, "Paris Climate Deal," *Guardian*, December 12, 2015 (secretary-general); Remarks by President Obama on the Paris Agreement, October 5, 2016; "Donald Trump Would 'Cancel' Paris Climate Deal," BBC News, May 27, 2016 (Trump).

7. Global Climate Project, *Global Climate Report 2019*; Daniel Yergin, *The Quest: Energy, Security, and the Remaking of the Modern World* (New York: Penguin, 2011), pp. 422–28.

8. Greta Thunberg, *No One Is Too Small to Make a Difference* (New York: Penguin, 2019), pp. 10, 62, 96–99; Charlotte Alter, Suyin Haynes, and Justin Worland, "2019 Person of the Year—Greta Thunberg," *Time*, December 11, 2019; Greta Thunberg, Luisa Neubauer, and Angela Valenzuela, "Why We Strike Again," *Project Syndicate*, November 29, 2019.

9. Mark Carney, "Breaking the Tragedy of the Horizons—Climate Change and Financial Stability," Speech, September 29, 2015.

10. "Final Report: Recommendations of the Task Force for Climate-related Financial Disclosures," Task Force for Climate-related Financial Disclosures, June 2017; Larry Fink's Letter to CEOs, January 14, 2020; Climate Bonds Initiative 2019; Andrew Edgecliffe-Johnson and Billy Nauman, "Fossil Fuel Divestment Had 'Zero' Climate Impact, Says Bill Gates," *Financial Times*, September 17, 2019.

11. Valerie Pavilonis and Matt Kristofferen, "Delay Second Half of The Game," *Yale Daily News*, November 23, 2019; Alan Murray and David Meyer, "Against Oil Divestment," *Fortune*, November 26, 2019 (Swensen).

12. Climate Accountability Institute and Union of Concerned Scientists, "Establishing Accountability for Climate Change Damages: Lessons from Tobacco Control," La Jolla, California, June 14–15, 2012, www.ucsusa.org/sites/default/files/attach/2016/04/establishing-accountability-climate-change-damages-lessons-tobacco-control.pdf; Amie Tsang and Stanley Reed, "*Guardian* Stops Accepting Fossil Fuel Ads," *New York Times*, January 30, 2020; Anna Bateson and Hamish Nicklin, "*Guardian* Will No Longer Accept Fossil Fuel Advertising," *Guardian*, January 29, 2020.

13. Hiroko Tabuchi and Nadja Popovich, "How Guilty Should You Feel About Flying?," *New York Times*, October 17, 2019; "Climate Confessions," NBC News, https://www.nbcnews.com/news/specials/climate-confessions-share-solutions-climate-change-n1054791; James Pickford, "RSC Brings Curtain Down," *Financial Times*, October 2, 2019.

Chapter 42: Green Deals

1. Opening Statement in the European Parliament Plenary Session by Ursula von der Leyen, Strasbourg, July 16, 2019; Ewa Krukowska and *Bloomberg*, "'A Quantum Leap in Its Ambition': Europe's Investment Bank," *Fortune*, November 15, 2019; Kelly Levin and Chantal Davis, "What Does 'Net-Zero Emissions' Mean? 6 Common Questions, Answered," WRI, September 17, 2019; European Commission, *Supplementary Report on Using the TEG Taxonomy*, June 2019; *Final Report on Financing a Sustainable European Economy* (March 2020).

2. Samuel Petrequin and Raf Casert, "EU Commission President Announces 'European Green Deal,'" *Christian Science Monitor* via Associated Press, December 11, 2019; Jean Pisani-Ferry, "A Credible Decarbonization Agenda Can Help Strengthen Europe's Economy," Peterson Institute for International Economics, December 9, 2019; European Commission, "Europe's Moment: Repair and Prepare for the Next Generation," May 27, 2020; EU press release, "Europe's Moment," May 27, 2020.

3. Christina Zhao, "Alexandria Ocasio-Cortez Warns, 'World Is Going to End in 12 Years'," *Newsweek*, January 22, 2019; Green New Deal Fact Sheet, February 7, 2019, https://assets.documentcloud.org/documents/5729035/Green-New-Deal-FAQ.pdf; "House Resolution 109, Recognizing the Duty of the Federal Government to Create a Green New Deal," February 2, 2019. For an analysis of a Green New Deal in terms of "how the economy has changed and government has evolved," see Jason Bordoff, "Getting Real About the Green New Deal," *Democracy Journal*, March 25, 2019.

Chapter 43: The Renewable Landscape

1. "The Father of Photovoltaics—Martin Green Profile," ABC, May 26, 2011.

2. John Fialka, "Why China Is Dominating the Solar Industry," *Scientific American*, September 19, 2016; Charlie Zhu and Bill Powell, "Special Report: The Rise and Fall of China's Sun King," Reuters, May 18, 2013.

3. Martin Green, "How Did Solar Cells Get So Cheap?," *Joule* 3, no. 3, March 20, 2019, pp. 631–33.

4. IHS Markit, *PV Module Supply Chain Tracker, PV Installation Tracker; PV Supplier Tracker*; Renewable Energy Policy Network for the 21st Century (REN21), *Renewables 2018: Global Status Report*, p. 97 ("cutthroat pricing").

5. IHS Markit, *Global Renewable Power Market Outlook*, March 2020.

6. Michael Stothard, "Isabelle Kocher," *Financial Times*, May 15, 2016 ("challenging"); REN 21, Renewables 2018, ("fierce competition"); Christopher Crane, CEO Plenary, Edison Electric Institute Annual Convention, June 11, 2019, transcript.

7. IHS Markit Global Energy Scenarios; Varun Sivaram, *Taming the Sun* (Cambridge, MA: MIT Press, 2018), pp. 43–45.

8. Sivaram, *Taming the Sun*, pp. 56, 71.

9. Ernest Moniz et al., *Optionality, Flexibility & Innovation: Pathways for Deep Decarbonization in California* (Washington, D.C.: Energy Futures Initiative, May 2019); Timothy P. Gardner, Michael Stoppard, Dan Clay, and Raul Timponi, "Exploring the Efficient Frontier: A Global Perspective on the Gas-Renewables Partnership," Global Gas Strategic Report, IHS Markit, September 2018.

Chapter 44: Breakthrough Technologies

1. IHS Markit and Energy Futures Initiative, *Advancing the Landscape of Clean Energy Innovation Breakthrough Energy Coalition*, February 2019, chapter 3; Third Way, "2019 Advanced Nuclear Map," October 2019.

2. IHS Markit launched a major research program on hydrogen in 2017 that focused on Europe, California, and China and continues in the Forum on Hydrogen and Renewable Gas.

3. Joeri Rogelj et al., "Mitigation Pathways Compatible with 1.5°C in the Context of Sustainable Development," in *Special Report: Global Warming of 1.5°C*, Intergovernmental Panel on Climate Change, Geneva, 2018; IPCC, B. Metz et al., eds., *IPCC Special Report on Carbon Dioxide Capture and Storage. Prepared by Working Group III of the Intergovernmental Panel on Climate Change* (Cambridge: Cambridge University Press, 2005); Francois Bastin et al., "The Global Tree Restoration Potential," *Science* 365, no. 6448 (July 5, 2019), pp. 76–79. For an overview of options, see U.S. National Petroleum Council, *Meeting the Dual Challenge: A Roadmap to At-Scale Deployment of Carbon Capture, Use, and Storage: Final Report*, December 12, 2019.

4. Salk Institute, Harnessing Plants Initiative, https://www.salk.edu/harnessing -plants-initiative/.

Chapter 45: What Does "Energy Transition" Mean in the Developing World?

1. Timipre Sylva, Nigerian minister of state and petroleum resources, Remarks at Russian Energy Week, October 2019; *WHO Guidelines for Indoor Air Quality: Household Fuel Combustion* (Geneva: World Health Organization, 2014), "Household Air Pollution and Health," May 8, 2018.

2. Government of India, *Economic Survey 2018–2019*, p. 180; Alex Thornton, "7 of the World's 10 Most Polluted Cities," World Economic Forum, March 5, 2019.

3. Shreerupa Mitra, ed. *Energizing India: Fueling a Billion Lives* (New Delhi: Rupa Publications, 2019), pp. 13–14, and for an overview of India's energy position.

4. Interview with Dharmendra Pradhan.

5. Mitra, *Energizing India*, p. 74 ("gas-based economy").

6. Interview with Dharmendra Pradhan.

Chapter 46: The Changing Mix

1. Atul Arya, "Whither Energy Transition?," Blog, December 13, 2019.

2. Remarks, International Energy Forum, Beijing, December 5, 2019; State Council, People's Republic of China, "Premier Calls for High Quality Energy Development," October 11, 2019, http://english.www.gov.cn/premier/news/201910/11/content _WS5da08a3fc6d0bcf8c4c14e92.html.

3. IHS Markit Mobility and Energy Future Scenarios 2019.

4. Tsvetana Paraskova, "IEA Chief: EVs Are Not the End of the Oil Era," OilPrice.com, January 22, 2019 (Fatih Birol); John Heimlich, "Tracking Impacts of COVID-19," Airlines for America, April 24, 2020.

5. Fenit Nirappil, "On Patrol with the Enforcer of DC's Plastic Straw Ban," *Washington Post*, January 28, 2019.

6. Prachi Patel, "Stemming the Plastic Tide: 10 Rivers Contribute Most of the Plastic in the Oceans," *Scientific American*, February 1, 2018; Source: Christian Schmidt, Tobias Krauth, and Stephan Wagner, "Export of Plastic Debris by Rivers into the Sea," *Environmental Science and Technology* 51, no. 21 (November 7, 2017); Jeremy Hess, Daniel Bednarz, Jaeyong Bae, and Jessica Pierce, "Petroleum and Health Care: Evaluating and Managing Health Care's Vulnerability to Petroleum Supply Shifts," *American Journal of Public Health* 101, no. 9 (2011), pp. 1568–79; *A Sea Change: Plastics Pathway to Sustainability, Multi-Client Study on Recycling & Sustainability*, IHS Markit, November 2018.

7. Mark P. Mills, "Testimony Before the U.S. Senate Energy Committee Sources and Uses of Minerals for a Clean Energy Economy," September 16, 2019.

Conclusion: The Disrupted Future

1. Daniel Yergin and Joseph Stanislaw, *The Commanding Heights: The Battle for the World Economy* (New York: Simon & Schuster, 2002).

2. Interview; Lee Hsien Loong, "The Endangered Asian Century," *Foreign Affairs*, July-August 2020, pp. 52, 61.

3. Chris Giles, "Richest Nations Face $17 Trillion Government Debt," *Financial Times*, May 25, 2020.

Epilogue: Net Zero

1. Lisa Friedman, Simoni Sengupta, and Coral Davenport, "Biden, Calling for Action, Commits U.S. to Halving Its Climate Emissions," *New York Times*, April 22, 2021.

2. Xi Jinping, Speech, Leaders Summit on Climate, April 22, 2021, http://www
   .xinhuanet.com/english/2021-04/22/c_139899289.htm.

3. European Parliament, "MEPs: Put a Carbon Price on Certain EU Imports to Raise
   Global Climate Ambition," press release, https://www.europarl.europa.eu/news
   /en/press-room/20210304IPR99208/meps-put-a-carbon-price-on-certain
   -eu-imports-to-raise-global-climate-ambition

4. Steven E. Koonin, *Unsettled: What Climate Science Tells Us, What It Doesn't, and
   Why It Matters* (Dallas: BenBella Books, 2021). Koonin was provost of Cal Tech
   and undersecretary for science in the U.S. Department of Energy in the Obama
   administration. Also see Michael Shellenberger, *Apocalypse Never: Why Environmental Alarmism Hurts Us All* (New York: Harper Collins, 2020).

5. See, for instance, Energy Transitions Commission, *Making Clean Electrification
   Possible: 30 Years to Electrify the World Economy*, April 2021. https://www.energy
   -transitions.org/publications/making-clean-electricity-possible.

6. Narendra Modi, Speech, CERAWeek 2021, March 5, 2021.

7. Interview with Noubar Afeyan by Walter Isaacson, CERAWeek 2021, March
   2021.

8. "The Hidden Cost of Oil," hearing before the Committee on Foreign Relations,
   United States Senate, March 3, 2006, https://www.govinfo.gov/content/pkg
   /CHRG-109shrg34739/html/CHRG-109shrg34739.htm.

9. *CERAWeek Conversation*, December 2020.

10. Jack Ewing, "Volvo Plans to Sell Only Electric Cars," *New York Times*, March 2,
    2021; Daniel Yergin, "How Electric, Self-Driving Cars and Ride-Hailing will
    Transform the Car Industry," *Wall Street Journal*, April 23, 2021.

11. White House, "Executive Order on America's Supply Chain," February 24, 2021;
    European Commission, *Critical Raw Materials Resilience: Charting a Path towards
    Greater Security and Sustainability*, September 3, 2020; Thierry Breton in Michael
    Peel and Henry Sanderson, "EU Sounds Alarm on Critical Raw Materials Short-
    ages," *Financial Times*, August 31, 2020.

12. *Hindustan Times*, September 4, 2020.

13. International Energy Agency, *The Role of Critical Minerals in Clean Energy Tran-
    sitions*, May 2021, pp. 5, 8, 11–12, 30, 50, 122, 235, https://www.iea.org/reports
    /the-role-of-critical-minerals-in-clean-energy-transitions; Mark Mills, "Biden's
    Not-So-Clean Energy Transition," *Wall Street Journal*, May 11, 2021.

14. International Energy Agency, "Fuel Supply," in *World Energy Investment 2020*, De-
    cember 2020, https://www.iea.org/reports/world-energy-investment-2020/fuel
    -supply#abstract; International Energy Agency, *Net Zero by 2020: A Roadmap for
    the Global Energy Sector*, May 2021, https://www.iea.org/reports/net-zero-by-2050;
    Fatih Birol, Columbia University Center for Global Energy Policy, May 19, 2021;
    Yuka Obayashi and Sonali Paul, "Asia Snubs IEA's Call," Reuters, May 19, 2021
    (Japan); "Nations Dispute IEA's road map," *Financial Times*, May 24, 2021.

15. White House, "Executive Order on Climate-Related Financial Risk," May 20,
    2021; White House, "FACT SHEET: President Biden Directs Agencies to Analyze
    and Mitigate the Risk Climate Change Poses to Homeowners and Consumers,
    Businesses and Workers, and the Financial System and Federal Government Itself,"

May 20, 2021; Austin Landis, "President Biden Signs Order to Address Financial Risks of Climate Change," NY1, May 20, 2021, https://www.ny1.com/nyc/all-boroughs/news/2021/05/20/biden-signs-order-financial-risk-climate-change.

16. White House, *Interim National Security Strategy Guidance*, March 3, 2021; Reuters, January 19, 2021 (Blinken).

17. White House, *National Security Strategy*, February 2015.

18. Wang Yi in Ryan Hass, "How China is Responding to Escalating Strategic Competition with the U.S.," *China Leadership Monitor*, March 1, 2021, https://www.brookings.edu/articles/how-china-is-responding-to-escalating-strategic-competition-with-the-us; Xi Jinping, speech, World Economic Forum, January 25, 2021, and speech, Boao Forum, April 20,2021, XinhuaNet.; Chris Buckley, "The East is Rising," *New York Times*, March 3, 2021 ("biggest threat").

Illustration Credits

Insert 2

page 1, top, left: https://commons.wikimedia.org/wiki/File:Mark_Sykes00.jpg.

page 1, top, right: Archives du Ministère de l'Europe et des Affaires étrangères—La Courneuve.

page 1, bottom: © VICE MEDIA LLC.

page 2, top: https://commons.wikimedia.org/wiki/File:Weizmann_and_feisal_1918 .jpg.

page 2, bottom: The Huntington Library, San Marino, California.

page 4, top: REUTERS/Social Media Website via Reuters TV.

page 4, bottom: U.S. Department of State.

page 5, top: © Office of the Iranian Supreme Leader/AP Photo.

page 5, bottom: Houthi Military Media Unit handout.

page 6, top: © The Economist Newspaper Limited, London, December 6th, 2014.

page 6, bottom: Staton R. Winter/Bloomberg via Getty Images.

page 7, top: Fars New Agency, CC BY 4.0., https://commons.wikimedia.org/wiki /File:Ali_Khamenei_meets_Bashar_al-Assad_in_Tehran_20190225_01.jpg.

page 7, middle: REUTERS/Hamad I Mohammed.

page 7, bottom: Jumbo Maritime.nl.

page 8, top: © Saudi Aramco—All Rights Reserved.

page 8, middle: Bandar Algaloud / Saudi Royal Council / Handout/Anadolu Agency /Getty Images.

page 9, top: U.S. Department of the Interior, National Park Service, Edison National Historic Site.

page 9, middle: Courtesy of Maurizio Pesce, https://flickr.com/photos/30364433@N05 /8765031426. Attribution 2.0 Generic (CC BY 2.0).

page 9, bottom: Felix Wong/South China Morning Post via Getty Images.

page 10, top: Stan Honda/AFP via Getty Images.

page 10, middle: REUTERS/Gene Blevins.

page 10, bottom: Courtesy of UBER.

page 11, top: Courtesy of Ford Motor Company.

page 11, bottom: Kiyoshi Ota/Bloomberg via Getty Images.

page 12, top: Copyright of The University of Manchester.

page 12, bottom: UNFCCC via Flickr.

page 13, top: Al Drago/Bloomberg via Getty Images.

page 13, bottom: Anders Hellberg, Wikipedia CC 4.0 Int'l license.

page 14, top: Graham Hely/Newspix.

page 14, middle: Kevin Frayer/Getty Images.

page 14, bottom: Alex Hofford/EPA/Shutterstock.

page 15, top: © Frans Lanting/lanting.com.

page 15, bottom: Courtesy of Siemens Gamesa.

page 16, top: Karim Sahib/AFP via Getty Images.

page 16, middle: Andrew Harnik/AP photo.

page 16, bottom: White House photo by Adam Schultz.

Index

Note: Page numbers in *italics* indicate maps and illustrations.

Abadi, Haider al-, 233–34
Abdulaziz bin Abdul Rahman (Ibn Saud), 60, 295
Abdulaziz bin Salman, 278, 294, 313, 321
Abqaiq oil facility, 286
Abraham Accords, 448
Abu Dhabi, 274, 300–302, 309
Advancing the Landscape of Clean Energy Innovation (study), 403
Afeyan, Noubar, 434
Afghanistan, 180, 210, 216, 246–47, 263–65, 297
Africa, 38, 152, 157, 178, 182, 228, 407, 416
Ahmadinejad, Mahmoud, 222–23
air pollution, 158, 341, 342–43, 407. *See also* carbon emissions; greenhouse gas emissions (GHG)
Aizenberg, Eitan, 256
Akerson, Daniel, 333
Alaska, 20, 33, 69
Alawites, 207, 242, 245
Alberta, Canada, 46–47, 48, 60
Alekperov, Vagit, 76
Al-Falih, Khalid, 279
Algeria, 87
Algiers Accord, 280
Alibaba, 362

Aliyev, Ilham, 121
Al Jazeera, 306
al-Jihad, 263, 264
Allison, Graham, 131
Al Qaeda, 216, 249, 264–66
   and the Gulf War, 237
Amerada, 18
American Journal of Public Health, 417
Angell, Norman, 461–464
Anglo-German naval race, 461, 463
Annam kingdom, 137
Ansar Allah, 249–50
Aphrodite field, 256
Apple, 358, 368
Arab-Israeli War (1967), 204–5
Arab League, 239
Arab nationalism, xv, 199–200, 203–5, 215
Arab oil embargo (1973), 53, 59
Arab Spring, 22, 91, 236–43, 248–52, 254, 429–30
Arak nuclear reactor, 226
Aramco. *See* Saudi Aramco
Arctic petroleum reserves, 111–14
artificial intelligence (AI), 174, 349, 354, 357, 364, 369, 424, 429
Arya, Atul, 412

Asia, 38, 163, 398, 402, 412, 416. *See also specific countries such as* China
Asia International Infrastructure Bank (AIIB), 182
Asian financial crisis of 1998, 75
Asquith, Herbert, 196
Assad, Bashar al-, 241–42, 243–45, 269
Assad, Hafez al-, 242
Association of Southeast Asian Nations (ASEAN), 148–50, 168–69, 169n
Astana International Financial Center, 177
Ataturk, Mustafa Kemal, 201–2
Australia, 38, 114, 167, 169, 189, 281
Austria, 30
Austro-Hungarian Empire, 196, 200
automobile industry
　automobile manufacturing, 42, 132
　and autonomous vehicles, 347–57, 368–69, 373
　Auto-Tech advances, xvi, 366–73, 415, 427
　Chinese market for U.S. cars, 171
　*See also e*lectric vehicles
Azerbaijan, 71, 74, 109, 120–23, 207, 281

Ba'athists, 207, 211–13, 216–17, 242, 268
Bab al-Mandeb, 251
Badri, Ibrahim Awad al-, 268
Baghdadi, Abu Bakr al-, 268–9, 271
Bahrain, 240–41, 306
Bai Meichu, 139–41, 150
Baker, James, 215
Bakken shale, 19–20, 21–22, 49
Balfour Declaration, 199
ballistic missiles, 226–27
Bank of England, 384
Banna, Hasan al-, 259–60
Barkindo, Mohammad Sanusi, 280–81, 319
Barnett Shale, 5–7, 14, 19, 23, *24*
Barra, Mary, 333–34, 370, 441
Bashneft, 76
battery technology, 327–34, 341, 344–46, 403, 429, 441–444
Bazhenov formation, 98
Beckman, Robert, 145, 146
"Beijing consensus," 116
Belarus, 74
Belt and Road Initiative (China), xiv, 134, 177–90, *185*
Ben Ali, Zine el Abidine, 236
Benghazi, Libya, 239
Bernanke, Ben, 26
Bhatt, Alia, 342
Biden, Hunter, 110
Biden, Joe, 110, 237, 392, 431, 436
Bin Laden, Osama, 263–64, 265

biomass and biofuels, 394, 408, 410
BlackRock, 385
Black Sea, 122
Blinken, Tony, 445–46
Blue Stream pipeline, 85
Boko Haram, 270
Bolshevik Revolution, 69, 72, 79
Bolsonaro, Jair, 44–45
Bosporus strait, 122
Boston, Massachusetts, 112–13
BP, 15, 76, 386
Brazil, 38, 44–45, 56–57, 320
Breakthrough Energy Coalition, 403
Breton, Thierry, 442
BRIC era, 56–57, 76–77, 273
Britain and the United Kingdom
　and Brexit, 343
　and carbon emission reduction, 388
　and Central Asian economic ties, 177
　and First Sino-Japanese War, 154
　and Iranian Revolution, 207
　and Iraqi oil infrastructure, 232
　and Middle East colonialism, 194–200, *197*
　and Nasser's Arab nationalism, 203–4
　and Syrian civil war, 247
　and the Thucydides Trap, 131, 154
　and Ukrainian independence, 80
　as "workshop of the world," 132
Brookings Institution, 59
Brouillette, Dan, 320
Brunei, 33, 143, 169n
Budapest Memorandum, 95
Burgan oil field, 24
Burisma, 110
Burkan missiles, 251–52
Burns, Larry, 351–52, 368
Burns, William, 225, 238
Bush, George H. W., 21–22, 172, 214–16, 237
Bush, George W., 172, 216–17

Cairo, Egypt, 237
California Air Resources Board (CARB), 329, 331, 336, 340
Caliph's Last Heritage, The (Sykes), 195
Camp, Garrett, 358–59
Camp Bucca, 268
Canada, 46–51, *50*, 57, 98, 281, 300, 316, 320, 322
carbon emissions
　and breakthrough energy technologies, 403
　and carbon capture technology, 404–6, 418–19, 439
　and carbon cycle, *383*
　and diesel vehicles, 335–37
　and electric vehicles, 328, 329, 331, 337, 346

and energy transition challenges, 377–78, 381–87, *383*

and global power politics, xi

and "green deal" proposals, 388–93, *390*

and impact of coronavirus, 411–12

impact of shale gas on, 13

and push for renewable energy sources, 394

by sector, *414*

See also net zero carbon

Carder, Daniel, 336

Carnegie Mellon University (CMU), 349–52

Carney, Mark, 384–85

Carpio, Antonio, 146

Carter, Jimmy, 53, 210–11

Caspian Pipeline, 121–22, 180

Caspian Sea, 71, 74

Caucasus, 120, 121

Central Asia, 120–26, 158, 177–80, 182, 184, 189. *See also specific countries*

Central Intelligence Agency (CIA), 103

Chan Heng Chee, 169

Chávez, Hugo, 272–73, 283

chemical weapons, 213, 245–47

Cheng Wei, 362–63

Cheniere, 32, 35–37, 38

Chesapeake, 31, 35–36

Chevrolet Volt, 332–34

Chevron, 65

Chiang Kai-shek, 139–41

China

and Auto-Tech advances, 368

Belt and Road Initiative, xiv, 134, 177–90, *185*

Century of Humiliation, xiv, 154, 460

and container shipping, 161–64

Cultural Revolution, 147–48, 166, 340

development of oil resources, 155–60

economic growth, 130–33, 161

and electric vehicles, 338, 340–42, 344–45

and energy transition challenges, 412–13, 416

financing for Russian infrastructure projects, 117–18

and global energy trends, 423

and global impact of coronavirus pandemic, 311–12, 322–23

impact of U.S. shale production on, 56–57

key challenges facing, 142–46

"Made in China" strategy, 174, 397

and Middle East oil resources, 223–24, 232

naval power in Chinese history, 152–54

and Northern Sea Route, 113

petroleum imports, 133

and push for renewable energy sources, 395–402, 432, 442

and ride-hailing services, 362–64

rising global position of, 150, 165–76, 188–89, 425–26

rivalry with U.S., xii, xiv–xv, 62, 129–35, 425–26

and Russian geopolitics, 77, 114, 115–19, 120–26

and South China Sea tensions, 136–41, 147–51, *149*, 152–53

and the Thucydides Trap, 131, 154, 425

trade and price wars, 26, 62, 315–16

and trade deficits, 39

and "unequal treaties," 119, 139

and U.S. manufacturing boom, 25–26

and U.S. transition to LNG exporter, 39–40

and varied approaches to climate change, 413

"workshop of the world," 132, xiv

China Dream, The (Liu), 132

China National Petroleum Company, 125

China Ocean Shipping Company, 163

China-Pakistan Economic Corridor, 186

"China's National Defense in the New Era," 173

Chinese Communist Party, 166–67, 174

Chinese Development Bank, 396

Chinese Military Council, 138

Chinese Ministry of Foreign Affairs, 144

Chinese Nationalists, 140, 146

chudo (Russian economic miracle), 74–75

Chukotka Peninsula, 69–70

chulha stoves, 408

Cleveland Clinic, 302

climate change

and electric vehicles, 330, 331

and energy transition challenges, xvi–xvii, 377–87

and European decarbonization efforts, 86

and global power politics, xi

and "green deal" proposals, 51, 64, 388–93

and pipeline battles in U.S., 48

politics of, xvi

variety of approaches to, 412

Clinton, Bill, 130, 153, 172, 216–17

Clinton, Hillary, 103, 150, 237–38

coal

and China's economic growth, 155–56

and electric vehicles, 339

and energy transition challenges, 378–79

and Germany's response to Fukushima disaster, 87

and hydrogen, 404

coal (cont.)
  impact of shale gas on, 12, 33
  and railroad transportation, 20
  and Russia's "pivot to the east," 116–17
  and varied approaches to climate
    change, 412
cobalt, 345, 442–43
Cold War, xii, xviii, 60, 69–70, 80, 90, 121,
    348, 426
Coldwell, Pedro Joaquín, 275
USS Cole bombing, 249, 264
colonialism, 193–96, 196–200, 197
Commanding Heights, The (Yergin and
    Stanislaw), xviii
commodity supercycle, 56–57, 77
ConocoPhillips, 65
container shipping, 161–64
Continental Resources, 19–20
COP 20 (Copenhagen) meeting, 381
COP 26 (Glasgow) meeting, 433
coronavirus pandemic, xi–xii, xv, 311–12,
    322–23, 434–35
  and Brazil, 45
  and carbon emissions, 411–12
  and China, xv
  and China Belt and Road Initiative,
    180–81
  and Chinese economy, 132
  and containerization, 163
  and current geopolitical challenges,
    424, 430
  and electric vehicle technology, 339,
    344, 373
  and energy transition challenges, xiii,
    xviii, 416–18, 420
  and global energy trends, 426, 427–28, 430
  and global oil market, 313, 314–15,
    434–436
  and green deal proposals, 393, 428
  impact on global geopolitics, 62, 65
  and Iraq, 235
  and oil price war, 318, 321–22
  origins of, 311–12
  and ride-hailing services, 365
  and Russian interests in Central Asia,
    125–26
  and Saudi oil industry, 310
  and U.S.–China relations, 124, 130–31, 174
  and Yemen, 251
corruption, 93, 167, 233–34, 292–93, 421
Council of Ministries (Soviet Union), 73
COVID-19. See coronavirus pandemic
Crane, Christopher, 400
Crimea, xiv, 93, 94–96, 96, 115,
    124–25, 246
Cultural Revolution, 147–48, 166, 340

cyber security and warfare, 70, 171, 355
Cyprus, 109, 256, 257

Dakota Access pipeline, 49–51, 50, 391
Damman #7 well, 295
Daqing oil field, 156
Darby, Abraham, 378, 429
Das Kapital (Marx), 148
decarbonization, xvii, 86, 377, 385, 390,
    390–91, 404
"decoupling," xv, xviii, 130–31
Defense Advanced Research Projects
    Agency (DARPA), 347–49
Democratic Party (U.S.), xv, 55, 64, 391
Democratic Republic of Congo, 345
Dempsey, Martin, 246
Deng Xiaoping, 147–48, 163, 167, 425
Denmark, 344, 400
dependency, 43, 232
developed countries, 382, 408, 412
developing countries, xvii, 56, 382, 407–10,
    412, 413
Devon Energy, 8, 9–10
DiDi, 362–64, 365, 369, 372
diesel fuel and vehicles, 54, 335–37
Diess, Herbert, 337
direct air capture, 405
distributed energy systems, 399–400
divestment movement, 385–86. See also
    shareholder activism
Doha, Qatar, 278–79
Donbas region, 79
Donilon, Thomas, 59
Dow, 29–30
drone technology, 251–52, 286
Drucker, Peter, 162
Dutch East Indies, 140
Dyukov, Alexander, 76

Eagle Ford Shale, 17, 24, 24, 42, 55
Earth Justice, 51
earthquakes, 28
East Asia, 33
East China Sea, 148
Eastern Europe, 88, 109
Eastern Mediterranean energy reserves,
    253–58
Eastern Mediterranean Gas Forum, 449
East Turkistan Islamic Movement, 180
Eberstadt, Nicholas, 132
Ebola epidemic, 315
"economic miracle" countries, 33, 74–75
Economic Survey (India), 408
Economist, The, 276, 286, 307–8
Edison, Thomas, 329
Egypt

and Arab nationalism, 203–5, 214
and Arab Spring protests, 237–38
and Eastern Mediterranean petroleum
    resources, 254, 256–57
and historical context of Middle East
    conflicts, 196
and Iranian Revolution, 209
and Islamic fundamentalism, 259–63,
    264, 270
and Qatar, 306
and Syrian civil war, 251
Einstein, Albert, 394–95
election interference, 70, 78, 81, 103–4
electric power and infrastructure, xvii,
    12–13, 184, 186, 234, 345–46, 404
electric vehicles (EVs), xvi, 327–46,
    368–71, 415, 427, 428, 430
embassy bombings in Kenya and Tanzania
    (1998), 264–65
emissions standards, 335–37, 339, 346
Energiewende ("energy turn"), 86–87,
    395–96
energy security and independence
    and Canadian imports to U.S., 47
    and China's development of petroleum
        resources, 160
    and current geopolitical challenges, 427
    and Eastern Mediterranean petroleum
        resources, 254–57
    and electric vehicles, 341
    and energy transition in the developing
        world, 408–9
    and energy transition in U.S., xiii
    and gas supplies to Europe, 84–89
    and Nixon administration, 53
    and opposition to Russian gas exports, 109
    and politics of U.S. shale production, 55
    and Russia-Europe relations, 83
    and South China Sea tensions, 171
    and varied approaches to climate change,
        412–13
"energy superpower" status, xiii, 57,
    70–71
Energy Transfer Partners, 49, 51
energy transition
    and breakthrough energy technologies,
        403–6
    and carbon capture technology, 419
    and current global challenges, xi–xviii,
        427–29
    and developing world, 407–10
    emerging consensus on climate issues,
        382–87
    and "green deal" proposals, 388–91,
        391–93
    historical perspective on, 377–79

and IPCC, 379–80
and Paris climate agreement, 380–82
and push for renewable energy sources,
    394, 400–401
and U.S. position, xiii
and varied approaches to climate
    change, 412, 434
Eni, 256
environmental issues and activism
    and American shale gas reserves, 113
    and Fukushima nuclear disaster, 87
    and global power politics, xi
    and hydraulic fracturing, 28–29
    and indoor air pollution in developing
        countries, 407–8
    and opposition to pipeline projects, 46–51
    and U.S. transition to LNG exporter, 37
    See also carbon emissions; climate change
Environmental Defense Fund, 28–29
EOG, 14–17
Erbil, Kurdistan, 232
Erdoğan, Recep Tayyip, 247, 305, 315
ESPO pipeline, 118
Estonia, 69
EU Council, 102
Eurasian Economic Union, 92, 93, 189
Europe
    and China Belt and Road Initiative,
        182, 184
    and Eastern Mediterranean petroleum
        resources, 258
    and impact of U.S. shale and LNG, 38, 55,
        61–62
    and push for renewable energy sources,
        398–99
    See also European Union (EU); *specific
        countries*
European Battery Alliance, 442
European Central Bank, 187
European Commission, 388–90
European Union (EU)
    and energy security issues in Europe,
        85–88
    and energy transition challenges, 381
    and "green deal" proposals, 388–91
    and Nord Stream 2 pipeline, 102, 104,
        108–9
    and Russian annexation of Crimea, 95
    and Russian gas supplies to Europe, 85
    and Russian geopolitical ambitions, 70, 115
    and Russia-Ukraine tensions, 93
    and Syrian refugees, 248
Europe's Optical Illusion (Angell), 462
Exclusive Economic Zones (EEZs) and
    territorial waters, 142–45, 148, 159,
    170, 257

extraterritoriality, 108, 139
ExxonMobil, 15, 65, 76, 395

Fabius, Laurent, 381
Fahd bin Abdulaziz Al Saud, 214
Faisal I, King of Iraq, 198–200, 202–3
Falcon rockets, 332
Farouk I, King of Egypt, 203
Federal Bureau of Investigation (FBI), 103
financial crisis of 2008, 26–27, 333, 429
Financial Stability Board, 385
Financial Times, 113, 273
financing for energy projects
    and China Belt and Road Initiative, 182–83
    and "green recovery" proposals, 428
    and push for renewable energy sources,
        397, 400–401
    and Russian interests in Central Asia,
        125–26
    and Russian LNG, 112
Fink, Larry, 385
First Opium War, 139
First Sino-Japanese War, 154, 460
5G technology, 175, 354
"flight shaming," 387, 415
Ford, Bill (and Ford Motor Company), 329,
    338, 346, 351, 369–70, 373
Ford, Henry, 372–73
Fort Laramie Treaty, 49
Fracking Debate, The (Raimi), 28
France, 138, 195–96, 201–2, 227, 232, 247,
    343
Freeport LNG facility, 24, 35–37, 38
Free Syrian Army (FSA), 244
Fukushima nuclear accident, 63, 87, 401, 430

G7, 129
G8, 129
G20, 129, 280, 319–20, 388, 426
Gadhafi, Muammar, 239
Gadkari, Nitin, 342
Gaidar, Yegor, 73
Galán, Ignacio, 437–38
gasoline
    and Auto-Tech advances, 368, 370–72
    and "clean diesel," 335
    and consumer behaviors, 421
    Mexican imports, 41, 43
    and oil embargo of 1973, 53–54
    and oil price war, 316–17, 323
    and pipeline battles in U.S., 47
Gates, Bill, 315, 385–86
Gates, Robert, 237–38
Gaza, 253
Gazprom, 76, 80, 86, 89, 105, 107–8,
    109, 125

Geely, 338
General Motors, 171, 329, 333–34, 369
Georges-Picot, François, 194–95, 196–98,
    201–2
Georgia (country), 82
Germany
    and "clean diesel," 336
    economic growth before World War I, 132
    and energy security issues in Europe,
        86–88
    and energy transition challenges, xvii
    and global order after First World
        War, 200
    and Iranian nuclear ambitions, 223, 227
    and Khashoggi affair, 305–6
    and Nord Stream 2 pipeline, 102, 104–5,
        107–8
    and push for renewable energy sources,
        395–96, 400–401
    and Russia's "pivot to the east," 117
    and Syrian refugees, 248
    and the Thucydides Trap, 131, 154
    and U.S.–China trade war, 175
Ghawar oil field, 24, 241
gig economy, 361
globalization, 56, 188, 314–15, 423–24
global oil market stresses, 272–278, 279–83,
    284–87, 288–90, 426
Global Times, 168
Google, 350–53, 357, 364, 368–69
Gorbachev, Mikhail, 73–74, 76
Gore, Al, 121, 379–80
GPS technology, 348, 353–54
Grand Challenge, 348–52
Grand Mosque attack, 261–63, 296
Great Depression, 4
Great Recession, 27
Greece, 184, 257
Green, Logan, 360
Green, Martin, 395, 397
greenhouse gas emissions (GHG), 47,
    383–84, 411–12. *See also* carbon
    emissions
"Green New Deal," 51, 64, 391–92
Green Party (U.S.), 391
Greenpeace, 50–51, 387
Grey, Edward, 463
Groner, Eli, 255
Groningen field, 33, 87, 108
Grotius, Hugo, 456–58, 464
Guaidó, Juan, 285
guar, 7
Guardian, The, 387
Guardian Council (Iran), 208
Gulf Cooperation Council, 240–41, 450
Gulf of Mexico, 10, 15, 22–23, 34–35

Gulf War, 121, 216–19, 237, 240
Gwadar, Pakistan, 186

Hainan Island, 158
Hajj pilgrimage, 303
Hamas, 253, 282
Hambantota, Sri Lanka, 187
Hamm, Harold, 18
Han dynasty, 178
Harnessing Plants Initiative, 405–6
Hashemites, 200
HD-981 (drill ship), 158–59
Hejaz, 198
Hess, John and Hess Corporation, 18, 19
Hezbollah, 228, 242, 245, 247, 250–53,
   256–57, 266, 289, 292
Hezbollah Al Hijaz, 213
Hidden Mahdi doctrine, 222
Himalayas, 189
Holder, Eric, 364
Hong Kong, 139, 163, 174, 180–81, 344
horizontal drilling, 8–10, 17, 19, 23, 98
hostage crises, 208, 226, 237, 261, 265
Houdina, Francis, 348
Houthi, Hussein al-, 249
Houthis, 207, 248–52, 286, 292, 294
Huawei, 175
Hu Jintao, 157
Hungary, 110
Hussein, Saddam, 211–19, 221, 229–30, 232
Hussein bin Ali, Sharif of Mecca, 198
hybrid cars, 330, 338–39
hydraulic fracturing ("fracking")
   and Barnett Shale, 5–7
   and Bazhenov formation, 98
   and "green deal" proposals, 392
   and horizontal drilling, 8, 10, 19, 98
   and net zero carbon, 435–36
   and origins of shale production, 15–16
   and Permian Basin, 22–23
hydrogen, 403–4, 419, 439–440
hydropower, 339, 394, 400, 402

Iberdrola, 437–38
Ibn Saud. *See* Abdulaziz bin Abdul Rahman
IHS Markit, 400, 414, 417
Ikhwan, 296
USS *Impeccable*, 148
improvised explosive devices (IEDs), 231,
   347–48
Independence gas terminal, 109
India
   "Act East," 189
   and Auto-Tech advances, 368
   and China's Belt and Road initiative, 189
   and electric vehicles, 342–43

and energy transition challenges, xvii,
   408–10
impact of shale revolution on, 56
and Iranian oil sanctions, 223–24
market reforms and natural gas, 409–10
and U.S. LNG, 39
and varied approaches to climate change,
   412–13
Indian Ocean, 138, 189
Indonesia, 33, 34, 169n, 179
Industrial Revolution, 132, 378–79
Influence of Sea Power Upon History, The
   (Mahan), 458–59
Institutional Revolutionary Party
   (Mexico), 42
intellectual property disputes, 171, 173, 364
Intergovernmental Panel on Climate Change
   (IPCC), 379–80, 382–83, 404, 438
intermittency issues, xvii, 398, 399,
   401, 403
International Energy Agency, 29, 404,
   415, 418
International Energy Forum, 280
International Institute for Strategic
   Studies, 165
international law
   and Exclusive Economic Zones (EEZs),
      144–45, 148, 159, 170, 257
   and freedom of navigation, 145, 150,
      165, 167
   and South China Sea tensions, 143,
      167–68
   See also sovereignty issues
International Monetary Fund (IMF), 187,
   312, 322
Iran
   and Arab Spring protests, 239–40
   and competition within Islam, xv
   and Eastern Mediterranean petroleum
      resources, 253–54, 256–57
   and Hezbollah, 228
   and historical context of Middle East
      conflicts, 194
   Iranian Revolution, 59, 206–10, 212–13,
      231, 252
   and Khatami's election, 220–22
   nuclear ambitions, 213, 217, 222–27, 274,
      285–86
   nuclear program, 61, 222–28
   regional strategies in Middle East,
      229–31, 233–35, 248–52, 449
   relationship with Qatar, 307
   and Russian geopolitical interests,
      121, 282
   sanctions imposed on, 55, 222–28,
      285–89

Iran (cont.)
  and sectarian conflict in Iraq, 229–31,
    233–35
  and Soviet invasion of Afghanistan, 210
  and Syrian civil war, 242, 247
  and Yemen conflict, 248–52
Iran-Arab Friendship Society, 204
Iran/Iraq War, 212–13
Iran nuclear negotiations, 224–26
Iraq
  and Arab Spring protests, 242
  Iraq War, 217
  and ISIS, 266–67, 270
  oil resources, 57, 203, 231–32
  and Saddam Hussein regime, 211–19
  sectarian conflict over, 229–35
  and "Sykes-Picot," 198
  and Syrian civil war, 246–47
Iraq Petroleum Company, 243
Ischinger, Wolfgang, 105
Islamic Dawa Party, 230, 233
Islamic fundamentalism and jihadism,
    259–65
  and China Belt and Road Initiative, 180
  and Grand Mosque attacks, 261–63
  and historical context of Middle East
    conflicts, 193–94
  and Houthi rebellion in Yemen, 249–52
  push to establish caliphate, xv, 194–95,
    266, 268–69
  and Saudi strategies for the future, 297,
    304, 306
  and social media, 269–70
  and Soviet invasion of Afghanistan, 211
  and Syrian civil war, 245
  See also terrorism
Islamic Revolutionary Guard Corps
    (Iran), 208–9, 213, 220–22, 230, 241,
    253, 289
ISIS (Islamic State in Iraq and Syria),
    193–94, 233–34, 239, 247, 259–61,
    265–71, 273, 289, 380
Israel
  and Eastern Mediterranean petroleum
    resources, 253–57
  and gas supplies to Europe, 109
  and historical context of Middle East
    conflicts, 203
  and Iran, 222–23, 225, 227
  and Mubarak regime in Egypt, 237
  and Nasser's Arab nationalism, 204–5
  and regional impact of Iranian
    Revolution, 209
  and Russia's growing influence in Middle
    East, 282
Italy, 109, 184, 257, 396

Jackson-Vanik amendment, 104
James Shoal, 140–41
Japan
  and China Belt and Road, 189
  and China's petroleum resources,
    156, 157
  and China's rise, 175
  and energy consumption, 413
  Fukushima nuclear accident, 401
  and push for renewable energy
    sources, 395
  and sanctions on Russia, 99
  and South China Sea tensions, 137–40,
    154, 167, 169, 171
  territorial disputes, 135, 142
  and U.S. transition to LNG
    exporter, 33
jet fuel, 54, 391–92
jihadists. See Islamic fundamentalism and
    jihadism
Jindal, Ajit, 342–43
USS John S. McCain, 150
Johnson, Boris, 432
Journal of Commerce, 32, 164

Kalanick, Travis, 359–61, 364
Kashagan field, 122
Kataib Hezbollah, 289
Kawaguchi, Yoriko, 171
Kazakhstan, 69, 71, 74, 120–23, 177–79,
    180, 184, 281
Keay, John, 455
Kenai, Alaska, 33
Kenya, 264
Kerry, John, 48, 431
Kettering, Charles, 329, 351
Keystone XL, 46–48, 50
KGB, 75, 103
Khamenei, Ali, 206, 221, 250, 289
Khashoggi, Jamal, 305
Khatami, Mohammad, 220–21, 222
Khobar Towers bombing, 221
Khodorkovsky, Mikhail, 76
Khomeini, Ayatollah Ruhollah, 59, 207–9,
    212–13, 220, 237, 242, 296
Khorgas, 184
Khosrowshahi, Dara, 364
Khrushchev, Nikita, 94
Khunjerab Pass, 186
Khurais processing facility, 286
Khuzestan, 212
Kissinger, Henry, 154
Koh, Tommy, 152, 168, 453
Kudrin, Alexi, 71, 100–101
Kurdistan and Kurdish population, 201–2,
    232, 245, 247, 270

Kuwait, 60–61, 121, 204–5, 213–15, 217, 274, 295
Kyivan Rus, 79
Kyrgyzstan, 120, 180, 184

Lagarde, Christine, 187
Laos, 169n, 184
Lawrence, T. E. ("Lawrence of Arabia"), 199
Leaders Climate Summit, 432
League of Nations, 201
Lebanon, 207, 228, 242, 247, 256, 292
Lee Hsien Loong, 426
Lee Kuan Yew, 123
Lenin, Vladimir, 72
Lenovo, 362
Leviathan field, 254–57
Lew, Jacob, 96–97
Libya, 238–39, 246, 270, 272
Limits of Growth, The (Club of Rome), 4
Lindmayer, Joseph, 395
lithium batteries, 327–28, 330, 332, 341, 344–45, 441
Lithuania, 62, 109
Little Ice Age, 378
Liu, Jean, 362–64
Liveris, Andrew, 29–30
López Obrador, Andrés Manuel, 43–44, 320
Louisiana, 32
Lubmin, Germany, 84–85, 89, 107
LUKOIL, 76
Lutz, Robert, 332
Lyft, 360–61, 369, 372

Mackinder, Halford, 120
Macron, Emmanuel, 107
Maduro, Nicolás, 273, 283, 285
Mahan, Alfred Thayer, 458–61, 463, 464
Mahathir, Mohammad, 187–88
Maidan Square protests, 81, 93
"Malacca Dilemma," 157, 182
Malacca Strait, 113, 157, 186
Malaysia, 33, 143, 169n, 232
Maliki, Nouri al-, 230, 233, 267
Manchuria, 139, 156
Mao Zedong, 117, 140, 147, 156, 167
Mapping the Global Future (National Intelligence Council), 314
Marcellus shale, 11, 27, 113
Mare Liberum (The Freedom of the Seas) (Grotius), 457
Maritime Silk Road, 179, 186
Markey, Edward, 392
Maronite Christians, 201, 228
Marshall Plan, 181
Marx, Karl, 148
Masdar, 302

Massachusetts Institute of Technology (MIT), 369
Mattis, James, 218
Ma Ying-jeou, 146
Mazrouei, Suhail al, 313
McClendon, Aubrey, 35
McLean, Malcom, 161–63
Medvedev, Dmitry, 84–85
Meesemaecker, Georges, 136–38
Menhall, James, 243
Merkel, Angela, 58, 84–85, 102, 105–6, 108, 115, 227, 336
Mesopotamia, 196, 198, 200, 201–2
methane emissions, 28–29, 383, 387, 392, 402
Methane Pioneer, 32, 33
Mexico, 41–44
Middle East
    and Arab nationalism, 203–5
    and Arab Spring, 22, 91, 236–43, 248–52, 254, 429–30
    and China Belt and Road initiative, 178, 182
    and current geopolitical challenges, 426
    and energy transition challenges, xv–xvi
    and global oil market, 52, 72, 200, 203, 231–35
    and Iranian designs on Iraq, 229–31, 233–35
    and Iranian nuclear ambitions, 222–28
    and Iranian Revolution, 206–9
    mandate system, 201–2, 203, 220
    and Persian Gulf wars, 210–19
    Russia's growing influence in, 282
    and struggle for Iraq, 229–35
    and "Sykes-Picot," 193–98, 197, 200–202, 215, 270
    and Syrian civil war, 241–48
    See also specific countries
Middle East respiratory syndrome (MERS), 279, 315
Mikhelson, Leonid, 111
Milestones (Qutb), 261
military power and technology
    and China-Russia strategic partnership, 117–19
    and China's rise, 167–71, 171–74
    and Chinese artificial islands, 143
    cruise missile attacks, 286
    defense spending, 77, 257
    modernization of U.S. military, 133–34
    and Russian geopolitical ambitions, 70
    and Russia-Ukraine tensions, 94–95
    and U.S.–China rivalry, 133–34
    See also naval power

Miller, Alexey, 76, 89, 125
Mitchell Energy, 5–6, 6–7, 8
Mitchell, George P., 4–5, 7, 15
mobility as a service (MaaS), 365–66,
    371–72, 424
modernization, xvi, 139, 167, 202, 207, 212,
    306, 404
Modi, Narendra, 39, 189, 409, 410, 434, 443
Mohammed bin Nayef, 294
Mohammed bin Salman (MBS)
    and Houthi rebellion in Yemen, 251
    and Iranian Revolution, 206
    and oil price war, 317–18
    and price war among petroleum
        producers, 316–17
    and Saudi strategies for the future, 291–94,
        297, 300–301, 303, 305–6, 307, 309
Mohammed bin Zayed (MBZ), 301
Mongols, 79, 153
Moniz, Ernest, 225, 402, 403
monopolies, 41–44
Moon Jae-in, 39
Morsi, Mohamed, 238
Motor Trend, 332
Mubadala initiative, 301–2
Mubarak, Hosni, 237–38, 254
mujahedeen fighters, 263–64, 297
Murkowski, Lisa, 55
Musk, Elon, 327–29, 332, 338, 340
Muslim Brotherhood, 238, 253, 260–61,
    261–62, 263, 306
"mutually assured ambiguity" (MAA), 170
Myanmar, 169n

Nahyan, Zayed al, 301
Naimi, Ali al-, 275–76, 278–79, 281
Najaf, Iraq, 234
Nansha Islands, 167–68
Napolitano, Janet, 94
Nasrallah, Hassan, 250–51, 257
Nasser, Gamal Abdel, 203–5, 207,
    260–61, 262
National Defense Strategy (U.S.), 172
National Intelligence Council, 314
nationalism and nationalist movements
    Arab, xv, 199–200, 203–5, 215, 261
    and China's territorial claims, 146
    and current geopolitical challenges,
        423, 425
    and historical context of Middle East
        conflicts, 194
    and Mexico, 43–44
    and Saudi strategies for the future, 300
    and South China Sea tensions, 137–41, 170
    and Syrian civil war, 248
    Ukrainian, 79

nationalization of petroleum industries, 41–43
National Security Agency (NSA), 90, 103
National Security Council (NSC), 237
National Security Strategy (U.S.), 172
Native Americans, 49–50
natural gas/liquefied natural gas, 31–40
    and the Barnett Shale, 6–7
    and China's development of resources,
        158, 160
    and energy transition in developing
        world, 410
    and expansion of U.S. production, 13
    liquefaction and regasification, 32–36
    partnership with renewable energy
        sources, 402
    prices. See oil and natural gas prices
    and Russian geopolitical ambitions, 71
    and South China Sea tensions, 138
    terminals for, 34–36, 86, 241
    and Trump administration, 39–40
    and U.S. exports to Mexico, 41–44
    and U.S. manufacturing boom, 26–28
    and varied approaches to climate
        change, 418
"Natural Resources and the Commodity
    Supercycle" (panel), 56
nature-based climate solutions, 405–6
naval power
    and China's Belt and Road initiative, 189
    and China's Malacca Dilemma, 182
    and Chinese history, 152–54
    and freedom of navigation, 145, 150,
        165, 167
    and "Petroleum Situation in the British
        Empire" (1918), 200
    and the Thucydides Trap, 131
    and U.S.–China rivalry, 134
    See also South China Sea
Nayef bin Saud, 221
Nazarbayev, Nursultan, 122–23
Nazarbayev University, 177–78
Nebraska Department of Roads, 348
Nenet people, 111
Netanyahu, Benjamin, 255–56
Netherlands, xvii, 33, 86, 87, 232, 344
net zero carbon, xi, xvii, 377, 388–402,
    403–4, 419, 431–34, 437–39
New China Construction Atlas
    (Bai Meichu), 140
Newcomen, Thomas, 378
New York Times, 208, 231–32, 348, 364, 387
Nguyen Phu Trong, 159
Nichols, Larry, 8, 10
Nigeria, 272, 278, 280
9-Dash Line, 141, 144–46, 149, 150, 157,
    453–54

Nissan, 42, 332, 340
nitrogen oxide (NOx) emissions, 335–36
Nixon, Richard, 53
Nobel Prizes, 379–80, 395
Noble Energy, 254, 256
Nord Stream pipelines, 84–85, 89, 102, 104–8, 106, 113, 126, 314
North Africa, 22, 87
North Atlantic Treaty Organization (NATO), 70, 80, 82, 88, 105–7, 135, 239
North Dakota, 18–20, 49–51, 50, 282–83
Northern Sea Route, 112–13
North Field, 34
North Korea, 39, 135
North Sea, 87, 398
Norway, 88, 99, 232, 281, 320, 339, 400
Novak, Alexander, 275, 280, 313
Novatek, 111–12
nuclear power, 63, 87, 155, 401, 403
Nuland, Victoria, 93

Obama, Barack
  and Arab Spring protests, 237, 238
  and China Belt and Road Initiative, 181, 183
  and electric car development, 340
  and energy transition challenges, 381–82
  and Iranian nuclear ambitions, 226
  and pipeline battles, 47 (Keystone), 51 (Dakota Access)
  and Russia-Ukraine tensions, 91–92, 94
  and Russian annexation of Crimea, 96
  and shale gas policies, 12
  and Snowden intelligence breach, 91–92
  and Syrian civil war, 245–47
  and U.S.–China trade relations, 172
Ocasio-Cortez, Alexandria, 50, 391
October War, 53
offshore energy reserves, 10, 15, 22–23, 42, 44, 277. See also Eastern Mediterranean energy reserves; wind power
Oil and Gas Climate Initiative, 29, 405
oil and natural gas prices
  American price controls, 53–54
  and Central Asian economic ties, 177
  and China's development of oil resources, 158
  and "commodity supercycle," 56–57
  and impact of collapse on Iraq, 233
  impact of hydraulic fracturing on, 9, 14–15
  and Iranian Revolution, 207
  and ISIS attacks in Iraq, 267–68
  and Mexican economy, 42
  and oil embargo of 1973, 53
  and peak oil/peak demand concerns, xvi, 156, 298–99, 413–14, 417

  and price shocks of 2014, 99
  and Russia's "pivot to the east," 116
  and U.S. manufacturing boom, 30
  and U.S. sanctions on Russia, 97, 99–101
oil sands, 46, 60
Oman, 224, 225, 281
On China (Kissinger), 154, 461
On the Law of War and Peace (Grotius), 457
OPEC-Plus ("Vienna Alliance"), 281–84, 312–15, 322–23, 435
Operation Decisive Storm, 251
Orange Revolution, 81–82
Organization of the Petroleum Exporting Countries (OPEC)
  and global impact of coronavirus pandemic, 312–13
  and Iranian Revolution, 206
  and oil embargo of 1973, 53
  and price war, 319–20
  and Soviet oil production, 72
Osirik reactor, 213
Ottoman Empire, xv, 195–96, 198, 202, 259
Overseas Private Investment Corporation (OPIC), 188

Pachauri, Rajendra, 380
Page, Larry, 350, 357
Pahlavi, Mohammed Reza Shah, 207–9
Pakistan, 180, 186–87, 207, 247
Palestine, 198–201, 203
Panama Canal, 62, 184
Papa, Mark, 14–17
Paracel Islands, 137, 140
Paraskevopoulos, Savvas (Mike Mitchell), 4
Paris Climate Agreement (2015), xvii, 380–82, 384–85, 405, 418, 446
Partido Revolucionario Institucional (PRI), 42
Pascual, Carlos, 223–24
Pasdaran (Iranian Revolutionary Guard), 213
Patriotism Under Three Flags (Angell), 462
peak oil/peak demand, xvi, 156, 298–99, 413–14, 417
Pemex, 41–44, 320
Peña Nieto, Enrique, 42
Pence, Mike, 173, 174
Pennsylvania, 113
People's Liberation Army Navy, 154
People's Republic of China, 140–41
Permian Basin, 21–24, 24, 63, 65, 399
Persian Gulf, 60, 111–12, 157, 210–19, 286, 287
Peterson Institute for International Economics, 390
Petrobras, 44

Philippines, 143, 145–46, 169n
photovoltaics, 394–97, 429. *See also* solar
  power
Pioneer Natural Resources, 22–23
pipelines
  and Arctic natural gas, 111–14
  and China Belt and Road Initiative, 179–80
  and environmental activism, 49–51, 391
  and gas supplies to Europe, 80–89, 102–10
  impact of shale production on U.S. foreign
    policy, 61
  and Mexican gas imports, 43–44
  and North Dakota oil production, 20
  and U.S. LNG, 32–34
  and U.S. political conflicts, 46–48,
    48–51, *50*
  See also specific pipeline names
Piraeus port, 184
plastics, 416–17, 429
Poland, 85, 110
Polar Silk Road, 113
pollution, 32–33, 341–43. *See also* carbon
    emissions; greenhouse gas emissions
    (GHG)
polysilicon, 397
Pompeo, Mike, 59, 168, 187
population growth, 4, 297, 368, 378, 413–14
populist politics, 43, 44–45, 423
Potential Gas Committee, 12
Power of Siberia pipeline, 117, 125, 126, 158
Pradhan, Dharmendra, 284, 342, 409–10
propane, 410
Public Investment Fund (PIF), 309
Putin, Vladimir, 70–71, 75–76, 125
  and Arctic gas reserves, 112, 114
  on gas negotiations, 83
  and impact of shale revolution on U.S.
    foreign policy, 61
  and opposition to Russian gas exports, 107
  and price war among petroleum
    producers, 315, 318–20
  relationship with Saudis, 311
  and Russia-Ukraine conflict, 82–83, 94–96
  and Russian gas supplies to Europe, 85
  and Russian interests in Central Asia,
    124–25
  and Russia's energy transition
    challenges, xiii
  and Russia's "pivot to the east," 115–19
  and Syrian civil war, 246–47
  and tensions with Ukraine, 91–93
  and U.S. politics, 103

Qatar, 34, 37–38, 114, 240, 278, 306–7
Quds Force, 228, 230–31, 234, 247–48,
    253, 268, 289

Quest, The (Yergin), xvii
Qutb, Sayyid, 260–64

Racketeer Influenced and Corrupt
    Organizations (RICO) Act, 51
railroads, 20, 48, 49, 179, 184, 187–88
Raimi, Daniel, 28
Ramírez, Rafael, 275
RAND Corporation, 133
Ras Tanura oil terminal, 241, *287*
Reagan, Ronald, 53–54, 88, 172
recycling, 416
Red Sea, 251, 303
Reema bint Bandar Al Saud, 317
refugee crises, 245, 247–48, 283
REN, 398, 399
Renault-Nissan, 332–33
renewable energy, 55, 323, 330, 394–402
Republic of China, 139
Republican Guard (Iraq), 216
Republican Party (U.S.), xv, 55
Rice, Condoleezza, 83
Richardson, Bill, 121
Richthofen, Ferdinand von, 178
ride-hailing services and taxis, 343, 347,
    358–65, 368, 370–71, 373
"Road to Rejuvenation, The" (exhibit), 166
Roman Empire, 178
Roosevelt, Theodore, 458
Rose Revolution, 82
Rosen, Harold, 327
Rosneft, 76, 104, 118, 159, 275, 314
RosUkrEnergo, 83
Rothschild, Lionel, 199
Rouhani, Hassan, 225, 226, 227
Rousseff, Dilma, 44
Royal Geographical Society, 120
Royal Shakespeare Company, 387
Rudd, Kevin, 169
Russia (Russian Federation), 69–75, 79,
    92–93
  Arctic gas reserves, 97–98
  and Central Asia, 120–26
  and China's Belt and Road initiative,
    182, 188–89
  and China's development of oil
    resources, 158
  and China's rise, 172
  and development of Arctic resources,
    110–14
  development of gas resources, 70–77
  and election interference, 70, 78, 81,
    103–4
  "energy superpower," 70–71
  and energy transition challenges, xiii–xiv
  and Gazprom, 80–82

and global impact of coronavirus
   pandemic, 312–13
and global oil market, 426
and historical context of Middle East
   conflicts, 196
impact of shale revolution on, 56–57
natural gas supplies to Europe, 78, 80–83,
   84–89, 102, 104–8, *106*, 113
oil and state power, 99–101
and OPEC-Plus deal, 321–23
"pivot to the east" strategy, 114, 115–19
and price war among petroleum
   producers, 317, 318–21
Russian-Georgian War, 91
sanctions on, 95–98, 99–101
and Syrian civil war, 246–47
and Ukraine conflict, 80, 95–98, *96*
and U.S. energy production, 63,
   65–66
and U.S. LNG, 40
wheat exports, 100

Sabetta, 111, 112
SABIC, 307
Sabine Pass LNG facility, 35–36, 38
Sadat, Anwar, 209, 262–63
Sakhalin Island, 111
Saleh, Ali Abdullah, 248–50, 252
Salk Institute, 405–6
Salman bin Abdulaziz Al Saud
   and Houthi rebellion in Yemen, 251
   and oil price war, 315, 317–18, 320
   and Russia relations, 311
   and Saudi strategies for the future, 293,
      299–300, 306
Samotlor oil field, 24
Samuelsson, Håkan, 441
Sanaa, Yemen, 248, 250
Sanders, Bernie, 392
Sangam, Urja, 409
Santa Rita 1 well, 21
SARS epidemic (2002), 132, 311–12, 315
Saudi Arabia
   and Arab Spring protests, 239–41
   and competition within Islam, xv
   conflict with Qatar, 306–7
   and energy transition challenges, xii
   and historical context of Middle East
      conflicts, 194
   impact of shale revolution on, 57
   and Iran, 206–7, 209, 221–28
   and Iran-Iraq War, 213–14
   and Khashoggi affair, 305–6
   and Nasser's Arab nationalism,
      204–5
   and OPEC-Plus deal, 321–23

origins of oil wealth, 295–96
and Permian Basin, 23–24
and price war among petroleum
   producers, 313–14, 316–23
and Russian geopolitical ambitions, 72,
   118, 303
and sectarian conflict in Iraq, 229
strategies for the future, 291–310
and Syrian civil war, 245
and U.S. energy production levels, 63
and Yemen conflict, 248–52
Saudi Aramco, 241, 275, 279, 307–10
Say's phoebe, 20
Scarborough Shoal, 145–46
Schroeder, Gerhard, 84
Scowcroft, Brent, 215
Sechin, Igor, 76, 104, 275, 314
Securities and Exchange Commission
   (SEC), 338
Šefčovič, Maroš, 57, 102
Sempra LNG facility, 37, 38
Senate Energy Committee, 55
September 11, 2001 terrorist attacks, 216,
   222, 249, 264–65
Sevastopol, Crimea, 95
shale oil and gas revolution
   and economic impact, 272–78, 279–83,
      284–290, 426
   and energy transition challenges, xviii
   geopolitical impact on Russia, 58–59
   impact on global geopolitics, 54–55,
      58–59, 60–66
   impact on Middle East oil, 224
   and net zero carbon, 436
   origins of industry, 14–17
   and Permian Basin, 21–24
   and Russian resources, 98
   in Texas, *24*
   and U.S. manufacturing boom, 26–28
   and U.S. transition to exporter, 31, 35–37,
      53–54
Shan Zhiqiang, 454
Shandong Yuhuang, 30
shareholder activism, 385–87, 419
sharia law, 225, 245, 268
Shattered Peace (Yergin), xviii
Sheffield, Scott, 23
Shell, 15, 36, 386
Shenzhen, China, 62
SH Griffin #4 natural gas well, 3–4, 7, 64
Shia/Shiites
   and Arab Spring protests, 240–41
   and Iranian goals in Middle East, 228
   and Middle East energy transition
      challenges, xv
   and Saddam Hussein regime in Iraq, 212

Shia/Shiites (*cont.*)
  and Saudi religious strategies, 297
  and sectarian conflict in Iraq, 229–31,
    233–34
Shultz, George, 88–89, 98
Siberia, 72, 74, 76, 88, 98, 121, 177
Sibneft, 76
Silicon Valley, 330, 334, 356–57
Silk Road, 178–79
Silk Road Fund, 182, 189
Silk Road International Film Festival, 183
Sinai Peninsula, 254
Singapore, 151, 152, 163, 165–66, 169,
  169n, 426
Sinopec, 62
Sisi, Abdel Fattah el-, 238
Sistani, Ali al-, 234
Sivaram, Varun, 401
Slovakia, 110
Smil, Vaclav, 379
Smith, Adam, 379, 457–58
Smith, Michael, 34–36
Snowden, Edward, 90–92, 94
Sochi Olympics, 94, 125
social media, 236, 237, 243–44, 269–70,
  294, 366–67
Society of Automotive Engineers, 353
solar power
  and China Belt and Road Initiative, 183
  and current geopolitical challenges,
    427, 429
  and energy transition challenges, xvii, 419
  and European decarbonization efforts, 86
  and global power politics, xi–xii
  and politics of U.S. crude oil exports, 55
  and push for renewable energy sources,
    394–98, 399–402
  and varied approaches to climate
    change, 412, 437–38
Soleimani, Qassem, 230–31, 233, 240–41,
  247, 248, 252, 289
Son, Masayoshi, 292
Sons of Iraq, 267
Sony, 330
Souki, Charif, 31–32, 34–36
South China Morning Post, 186
South China Sea
  and China's Belt and Road initiative, 188
  and China's energy transition
    challenges, xiv
  and China's political strategies, 147–51
  and China's territorial claims, 136–41,
    142–46
  and current geopolitical challenges, 426,
    452–53
  and fish resources, 138

historical context of current tensions,
  152–54, 452–64
  and navigation disputes, 165, 167
  and 9-Dash Line, 141, 144–46, *149*,
    150, 157
  oil and gas resources, 137, 157–60
  and trade relations, 161
  and U.S.–China rivalry, 134–35, 172,
    452–53
Southeast Asia, 153, 165–66, 189. *See also*
  specific countries
South Korea
  and China-Russia strategic
    partnership, 119
  and container shipping, 163
  and Iraqi oil infrastructure, 232
  and Northern Sea Route, 113
  and Russia's "pivot to the east," 119
  and South China Sea tensions, 154, 171
  and U.S. LNG exports, 33, 39
South Pars field, 34
South Sea Islands, 142
South Yemen, 248
sovereignty issues
  "absolute sovereignty" claims, xiv, 116,
    182, 189
  and China's development of petroleum
    resources, 157, 159
  and China's territorial claims, 136–37,
    139–40, 142–44, 146, 148–50
  and historical context of Middle East
    conflicts, 194, 195, 202, 203
  and South China Sea tensions, 167–68
sovereign wealth funds, 27, 37, 100, 301–2,
  309, 339, 421
Soviet Union
  collapse of, xiii, 56, 73–76, 78–79, 94,
    117, 120–21, 131, 179, 211, 281
  invasion of Afghanistan, 210–11,
    263–64, 297
  and Iranian Revolution, 207
  oil and gas exports to Europe, 87–89
SpaceX, 327, 332
Spain, 36, 70, 396
Spanish flu pandemic, 201
Spraberry and Wolfcamp, 24
Spratly Islands, 136–37, 140, 146, 167
Sri Lanka, 187
Stalin, Joseph, 72
Standard Oil, 295
Standing Rock Sioux tribe, 49
Stanford University, 275, 328, 334,
  349–51, 369
State Administration for Foreign Exchange
  (SAFE), 132
State Council (China), 351

State Grid Company (China), 184
Stavridis, James, 135
Stein, Jill, 391
Steinitz, Yuval, 257–58
Steinsberger, Nick, 6–7
Stevens, Chris, 239
Steward, Dan, 6
St. James Parish, Louisiana, 25–26
St. Petersburg International Economic Forum, 58, 89, 124
Strait of Hormuz, 62, 138, 157
Strata, 302
strategic partnerships, 117–18, 126, 184, 189, 311
Straubel, J. B., 327–31, 344
subsidies, 171, 339, 341–44, 373
Suez Canal, 113, 186, 196, 199, 204, 251, 259
Sunni Awakening, 230, 267
Sunnis
    and Arab Spring protests, 240, 242
    and Hamas, 253
    and historical context of Middle East conflicts, 202
    and ISIS attacks in Iraq, 267
    and Middle East energy transition challenges, xv
    and Saddam Hussein regime in Iraq, 212
    and sectarian conflict in Iraq, 229–30, 233
supply chains, 56, 130, 425, 429
Supreme Shia Council (Lebanon), 242
sustainability, 188, 351–52, 385, 389, 409
Sweden, 400, 415
Swensen, David, 386
Sykes, Mark, 195–96, 196–98, 201
Sykes-Picot line, 193–98, 197, 200–202, 215, 270
Sylva, Timipre, 407
Syria
    civil war, 241–48
    and historical context of Middle East conflicts, 201–3
    and Iranian Revolution, 207
    and ISIS, 266–71
    Kurds, 245–47
    and Nasser's Arab nationalism, 204
    and Russia's growing influence in Middle East, 282
    and "Sykes-Picot," 198
Syrian Democratic Forces, 247

Tahrir Square, Cairo, 237, 238
Taiwan
    and container shipping, 163
    and People's Republic of China, 134–35, 157, 173

    and South China Sea tensions, 140–41, 143, 146, 150, 153, 154, 460
    and U.S.–China rivalry, 134–35
    and U.S. LNG, 33
Taiwan Strait, 153
Tajikistan, 120, 180
takfir, 261, 264
Taliban, 216, 265
Tamar field, 254
tanker ships, 33–34, 157, 251, 391
Tanzania, 264
tariffs, 81, 92, 286, 389, 395–96, 410. See also taxation and tax policy
Tarim Basin, 180
Tata Motors, 343
Tavares, Carlos, 339
taxation and tax policy, 55, 339–40, 346, 388, 421. See also subsidies; tariffs
taxis. See ride-hailing services and taxis
Taxonomy (EU report), 389
TC Energy (Trans Canada), 47
technological advances
    and Arctic gas reserves, 111
    and automobile industry, xvi, 366–73, 415, 427
    and autonomous vehicles, 347–57, 368–69, 373
    China, 174–75
    and container shipping, 161–64
    and current geopolitical challenges, 425
    and electric vehicles, xvi, 327–46, 368–71, 415, 427, 430
    and hydraulic fracturing techniques, 7
    and "low-carbon energy," 418
    and oil sands production, 46
    and pace of innovation, 429
    and sanctions against Russia, 97–98
    and U.S. oil production levels, 64
Tengiz field, 122
terrorism, 216, 221–22, 228, 230–31, 241, 249, 253, 261–65, 270, 286–88, 296
Tesla, 330–34, 338–40, 344
Texas, 3–6, 7, 9–10, 20, 23, 24, 29–30, 399
Thatcher, Margaret, 14
USS Theodore Roosevelt, 447
3D printing, 404, 425
Three Forks stratum, 19
Three Mile Island nuclear accident, 349
Thrun, Sebastian, 349–51, 352–53, 357
Thucydides Trap, 131, 154, 425
Thunberg, Greta, 384
Tibet, 150
Time, 18, 20
TNK-BP, 76
Tokayev, Kassym-Jomart, 123
Total, 112

Toyoda, Akio, 369
Toyota, 338, 369
trade wars, 26, 130, 171, 174–75, 286
Trans-Pacific Partnership, 134
Treaty of Lausanne, 201
Treaty of Nanking, 139
Truman, Harry, 60, 211
Trump, Donald
    impeachment, xiv, 110
    and influence of U.S., 188
    and Iranian nuclear ambitions, 227
    and Nord Stream 2 pipeline project,
        105–7, 108
    and pipeline battles in U.S., 51
    and price war among petroleum
        producers, 317–20
    relationship with Russia, 103
    and Russian election interference, 78
    and Saudi relations, 306
    and Syrian civil war, 247
    and U.S.–China rivalry, 134, 171–72
    and U.S. LNG, 39–40
    withdrawal from Paris Agreement, 382
Tunisia, 236
Turkey
    and Eastern Mediterranean petroleum
        resources, 257
    and global order after First World War,
        200–202
    and Iraqi oil infrastructure, 232
    and ISIS, 269
    and Khashoggi affair, 305
    and Qatar
    and Russian gas pipelines, 85, 104
    and Saudi Arabia, 306–7
    and Syrian civil war, 245
Turkmenistan, 120, 179–80
Turkomans, 202
Tusk, Donald, 102
Twitter, 237, 382

Uber, 358–65, 367, 369, 372
Uchiyamada, Takeshi, 338
Uighurs, 180
Ukraine
    and collapse of Soviet Union, 73–74
    and current geopolitical challenges, 426
    and East-West tensions, 90–93
    and gas supplies to Europe, 110
    and Gazprom, 80–82
    and politics of U.S. shale production, 55
    and Russian annexation of Crimea, 94–98
    and Russian gas exports, 102, 104–5,
        107–8, 110, 112, 113
    and Russian geopolitical strategy, 78–80
    and Russian political isolation, 246

United Arab Emirates (UAE)
    and Arab Spring protests, 241
    and Iranian nuclear ambitions, 225, 228
    and price war among petroleum
        producers, 313
    and Saudi Arabia, 300–302, 306
    and Yemen conflict, 251–52
United Arab Republic, 243
United Kingdom. See Britain and the United
    Kingdom
United Nations, 148, 203, 226, 239, 251–52,
    284, 379–80
United Nations Convention on the Law of
    the Sea (UNCLOS), 144–46, 152,
    453, 458
United Nations Security Council, 223
United States
    abundance of natural gas, 420
    and China, 150, 165–76, 188–89, 425–26
    and first oil imports from the Middle
        East, 52
    as gas exporter, 31–40
    and green deal proposals, 321–23, 391–93
    and Iran, 209, 220–28
    and Iraq War, 217
    and ISIS, 267, 270–71
    and Latin American energy markets,
        41–45
    manufacturing renaissance, 25–28, 29–30
    and mujahedeen fighters in
        Afghanistan, 297
    and Nasser's Arab nationalism, 203, 204
    oil and gas production leadership,
        xii, 114
    and Persian Gulf conflicts, 210–19
    and pipeline battles, 46–48, 49–51, 50
    pivot to Asia, 181
    and price war among petroleum
        producers, 314, 316
    and push for renewable energy sources,
        398–99, 432, 442
    and Russian geopolitical strategy, 78–80,
        95–96, 114, 115–16, 118, 122
    and Russian interests in Central Asia,
        124–25
    and Saudi Arabia, 303, 308
    Security Council, 223
    and shale gas and oil production, xi–xii,
        3–6, 9–10, 14–17, 24, 52–57, 58–66
    and Syrian civil war, 246
    terrorist attacks against, 208, 216, 221,
        222, 226, 228, 237, 249, 264–65
    as top oil producer, 65
    and Ukrainian independence, 80
    and varied approaches to climate change,
        413–14

and Yemen conflict, 249, 251
See also specific U.S. institutions
U.S. Air Force, 307
U.S. Army Corps of Engineers, 49, 51
U.S. Central Command, 307
U.S. Congress, 103, 306
U.S. Department of Defense, 172
U.S. Department of Energy, 37
U.S. Department of Justice, 20–21, 336
U.S. Department of State, 48, 223–24, 225
U.S. Energy Information Administration, 98, 160
U.S. Environmental Protection Agency (EPA), 336
U.S. Fifth Fleet, 240
U.S. Fish and Wildlife Service, 20–21
U.S. Geological Survey, 97
U.S. House of Representatives, xiv, 180
U.S. International Development Finance Corporation, 188–89
U.S. Marines, 133
U.S. Navy, 145, 157
U.S. Senate, xiv, 107, 145, 252, 305–6, 340
University of Texas, 9, 21
Uteybi, Juhayman al-, 261–62, 263
Utica shale, 11
Uzbekistan, 120

Varadi, Peter, 395
Venezuela, 72, 272–75, 282–83, 285
Versailles Peace Conference, 200–201
Vietnam, 136–38, 143, 145–46, 150–51, 158–59, 163, 168
Vision 2030 (Abu Dhabi)
Vision 2030 (Saudi Arabia), 294, 300, 302, 303–4, 309, 310, 319
Vision Realization Programs, 302
Voice of the Arabs radio, 204
Voice of the Islamic Republic of Iran, 297
Volkswagen, 335–37, 340, 369–70
Volvo, 337–38, 441
Von der Leyen, Ursula, 388, 390–91

Wagoner, Rick, 333, 351
Wahhab, Muhammad ibn Abd al-, 296
Wahhabism, 296–97
Wall Street Bar and Grill (Midland, Texas), 22
Wall Street Journal, 11
Wan Gang, 340–41
Wang Jinshu, 25–26
Wang Yi, 447
Wang Yilin, 125
Warren, Elizabeth, 392
Washington Post, 305
Watt, James, 379

weapons of mass destruction (WMD), 216, 218–19, 239. *See also* chemical weapons
Weihai, China, 154, 460
Weizmann, Chaim, 199
West Siberia, 72, 74, 98
West Texas, 21
West Virginia University, 336
Whittaker, William "Red," 349–50, 352
Williston Basin, 18, 19
Wilson, Woodrow, 201
wind power
and Britain's Brexit process, 343
and China Belt and Road Initiative, 183
and China's economic growth, 132
and energy transition challenges, xii, xvii, 419, 437–38
and European decarbonization efforts, 86
and global energy trends, xi, 394, 395–96, 398–99, 412, 422, 427, 429
impact on wildlife, 20
and politics of U.S. shale production, 55
Winter Olympics, Sochi, 94
Wisner, Frank, 237–38
The Woodlands, Texas, 5–6
Wordsworth, William, 236
World Bank, 232
World Health Organization (WHO), 407–8
World Resources Institute, 389
World Trade Center attack (9/11), 216, 264–65
World Trade Organization (WTO), xv, 130, 156, 171–72, 425
World War I
and German economic growth, 132
global order after, 198–200, 200–203
and historical context of Middle East conflicts, 195
and origins of Arab nationalism, 204
and the Thucydides Trap, 131, 154
and Ukrainian nationalism, 79
World War II
and Permian Basin oil production, 21
and Soviet expansion of oil production, 72
Wright, Lawrence, 261
Wu Qian, 173
Wu Shengli, 154, 453, 460

Xi Jinping
and China Belt and Road Initiative, 177–79, 184, 188, 189
and Chinese investment in United States, 26
and energy transition challenges, 381
political background of, 166–67
and Russian interests in Central Asia, 124–25

Xi Jinping (cont.)
    and Russia's "pivot to the east," 115,
        117, 119
    and Russia-Ukraine tensions, 94
    and the Thucydides Trap, 131
    and U.S.–China rivalry, 133–34, 447
Xinjiang region, 180

Yamal-Europe pipeline, 85
Yamal Peninsula, 111, 113–14
Yang Jiechi, 150–51
Yanukovych, Viktor, 81, 93
Yazidis, 202, 270
Yeltsin, Boris, 73–75, 75–76
Yemen, 204, 248–52, 264, 286–88, 292
Yuhuang Chemical Company, 25–26

Yukos, 76
Yushchenko, Viktor, 81–82

Zanganeh, Bijan, 278, 280
Zawahiri, Ayman al-, 264, 265
Zaydis, 249
Zelensky, Volodymyr, 107
Zeng Ansha (James Shoal), 141
zero-emissions vehicles (ZEVs), 329, 331
Zhang Qian, 178
Zheng He, 152–53, 164, 179, 454–57, 464
Zimmer, John, 360–61
Zimride, 360
Zionism, 199, 222
Zoellick, Robert, 175
Zohr field, 256

ALLEN LANE
an imprint of
PENGUIN BOOKS

Also Published

Clare Jackson, *Devil-Land: England Under Siege, 1588-1688*

Steven Pinker, *Rationality: Why It Is, Why It Seems Scarce, Why It Matters*

Volker Ullrich, *Eight Days in May: How Germany's War Ended*

Adam Tooze, *Shutdown: How Covide Shook the World's Economy*

Tristram Hunt, *The Radical Potter: Josiah Wedgwood and the Transformation of Britain*

Paul Davies, *What's Eating the Universe: And Other Cosmic Questions*

Shon Faye, *The Transgender Issue: An Argument for Justice*

Dennis Duncan, *Index, A History of the*

Richard Overy, *Blood and Ruins: The Great Imperial War, 1931-1945*

Paul Mason, *How to Stop Fascism: History, Ideology, Resistance*

Cass R. Sunstein and Richard H. Thaler, *Nudge: Improving Decisions About Health, Wealth and Happiness*

Lisa Miller, *The Awakened Brain: The Psychology of Spirituality and Our Search for Meaning*

Michael Pye, *Antwerp: The Glory Years*

Christopher Clark, *Prisoners of Time: Prussians, Germans and Other Humans*

Rupa Marya and Raj Patel, *Inflamed: Deep Medicine and the Anatomy of Injustice*

Richard Zenith, *Pessoa: An Experimental Life*

Michael Pollan, *This Is Your Mind On Plants: Opium—Caffeine—Mescaline*

Amartya Sen, *Home in the World: A Memoir*

Jan-Werner Müller, *Democracy Rules*

Robin DiAngelo, *Nice Racism: How Progressive White People Perpetuate Racial Harm*

Rosemary Hill, *Time's Witness: History in the Age of Romanticism*

Lawrence Wright, *The Plague Year: America in the Time of Covid*

Adrian Wooldridge, *The Aristocracy of Talent: How Meritocracy Made the Modern World*

Julian Hoppit, *The Dreadful Monster and its Poor Relations: Taxing, Spending and the United Kingdom, 1707-2021*

Jordan Ellenberg, *Shape: The Hidden Geometry of Absolutely Everything*

Duncan Campbell-Smith, *Crossing Continents: A History of Standard Chartered Bank*

Jemma Wadham, *Ice Rivers*

Niall Ferguson, *Doom: The Politics of Catastrophe*

Michael Lewis, *The Premonition: A Pandemic Story*